Modern Trends in Structural and Solid Mechanics 1

Series Editor
Noël Challamel

Modern Trends in Structural and Solid Mechanics 1

Statics and Stability

Edited by

Noël Challamel
Julius Kaplunov
Izuru Takewaki

iSTE

WILEY

First published 2021 in Great Britain and the United States by ISTE Ltd and John Wiley & Sons, Inc.

ISTE Ltd
27-37 St George's Road
London SW19 4EU
UK

www.iste.co.uk

John Wiley & Sons, Inc.
111 River Street
Hoboken, NJ 07030
USA

www.wiley.com

Library of Congress Control Number: 2020952688

British Library Cataloguing-in-Publication Data
A CIP record for this book is available from the British Library
ISBN 978-1-78630-714-9

MIX
Paper from responsible sources
FSC® C013604

Contents

Preface

Short Bibliographical Presentation of Prof. Isaac Elishakoff

This book is dedicated to Prof. Isaac Elishakoff by his colleagues, friends and former students, on the occasion of his seventy-fifth birthday.

Figure P.1. *Prof. Isaac Elishakoff*

For a color version of all the figures in this chapter, see www.iste.co.uk/challamel/mechanics1.zip.

Prof. Isaac Elishakoff is an international leading authority across a broad area of structural mechanics, including dynamics and stability, optimization and anti-optimization, probabilistic methods, analysis of structures with uncertainty, refined theories, functionally graded material structures, and nanostructures. He was born in Kutaisi, Republic of Georgia, on February 9, 1944.

Figure P.2. *Elishakoff in middle school in the city of Sukhumi, Georgia*

Elishakoff holds a PhD in Dynamics and Strength of Machines from the Power Engineering Institute and Technical University in Moscow, Russia (Figure P.4 depicts the PhD defense of Prof. Isaac Elishakoff).

Figure P.3. *Elishakoff just before acceptance to university. Photo taken in Sukhumi, Georgia*

Figure P.4. *Public PhD defense, Moscow Power Engineering Institute and State University; topic "Vibrational and Acoustical Fields in the Circular Cylindrical Shells Excited by Random Loadings", and dedicated to the evaluation of noise levels in TU-144 supersonic aircraft*

His supervisor was Prof. V. V. Bolotin (1926–2008), a member of the Russian Academy of Sciences (Figure P.5 shows Elishakoff with Bolotin some years later).

Figure P.5. *Elishakoff with Bolotin (middle), member of the Russian Academy of Sciences, and Prof. Yukweng (Mike) Lin (left), member of the US National Academy of Engineering. Photo taken at Florida Atlantic University during a visit from Bolotin*

Currently, Elishakoff is a Distinguished Research Professor in the Department of Ocean and Mechanical Engineering at Florida Atlantic University. Before joining the university, he taught for one year at Abkhazian University, Sukhumi in the

Republic of Georgia, and 18 years at the Technion – Israel Institute of Technology in Haifa, where he became the youngest full professor at the time of his promotion (Figure P.6 shows Elishakoff presenting a book to Prof. Josef Singer, Technion's former president).

Figure P.6. *Prof. Elishakoff presenting a book to Prof. J. Singer, Technion's President; right: Prof. A. Libai, Aerospace Engineering Department, Technion*

Elishakoff has lectured at about 200 meetings and seminars, including about 60 invited, plenary or keynote lectures, across Europe, North and South America, the Middle East and the Far East.

Prof. Elishakoff has made vital and outstanding contributions in a number of areas in structural mechanics. In particular, he has analyzed random vibrations of homogeneous and composite beams, plates and shells, with special emphasis on the effects of refinements in structural theories and cross-correlations. Free structural vibrations have been tackled using a non-trivial generalization of Bolotin's dynamic edge effect method. Nonlinear buckling has been investigated using a novel method, incorporating experimental analysis of imperfections. As a result, the fundamental concept of closing the gap – spanning the entire 20th century – between theory and practice in imperfection-sensitive structures has been proposed. Novel methods of evaluating structural reliability have been proposed, taking into account the error associated with various low-order approximations, as well as human error; innovative generalization of the stochastic linearization method has been advanced.

A non-probabilistic theory for treating uncertainty in structural mechanics has been established. Dynamic stability of elastic and viscoelastic structures with imperfections has been studied. An improved, non-perturbative stochastic finite element method for structures has been developed. The list of Elishakoff's remarkable research achievements goes on.

His research has been acknowledged by many awards and prizes. He is a member of the European Academy of Sciences and Arts, a Fellow of the American Academy of Mechanics and ASME, and a Foreign Member of the Georgian National Academy of Sciences. Elishakoff is also a recipient of the Bathsheva de Rothschild prize (1973) and the Worcester Reed Warner Medal of the American Society of Mechanical Engineers (2016).

Figure P.7. *Elishakoff having received the William B. Johnson Inter- Professional Founders Award*

Elishakoff is directly involved in numerous editorial activities. He serves as the book review editor of the "Journal of Shock and Vibration" and is currently, or has previously been an associate editor of the *International Journal of Mechanics of Machines and Structures, Applied Mechanics Reviews*, and *Chaos, Solitons & Fractals*. In addition, he is or has been on the editorial boards of numerous journals, for example *Journal of Sound and Vibration*, *International Journal of Structural Stability and Dynamics*, *International Applied Mechanics* and *Computers & Structures*. He also acts as a book series editor for Elsevier, Springer and Wiley.

Figure P.8. *Inauguration as the Frank Freimann Visiting Professor of Aerospace and Mechanical Engineering; left: Rev. Theodore M. Hesburgh, President of the University of Notre Dame; right: Prof. Timothy O'Meara, Provost*

Prof. Elishakoff has held prestigious visiting positions at top universities all over the world. Among them are Stanford University (S. P. Timoshenko Scholar); University of Notre Dame, USA (Frank M. Freimann Chair Professorship of Aerospace and Mechanical Engineering and Henry J. Massman, Jr. Chair Professorship of Civil Engineering); University of Palermo, Italy (Visiting Castigliano Distinguished Professor); Delft University of Technology, Netherlands (multiple appointments, including the W. T. Koiter Chair Professorship of the Mechanical Engineering Department – see Figure P.9); Universities of Tokyo and Kyoto, Japan (Fellow of the Japan Society for the Promotion of Science); Beijing University of Aeronautics and Astronautics, People's Republic of China (Visiting Eminent Scholar); Technion, Haifa, Israel (Visiting Distinguished Professor); University of Southampton, UK (Distinguished Visiting Fellow of the Royal Academy of Engineering).

Figure P.9. *Prof. Elishakoff with Prof. Warner Tjardus Koiter, Delft University of Technology (center), and Dr. V. Grishchak, of Ukraine (right)*

Figure P.10. *Elishakoff and his colleagues during the AIAA SDM Conference at Palm Springs, California in 2004; Standing, from right to left, are Prof. Elishakoff, the late Prof. Josef Singer and Dr. Giora Maymon of RAFAEL. Sitting is the late Prof. Avinoam Libai*

Elishakoff has made a substantial contribution to conference organization. In particular, he participated in the organization of the Euro-Mech Colloquium on "Refined Dynamical Theories of Beams, Plates and Shells, and Their Applications" in Kassel, Germany (1986); the *Second International Conference on Stochastic Structural Dynamics*, in Boca Raton, USA (1990); "International Conference on Uncertain Structures" in Miami, USA and Western Caribbean (1996). He also coordinated four special courses at the International Centre for Mechanical Sciences (CISM), in Udine, Italy (1997, 2001, 2005, 2011).

Prof. Elishakoff has published over 540 original papers in leading journals and conference proceedings. He championed authoring, co-authoring or editing of 31 influential and extremely well-received books and edited volumes.

Here follows some praise of his work and books:

– "It was not until 1979, when Elishakoff published his reliability study … that a method has been proposed, which made it possible to introduce the results of imperfection surveys … into the analysis …" (Prof. Johann Arbocz, Delft University of Technology, The Netherlands, *Zeitschrift für Flugwissenschaften und Weltraumforschung*).

– "He has achieved world renown … His research is characterized by its originality and a combination of mathematical maturity and physical understanding which is reminiscent of von Kármán …" (Prof. Charles W. Bert, University of Oklahoma).

– "It is clear that Elishakoff is a world leader in his field … His outstanding reputation is very well deserved …" (Prof. Bernard Budiansky, Harvard University).

– "Professor Isaac Elishakoff … is subject-wise very much an all-round vibrationalist" (P. E. Doak, Editor in Chief, *Journal of Sound and Vibration*, University of Southampton, UK).

– "This is a beautiful book …" (Dr. Stephen H. Crandall, Ford Professor of Engineering, M.I.T.).

– "Das Buch ist in seiner Aufmachunghervorragendgestaltet und kannalsäusserstwertvolleErganzung … wäzmstensempfohlenwerden …" [The book's appearance is perfectly designed and can be highly recommended as a valuable addition.] (Prof. Horst Försching, Institute of Aeroelasticity, Federal Republic of Germany, *Zeitschrift für Flugwissenschaften und Weltraumforschung*).

– "Because of you, Notre Dame is an even better place, a more distinguished University" (Prof. Rev. Theodore M. Hesburgh, President, University of Notre Dame).

– "It is an impressive volume …" (Prof. Warner T. Koiter, Delft University of Technology, The Netherlands).

– "This extremely well-written text, authored by one of the leaders in the field, incorporates many of these new applications … Professor Elishakoff's techniques for developing the material are accomplished in a way that illustrates his deep insight into the topic as well as his expertise as an educator … Clearly, the second half of the text provides the basis for an excellent graduate course in random vibrations and buckling … Professor Elishakoff has presented us with an outstanding instrument for teaching" (Prof. Frank Kozin, Polytechnic Institute of New York, *American Institute of Aeronautics and Astronautics Journal*).

– "By far the best book on the market today …" (Prof. Niels C. Lind, University of Waterloo, Canada).

– "The book develops a novel idea … Elegant, exhaustive discussion … The study can be an inspiration for further research, and provides excellent applications in design …" (Prof. G. A. Nariboli, *Applied Mechanics Reviews*).

– "This volume is regarded as an advanced encyclopedia on random vibration and serves aeronautical, civil and mechanical engineers …" (Prof. Rauf Ibrahim, Wayne State University, *Shock and Vibration Digest*).

– "The book deals with a fundamental problem in Applied Mechanics and in Engineering Sciences: How the uncertainties of the data of a problem influence its solution. The authors follow a novel approach for the treatment of these problems … The book is written with clarity and contains original and important results for the engineering sciences …" (Prof. P. D. Panagiotopoulos, University of Thessaloniki, Greece and University of Aachen, Germany, *SIAM Review*).

– "The content should be of great interest to all engineers involved with vibration problems, placing the book well and truly in the category of an essential reference book …" (Prof. I. Pole, *Journal of the British Society for Strain Measurement*).

– "A good book; a different book … It is hoped that the success of this book will encourage the author to provide a sequel in due course …" (Prof. John D. Robson, University of Glasgow, Scotland, UK, *Journal of Sound and Vibration*).

– "The book certainly satisfies the need that now exists for a readable textbook and reference book …" (Prof. Masanobu Shinozuka, Columbia University).

– "[the] author ties together reliability, random vibration and random buckling … Well written … useful book …" (Dr. H. Saunders, *Shock and Vibration Digest*).

– "A very useful text that includes a broad spectrum of theory and application" (*Mechanical Vibration*, Prof. Haym Benaroya, Rutgers University).

– "A treatise on random vibration and buckling … The reviewer wishes to compliment the author for the completion of a difficult task in preparing this book on a subject matter, which is still developing on many fronts …" (Prof. James T. P. Yao, Texas A&M University, *Journal of Applied Mechanics*).

– "It seems to me a hard work with great result …" (Prof. Hans G. Natke, University of Hannover, Federal Republic of Germany).

– "The approach is novel and could dominate the future practice of engineering" (*The Structural Engineer*).

– "An excellent presentation … well written … all readers, students, and certainly reviewers should read this preface for its excellent presentation of the philosophy and raison d'être for this book. It is well written, with the material presented in an informational fashion as well as to raise questions related to unresolved … challenges; in the vernacular of film critics, 'thumbs up'" (Dr. R. L. Sierakowski, U.S. Air Force Research Laboratory, *AIAA Journal*).

– "This substantial and attractive volume is a well-organized and superbly written one that should be warmly welcomed by both theorists and practitioners … Prof. Elishakoff, Li, and Starnes, Jr. have given us a jewel of a book, one done with care and understanding of a complex and essential subject and one that seems to have ably filled a gap existing in the present-day literature and practice" (*Current Engineering Practice*).

– "Most of the subjects covered in this outstanding book have never been discussed exclusively in the existing treatises … (*Ocean Engineering*).

– "The treatment is scholarly, having about 900 items in the bibliography and additional contributors in the writing of almost every chapter … This reviewer believes that *Non-Classical Problems in the Theory of Elastic Stability* should be a useful reference for researchers, engineers, and graduate students in aeronautical, mechanical, civil, nuclear, and marine engineering, and in applied mechanics" (*Applied Mechanics Reviews*).

– "What more can be said about this monumental work, other than to express admiration? … The study is of great academic interest, and is clearly a labor of love. The author is to be congratulated on this work …" (Prof. H. D. Conway, Department of Theoretical and Applied Mechanics, Cornell University).

– "This book … is prepared by Isaac Elishakoff, one of the eminent solid mechanics experts of the 20th century and the present one, and his distinguished coauthors, will be of enormous use to researchers, graduate students and professionals in the fields of ocean, naval, aerospace and mechanical engineers as well as other fields" (Prof. Patricio A. A. Laura, Prof. Carlos A. Rossit, Prof. Diana V. Bambill, Universidad Nacional del Sur, Argentina, *Ocean Engineering*).

– "This book is an outstanding research monograph … extremely well written, informative, highly original … great scholarly contribution …. There is no comparable book discussing the combination of optimization and anti-optimization … magnificent monograph …. This book, which certainly is written with love and passion, is the first of its kind in applied mechanics literature, and has the potential of having a revolutionary impact on both uncertainty analysis and optimization" (Prof. Izuru Takewaki, Kyoto University, *Engineering Structures*).

– "This book is a collection of a surprisingly large number of closed form solutions, by the author and by others, involving the buckling of columns and beams, and the vibration of rods, beams and circular plates. The structures are, in general, inhomogeneous. Many solutions are published here for the first time. The text starts with an instructive review of direct, semi-inverse, and inverse eigenvalue problems. Unusual closed form solutions of column buckling are presented first, followed by closed form solutions of the vibrations of rods. Unusual closed form solutions for vibrating beams follow. The influence of boundary conditions on eigenvalues is discussed. An entire chapter is devoted to boundary conditions involving guided ends. Effects of axial loads and of elastic foundations are presented in two separate chapters. The closed form solutions of circular plates concentrate on axisymmetric vibrations. The scholarly effort that produced this book is remarkable" (Prof. Werner Soedel, then Editor-in-Chief of *Journal Sound and Vibration*).

– "The field has been brilliantly presented in book form …" (Prof. Luis A. Godoy *et al.*, Institute of Advanced Studies in Engineering and Technology, Science Research Council of Argentina and National University of Cordoba, Argentina, *Thin-Walled Structures*).

– "Elishakoff is one of the pioneers in the use of the probabilistic approach for studying imperfection-sensitive structures" (Prof. Chiara Bisagni and Dr. Michela Alfano, Delft University of Technology; *AIAA Journal*).

– "Recently, Elishakoff *et al.* presented an excellent literature review on the historical development of Timoshenko's beam theory" (Prof. Zhenlei Chen *et al.*, *Journal of Building Engineering*).

Professor Isaac Elishakoff is the author or co-author of an impressive list of seminal books in the field of deterministic and non-deterministic mechanics, presented below.

Books by Elishakoff

Ben-Haim, Y. and Elishakoff, I. (1990). *Convex Models of Uncertainty in Applied Mechanics.* Elsevier, Amsterdam.

Cederbaum, G., Elishakoff, I., Aboudi, J., Librescu, L. (n.d.). *Random Vibration and Reliability of Composite Structures.* Technomic, Lancaster.

Elishakoff, I. (1983). *Probabilistic Methods in the Theory of Structures.* Wiley, New York.

Elishakoff, I. (1999). *Probabilistic Theory of Structures.* Dover Publications, New York.

Elishakoff, I. (2004). *Safety Factors and Reliability: Friends or Foes?* Kluwer Academic Publishers, Dordrecht.

Elishakoff, I. (2005). *Eigenvalues of Inhomogeneous Structures: Unusual Closed-Form Solutions of Semi-Inverse Problems.* CRC Press, Boca Raton.

Elishakoff, I. (2014). *Resolution of the Twentieth Century Conundrum in Elastic Stability.* World Scientific/Imperial College Press, Singapore.

Elishakoff, I. (2017). *Probabilistic Methods in the Theory of Structures: Random Strength of Materials, Random Vibration, and Buckling.* World Scientific, Singapore.

Elishakoff, I. (2018). *Probabilistic Methods in the Theory of Structures: Solution Manual to Accompany Probabilistic Methods in the Theory of Structures: Problems with Complete, Worked Through Solutions.* World Scientific, Singapore.

Elishakoff, I. (2020). *Dramatic Effect of Cross-Correlations in Random Vibrations of Discrete Systems, Beams, Plates, and Shells.* Springer Nature, Switzerland.

Elishakoff, I. (2020). *Handbook on Timoshenko-Ehrenfest Beam and Uflyand-Mindlin Plate Theories.* World Scientific, Singapore.

Elishakoff, I. and Ohsaki, M. (2010). *Optimization and Anti-Optimization of Structures under Uncertainty.* Imperial College Press, London.

Elishakoff, I. and Ren, Y. (2003). *Finite Element Methods for Structures with Large Stochastic Variations*. Oxford University Press, Oxford.

Elishakoff, I., Lin, Y.K., Zhu, L.P. (1994). *Probabilistic and Convex Modeling of Acoustically Excited Structures*. Elsevier, Amsterdam.

Elishakoff, I., Li, Y., Starnes Jr., J.H. (2001). *Non-Classical Problems in the Theory of Elastic Stability*. Cambridge University Press, Cambridge.

Elishakoff, I., Pentaras, D., Dujat, K., Versaci, C., Muscolino, G., Storch, J., Bucas, S., Challamel, N., Natsuki, T., Zhang, Y., Ming Wang, C., Ghyselinck, G. (2012). *Carbon Nanotubes and Nano Sensors: Vibrations, Buckling, and Ballistic Impact*. ISTE Ltd, London, and John Wiley & Sons, New York.

Elishakoff, I., Pentaras, D., Gentilini, C., Cristina, G. (2015). *Mechanics of Functionally Graded Material Structures*. World Scientific/Imperial College Press, Singapore.

Books edited or co-edited by Elishakoff

Ariaratnam, S.T., Schuëller, G.I., Elishakoff, I. (1988). *Stochastic Structural Dynamics – Progress in Theory and Applications*. Elsevier, London.

Casciati, F., Elishakoff, I., Roberts, J.B. (1990). *Nonlinear Structural Systems under Random Conditions*. Elsevier, Amsterdam.

Chuh, M., Wolfe, H.F., Elishakoff, I. (1989). *Vibration and Behavior of Composite Structures*. ASME Press, New York.

David, H. and Elishakoff, I. (1990). *Impact and Buckling of Structures*. ASME Press, New York.

Elishakoff, I. (1999). *Whys and Hows in Uncertainty Modeling*. Springer, Vienna.

Elishakoff, I. (2007). *Mechanical Vibration: Where Do We Stand?* Springer, Vienna.

Elishakoff, I. and Horst, I. (1987). *Refined Dynamical Theories of Beams, Plates and Shells and Their Applications*. Springer Verlag, Berlin.

Elishakoff, I. and Lin, Y.K. (1991). *Stochastic Structural Dynamics 2 – New Applications*. Springer, Berlin.

Elishakoff, I. and Lyon, R.H. (1986), *Random Vibration-Status and Recent Developments*. Elsevier, Amsterdam.

Elishakoff, I. and Seyranian, A.P. (2002). *Modern Problems of Structural Stability*. Springer, Vienna.

Elishakoff, I. and Soize, C. (2012). *Non-Deterministic Mechanics*. Springer, Vienna.

Elishakoff, I., Arbocz, J., Babcock Jr., C.D., Libai, A. (1988). *Buckling of Structures: Theory and Experiment*. Elsevier, Amsterdam.

Lin, Y.K. and Elishakoff, I. (1991). *Stochastic Structural Dynamics 1 – New Theoretical Developments*. Springer, Berlin.

Noor, A.K., Elishakoff, I., Hulbert, G. (1990). *Symbolic Computations and Their Impact on Mechanics*. ASME Press, New York.

Figure P.11. *Elishakoff with his wife, Esther Elisha, M.D., during an ASME awards ceremony*

On behalf of all the authors of this book, including those friends who were unable to contribute, we wish Prof. Isaac Elishakoff many more decades of fruitful works and collaborations for the benefit of world mechanics, in particular.

Modern Trends in Structural and Solid Mechanics 1 – the first of three separate volumes that comprise this book – presents recent developments and research discoveries in structural and solid mechanics, with a focus on the statics and stability of solid and structural members.

The book is centered around theoretical analysis and numerical phenomena and has broad scope, covering topics such as: buckling of discrete systems (elastic chains, lattices with short and long range interactions, and discrete arches), buckling of continuous structural elements including beams, arches and plates, static investigation of composite plates, exact solutions of plate problems, elastic and inelastic buckling, dynamic buckling under impulsive loading, buckling and post-buckling investigations, buckling of conservative and non-conservative systems, buckling of micro and macro-systems. The engineering applications concern both small-scale phenomena with micro and nano-buckling up to large-scale structures, including the buckling of drillstring systems.

Each of the three volumes is intended for graduate students and researchers in the field of theoretical and applied mechanics.

Prof. Noël CHALLAMEL
Lorient, France
Prof. Julius KAPLUNOV
Keele, UK
Prof. Izuru TAKEWAKI
Kyoto, Japan
February 2021

1

Static Deformations of Fiber-Reinforced Composite Laminates by the Least-Squares Method

Accurate solutions of the linear elasticity equations governing three-dimensional static deformations of fiber-reinforced composite laminates are needed to efficiently design them for structural applications, by considering failure modes in them. In this chapter, the governing equations are written as first-order partial differential equations for three displacements, three transverse stresses and three in-plane strains. The mixed formulation facilitates the satisfaction of continuity conditions at an interface between two adjacent plies. These nine equations are numerically solved by minimizing residuals in them and in the boundary conditions by using the least-squares method. The functional of the residuals is evaluated by expressing the unknown quantities as the product of complete polynomials of different orders in the three independent coordinates and using appropriate quadrature rules. Minimization of the functional, with respect to coefficients appearing in the polynomials for the solution variables, provides a system of simultaneous linear algebraic equations that are numerically solved. It is shown that polynomial functions of degree, at most, 8 in the in-plane coordinates and 3 in the thickness coordinate for each layer of a laminate, provide accurate solutions for stresses and displacements when compared with the analytical solutions of problems. Through-the-thickness stress distributions are found to agree very well with those found analytically. It provides an efficient numerical scheme for accurately finding stresses and displacements in a laminate.

Chapter written by Devin BURNS and Romesh C. BATRA.

1.1. Introduction

Composites are being increasingly used in transportation, aviation and defense industries because of their high specific strength and our ability to tailor their elastic moduli and ultimate strengths in desired directions and at critical locations in the structure. They are usually in the form of a laminate composed of numerous plies, with each ply having unidirectional fibers. The volume fraction of fibers in different plies can be varied to suit the intended application. Recognizing that design is an iterative process, it is imperative that we can easily and accurately analyze a structure's response to external stimuli.

Closed-form solutions for three-dimensional linear elasticity equations for composite laminates are scarce. Assuming that the Euler–Bernoulli beam theory applies to a composite laminated beam and the Kirchhoff–Love theory is relevant for a composite laminated plate, enables us to analytically solve some boundary value problems (see, for example, Jones (1998) and Hyer (2009)). Elishakoff has reviewed various theories for plates and laminates and has provided their historical perspective (Elishakoff 2018). However, for general structures, we resort to numerical methods for solving the governing equations.

A challenge in solving problems for laminates is accurately finding transverse stresses that can cause delamination between adjacent plies. For simply supported edges, the linear elasticity equations have been solved by expressing the three displacements as double Fourier series in the in-plane coordinates and deducing ordinary differential equations for them in the thickness direction (Vlasov 1957; Pagano 1969; Srinivas and Rao 1970). For other boundary conditions at the edges, Vel and Batra employed the Eshelby–Stroh formalism and satisfied boundary conditions at the edges in the sense of Fourier series (Vel and Batra 1999; Vel and Batra 2000) that rely on St. Venant's principle. This principle states that the solution at points away from the edges is unaffected by the boundary conditions, provided that they are equipollent to the same resultant force and moment (Toupin 1965). The distance from the edges, usually called the decay length, where the solution is unaffected, depends on the lowest frequency of free vibrations of a slice of the structure of characteristic length l and the maximum eigenvalue of the elasticity matrix. For composite laminates, the decay length has been estimated in Horgan and Baxter (1998).

A common approach, called the state space method, is to take the three transverse stresses and the three displacements as variables, cast governing equations for them as first-order partial differential equations (PDEs) in the thickness coordinate, z, and find their values simultaneously. The boundary value problem is thus reduced to one that looks like an initial value problem. One way to solve these PDEs is to find a propagation matrix that relates variables for two different values of z (see, for example, Bahar (1975)). Kant and Ramesh have presented Goldberg and Bogdanoff's method of numerically solving these six first-order equations (Goldberg and Bogdanoff 1957; Kant and Ramesh 1981) that satisfy the prescribed boundary conditions at the top and the bottom surfaces of a plate. Here, we also solve for the three in-plane strains, along with the aforementioned six unknowns, use the least-squares method, polynomial basis functions of possibly different orders in each one of the three directions defined on the entire structure's domain, include in the functional to be minimized boundary conditions at the six edges of a laminate and derive a set of simultaneous algebraic equations; see, for example, Moleiro *et al.* (2011) who have analyzed numerous problems. The contribution of this work is in demonstrating that polynomial basis functions defined on the entire domain enable us to find a reasonably accurate solution of the linear elasticity equations, for rectangular composite laminates of side length equal to twice the thickness and sandwich structures with the ratio of the axial elastic modulus of a facesheet to that of the core less than 1000.

The rest of the chapter is organized as follows. The problem formulation is described in section 1.2. In section 1.3, the developed algorithm is verified by comparing predictions from it with the analytical solutions for a simply supported four-layer laminate and a sandwich structure, which also has results for a three-layer laminate with two opposite edges simply supported and the other two edges either free or clamped. Significant outcomes of the work are briefly summarized as conclusions.

1.2. Formulation of the problem

We formulate the problem for a rectangular $a \times b \times h$ lamina that can be easily extended to a laminate made of p layers. We employ rectangular Cartesian coordinates (x_1, x_2, x_3), shown in Figure 1.1, to describe the position of a material particle in the unstressed reference configuration. In the absence of body forces, infinitesimal deformations of the lamina are governed by the following equilibrium equations [1.1], stress–strain relations [1.2] and strain–displacement relations [1.3]:

$$\sigma_{ij,j} \equiv \frac{\partial \sigma_{ij}}{\partial x_j} = 0 \qquad [1.1]$$

$$\sigma_{ij} = C_{ijkl} e_{kl}, \qquad C_{ijkl} = C_{klij} = C_{jikl} \qquad [1.2]$$

$$e_{kl} = \frac{1}{2}\left(u_{k,l} + u_{l,k}\right), \quad i,j,k,l = 1,2,3. \qquad [1.3]$$

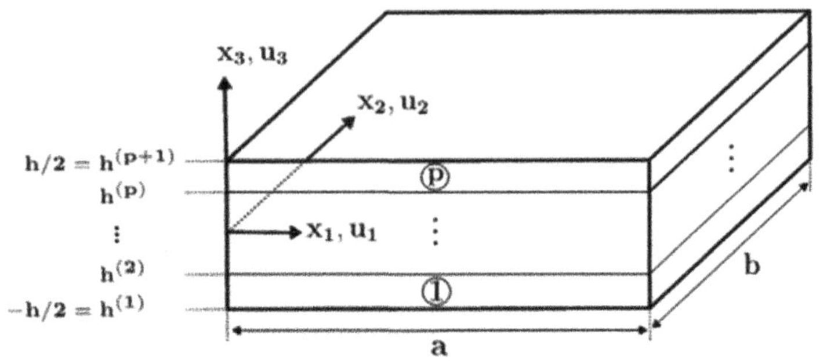

Figure 1.1. *Geometry and coordinate-axes of a laminated plate*

In equations [1.1]–[1.3], **σ** is the stress tensor, **e** is the infinitesimal strain tensor, **u** is the displacement vector and **C** is the matrix of elasticities. Here, a layer is modeled as a transversely isotropic material with the axis of transverse isotropy along the fiber. A repeated index implies summation over the range of the index. The substitution from equation [1.3] into [1.2], and the result into equation [1.1] gives three second-order partial differential equations for the three displacement components that are to be solved under the prescribed boundary conditions of surface tractions on one part of the boundary and displacements on the other part. Of course, linearly independent components of **u** and the surface traction vector, $f_i = \sigma_{ij} n_j$, can alternatively be prescribed at a point of the boundary. Here, n_j is the jth component of the outward unit normal to the boundary.

As is often done, we use the Voigt notation to express **σ** and **e** as six-dimensional vectors and write equation [1.2] as

$$\sigma_\alpha = C_{\alpha\beta} e_\beta, \quad \alpha, \beta = 1,2,3,4,5,6. \qquad [1.4]$$

In equation [1.4], **C** is a 6 x 6 symmetric matrix of elasticities of the layer material, and $(\sigma_1, \sigma_2, \sigma_3, \sigma_4, \sigma_5, \sigma_6) = (\sigma_{11}, \sigma_{22}, \sigma_{12}, \sigma_{13}, \sigma_{23}, \sigma_{33})$. A similar notation is used for **e.**

We use a mixed formulation and take $\mathbf{s}^k = (u_1^k, u_2^k, u_3^k, \sigma_4^k, \sigma_5^k, \sigma_6^k, e_1^k, e_2^k, e_3^k)$ as unknowns at a point in each plate layer. In order to solve for $\mathbf{s}^k$, we define the following residuals on the kth layer that only involve first-order derivatives for the elements of $\mathbf{s}^k$:

$$R_1{}^k \equiv u_{1,1}{}^k - e_{11}{}^k$$

$$R_2{}^k \equiv u_{2,2}{}^k - e_{22}{}^k \qquad [1.5a]$$

$$R_3{}^k \equiv u_{1,2}{}^k + u_{2,1}{}^k - 2e_{12}{}^k$$

$$R_4{}^k \equiv \sigma_4{}^k - A_{55}{}^k u_{1,3}{}^k - A_{55}{}^k u_{3,1}{}^k - A_{45}{}^k u_{2,3}{}^k - A_{45}{}^k u_{3,2}{}^k$$

$$R_5{}^k \equiv \sigma_5{}^k - A_{45}{}^k u_{1,3}{}^k - A_{45}{}^k u_{3,1}{}^k - A_{44}{}^k u_{2,3}{}^k - A_{44}{}^k u_{3,2}{}^k \qquad [1.5b]$$

$$R_6{}^k \equiv \sigma_6{}^k - A_{13}{}^k e_1{}^k - A_{23}{}^k e_3{}^k - A_{36}{}^k e_3{}^k - A_{33}{}^k u_{3,3}{}^k$$

$$\begin{aligned} R_7{}^k = {} & \hat{C}_{11}{}^k e_{1,1}{}^k + \hat{C}_{12}{}^k e_{2,1}{}^k + \hat{C}_{16}{}^k e_{3,1}{}^k + \hat{C}_{13}{}^k \sigma_{6,1}{}^k + \hat{C}_{16}{}^k e_{1,2}{}^k \\ & + \hat{C}_{26}{}^k e_{2,2}{}^k + \hat{C}_{66}{}^k e_{3,2}{}^k + \hat{C}_{36}{}^k \sigma_{6,2}{}^k + \sigma_{4,3}{}^k \end{aligned} \qquad [1.5c]$$

$$\begin{aligned} R_8{}^k = {} & \hat{C}_{16}{}^k e_{1,1}{}^k + \hat{C}_{26}{}^k e_{2,1}{}^k + \hat{C}_{66}{}^k e_{3,1}{}^k + \hat{C}_{36}{}^k \sigma_{6,1}{}^k + \hat{C}_{12}{}^k e_{1,2}{}^k \\ & + \hat{C}_{22}{}^k e_{2,2}{}^k + \hat{C}_{26}{}^k e_{3,2}{}^k + \hat{C}_{23}{}^k \sigma_{6,2}{}^k + \sigma_{5,3}{}^k \end{aligned}$$

$$R_9{}^k = \sigma_{4,1}{}^k + \sigma_{5,2}{}^k + \sigma_{6,3}{}^k$$

In equations [1.5b] and [1.5c], the coefficients $A_{ij}{}^k$ represent the elasticities of the kth layer material, with respect to the layer material principal axes rotated with respect to the analysis coordinate system (see Figure 1.1). The $\hat{C}_{ij}{}^k$ quantities are found by expressing the in-plane stresses and transverse strains in terms of the transverse stresses and in-plane strains (i.e. the layerwise continuous variables) from equations [1.2] and [1.3] (see Moleiro *et al.* 2011). It should be highlighted that in-plane strains have been incorporated into the mixed formulation, in order to recast the governing equations into first-order form, which allows the use of C^0 basis functions for the interpolation of unknown variables, discussed later. While this increases the number of unknowns to solve for, the advantage is that continuity of surface tractions and displacements at layer interfaces can now be enforced.

We note that $R^k{}_a = 0$, a = 1, 2, …, 9 are the nine equations for the nine unknowns in $\mathbf{s}^k$. Recall that there are three boundary conditions prescribed at each bounding face of the lamina. Generally, the top and the bottom faces of a lamina have prescribed tractions, $f_i = f_i^p$, on them, and edges x = 0, a and y = 0, b have a combination of f_i and u_i, where f_i^p and u_i^p are the known function of the in-plane coordinates. For example, at a clamped edge, $u_i^p = 0$, and at a free edge, $f_i^p = 0$. At a simply supported edge x = 0, we set

$$u_2^p = 0, u_3^p = 0, f_1^p = 0. \qquad [1.6]$$

Equation [1.6]$_3$ is equivalent to a null in-plane normal stress at the edge x = 0. Since in-plane stresses are not directly computed in the present formulation, we write this boundary condition residual in terms of the variables in **s** as

$$R^k = \hat{C}_{11}{}^k e_1{}^k + \hat{C}_{12}{}^k e_2{}^k + \hat{C}_{16}{}^k e_3{}^k + \hat{C}_{13}{}^k \sigma_6{}^k \qquad [1.7]$$

which also uses the constitutive relation in equation [1.4]. Thus, for each one of the six bounding faces, we will have three residuals that we denote by R_{10} through R_{27}.

We define the functional, J, in terms of the residuals, with contributions from each material layer as

$$J = \frac{1}{2}\left(\sum_{k=1}^{p} \int_{\Omega} \int_{h^k} R_a{}^k \cdot R_a{}^k dx_3 d\Omega + \sum_{k=1}^{p} \int_{\Omega} R_b{}^k \cdot R_b{}^k d\Omega\right) \qquad [1.8]$$

We note that the repeated index *a* goes from 1 to 9, and the second surface integral is for the six bounding surfaces of a lamina, and in it, $R_b{}^k$ (b = 10, 11, …, 27) are residuals for the boundary conditions of the kth layer. In equation [1.8], Ω denotes either the in-plane domain of the plate or one of its edge surfaces, h^k is the thickness of the kth layer. In evaluating integrals with respect to x_3, elasticities of the individual layer are considered.

Each element of the nine-dimensional vector $\mathbf{s}^k$ is expressed as the product of complete Lagrange polynomials of degrees N_1, N_2 and N_3 in x_1, x_2 and x_3 defined as follows:

$$s_a = \sum_{i=1}^{N_1} \sum_{j=1}^{N_2} \sum_{k=1}^{N_3} S_{aijk} \psi_i(x_1) \psi_j(x_2) \psi_k(x_3) \qquad [1.9]$$

$$\psi_i(\xi) = \frac{(\xi-1)(\xi+1)L_P{}'(\xi)}{p(p+1)L_p(\xi_i)(\xi-\xi_i)} \qquad [1.10]$$

Note that equation [1.9] has 9 x N_1 x N_2 x N_3 unknowns, S_{aijk}, for each layer. In equation [1.10], the basis function ψ_i is written in natural coordinates, ξ, and in terms of the Pth-order Lagrange polynomial $L_P(\xi)$ and its derivative, indicated by the prime symbol. The quantity ξ_i is a root of the equation $P_n(\xi) = 0$, where P_n is a Legendre polynomial of order *n*. Basis functions given in [1.10] are associated with Gauss–Lobatto points. Substitution from equation [1.10] into equation [1.9], the result into equation [1.8], and the numerical evaluation of the integral by using the Gauss–Lobatto quadrature rule of order P_n in the three directions, gives J as a function of S_{aijk}. We deduce the needed linear algebraic equations by setting

$$\frac{\partial J}{\partial S_{aijk}} = 0. \qquad [1.11]$$

We realize that expressions for the residuals have different units. When equation [1.11] is written as **KA** = **F**, it is likely that the use of non-dimensional variables throughout the chapter will improve the condition number of the matrix **K** and reduce error. However, we have not tried this. A feature of the equations derived from [1.11] using basis functions of type [1.10] is that they are insensitive to shear-locking effects, which means that reduced integration is not needed in the thickness direction.

We note that the above formulation holds for a laminate, when the continuity of variables u_1, u_2, u_3, σ_4, σ_5, σ_6, e_1, e_2, e_3 is enforced by adding the appropriate residuals in equation [1.8] or using a layerwise theory. Here, we use a layerwise theory, where the contribution from each layer is included in the summation in equation [1.8] and the continuity of the variables in **s** at each layer interface is ensured.

1.3. Results and discussion

1.3.1. *Verification of the numerical algorithm*

To verify the algorithm and to establish the accuracy of computed results, we study the problem analytically analyzed by Pagano (1969). It involves a four-layered [0/90/90/0] simply supported square laminate of side length a, with the sinusoidal surface traction

$$f^p = [0,0,-q_0 \sin\left(\frac{m\pi x}{a}\right)\sin\left(\frac{n\pi y}{a}\right)] \qquad [1.12]$$

applied only on the top surface. The material of the layers has the following values of the moduli:

$$\frac{E_L}{E_T} = 25, \frac{G_{LT}}{E_T} = 0.5, \frac{G_{TT}}{E_T} = 0.2, \upsilon_{LT} = \upsilon_{TT} = 0.25 \qquad [1.13]$$

Here, E, G and υ denote Young's modulus, shear modulus and Poisson's ratio, respectively, and subscripts L and T indicate directions parallel and transverse to the fiber direction. Following Pagano, we express the results in terms of the non-dimensionalized quantities defined in equation [1.4] and employ (x, y, z) = (x_1, x_2, x_3) as the coordinate axes and (u, v, w) = (u_1, u_2, u_3)

$$\left(\bar{\sigma}_{xx}, \bar{\sigma}_{yy}, \bar{\sigma}_{xy}\right) = \frac{1}{q_0(a/h)^2}\left(\sigma_{xx}, \sigma_{yy}, \sigma_{xy}\right)$$

$$\left(\bar{\sigma}_{xz}, \bar{\sigma}_{yz}\right) = \frac{1}{q_0(a/h)}\left(\sigma_{xz}, \sigma_{yz}\right), \qquad \bar{\sigma}_{zz} = \frac{\sigma_{zz}}{q_0} \qquad [1.14]$$

$$(\bar{u},\bar{v}) = \frac{E_2}{q_0\, h(a/h)^3}(u,v), \qquad \bar{w} = \frac{100E_2}{q_0\, h(a/h)^3}w.$$

In Table 1.1, we compare our results of select quantities with those in Pagano (1969) for the plate aspect ratio a/h = 100, 10, 4 and 2. It is clear that the developed least-squares method algorithm yields highly accurate results for the simply supported laminate.

a/h	-	$\bar{\sigma}_{xx}$ [1]	$\bar{\sigma}_{yy}$ [2]	$\bar{\sigma}_{xy}$ [3]	$\bar{\sigma}_{xz}$ [4]	$\bar{\sigma}_{yz}$ [5]	$\bar{w}$ [6]
2	Pagano	1.38841 -0.91165	0.83508 -0.79465	-0.08630 0.06732	0.15300	0.29458	5.0745
	Present	1.38020 -0.90607	0.83038 -0.79049	-0.08599 0.06711	0.15311	0.29428	5.0643
4	Pagano	0.72026 -0.68434	0.66255 -0.66551	-0.04666 0.04581	0.21933	0.29152	1.93672
	Present	0.72020 -0.68427	0.66246 -0.66541	-0.04665 0.04575	0.21939	0.29154	1.93660
10	Pagano	0.55861 -0.55909	0.40095 -0.40257	-0.02750 0.02764	0.30137	0.19595	0.73698
	Present	0.55862 -0.55910	0.40096 -0.40257	-0.02747 0.02761	0.30140	0.19597	0.73698
100	Pagano	0.53885 -0.53887	0.27101 -0.27103	-0.02135 0.02136	0.33880	0.13894	0.43460
	Present	0.53883 -0.53885	0.27100 -0.27102	0.021353 -0.021355	0.33880	0.13894	0.43460

Table 1.1. *Comparison of the results with the 3D exact solution of Pagano for the [0/90/90/0] laminate*

1 – (a/2, a/2, ±h/2) 2 – (a/2, a/2, ±h/4) 3 – (0, 0, ±h/2) 4 – (0, a/2, 0)

5 – (a/2, 0, 0) 6 – (a/2, a/2, 0)

For a/h = 100, the maximum error in the computed quantities equals 0.023% for the in-plane shear stress at point $(0,0,-h/2)$, and for a/h = 2, the maximum error is 0.612% for the in-plane axial stress at point $(a/2,a/2,-h/2)$. The errors for a/h = 10 and 4 are between those for a/h = 100 and 2. The through-the-thickness plots of $\bar{u}(0,a/2,z)$, $\bar{\sigma}_{xx}(a/2,a/2,z)$, $\bar{\sigma}_{xz}(0,a/2,z)$, $\bar{\sigma}_{yz}(a/2,0,z)$ and $\bar{\sigma}_{zz}(a/2,a/2,z)$ coincide well with those from Pagano's solution and are omitted.

For a/h = 10 and 2, we conducted the following five numerical experiments, E_1, E_2, ..., E_5, by varying N_1, N_2 and N_3. In E_1 and E_2, N_3 was set equal to either 3 or 4 and $N_1 = N_2$ was assigned values 4, 6 and 8; in E_3 and E_4, $N_1 = N_2 = 4$ and N_3 was given values 3, 4, 6 and 8; and in E_5, $N_1 = N_2 = N_3$ were assigned values 4, 6 and 8. Thus, a total of 13 combinations of N_1, N_2 and N_3 were evaluated for each value of a/h, and values of u, v, w, σ_{xx}, σ_{yy}, σ_{xy} and σ_{xz} at typical points were compared with their analytical values. For a/h = 10, $N_1 = N_2 = 8$, $N_3 = 3$ (total degrees of freedom (DOFs) = 9,477) provided an error of less than 0.1% in the value of each of these seven variables. Keeping $N_1 = N_2 = 8$ and increasing $N_3 = 4$, 5 marginally decreased the error, suggesting that there is no benefit derived from the increased computation cost. For a/h = 4, all 13 combinations of values of N_1, N_2 and N_3 resulted in an error of 7.25% in w and the error in σ_{xz} decreased from 37.7% for $N_1=N_2=N_3 = 4$ to 0.03% for $N_1 = N_2 = N_3 = 8$ (DOFs = 24,057).

1.3.2. *Simply supported sandwich plate*

We now study deformations of a sandwich plate with the top and bottom facesheets made of an orthotropic Aragonite crystal and the core made of a softer material. Material properties of the facesheets and the core are the same as those in Srinivas and Rao (1970). The top surface of the plate is loaded by sinusoidal tractions given by equation [1.12] and the bottom surface is traction-free. Since the load on the top surface given by equation [1.12] is different from that in Srinivas and Rao (1970), we first solved the problem analytically using their methodology, then verified our software by ensuring that values of displacements and stresses at several points agreed with those reported in Srinivas and Rao (1970). The thickness of each facesheet equals 0.1h and that of the core 0.8h, where h equals the sandwich plate thickness.

Defining $\beta = E_{x1}/E_{x2}$, where E_{x1} and E_{x2} are, respectively, elastic moduli of the facesheets and the core in the x direction, we compare the results from the least-squares method and the analytical solution for β = 1, 10, 100, 1000, 10,000, and aspect ratio a/h = 10, 5 and 2 in Tables 1.2–1.4. Quantities compared are the centroidal deflection, stresses σ_{xx}, σ_{yy} and σ_{xz}, normalized as wE_{x2}/q_0, σ_{xx}/q_0, σ_{yy}/q_0 and σ_{xz}/q_0. Note that $z = 0.4h^+$ and $0.4h^-$, respectively, represent interfaces between the top facesheet and the core. For the three aspect ratios and for all values of β except β = 10,000, the least-squares methodology gives results that differ from their analytical values by less than 0.4%. Even for β = 10,000, the maximum error in all quantities is less than 1.3%, except for that in σ_{xz}. Thus, we should not use the least-squares method for β = 10,000. We have not investigated if increasing values of N_1, N_2 and N_3 in equation [1.9] will reduce this error.

For a/h = 2 and β = 1000, the error in the 15 variables listed in Table 1.4 was less than 1% for $N_1 = N_2 = 6$ and $N_3 = 4$ (DOFs = 7,497) and less than 0.6% for $N_1 = N_2 = 8$ and $N_3 = 4$ (DOFs = 12,393). Thus, the present method provides accurate values of the variables at critical points for a rather modest computational cost.

1.3.3. *Laminate with arbitrary boundary conditions*

We now present results for a [0/90/0] laminate simply supported on edges y = 0, b, with the other two edges either clamped or traction-free, and loaded on the top surface by the surface tractions listed in equation [1.12]. For the plate with two edges free, the present results have been computed with $N_1 = N_2 = 11$ and $N_3 = 5$ in equation [1.9] for a total of 14,157 DOFs. These values could not be increased due to memory limitation of the Dell laptop used for the computational work. We are now converting the code from MATLAB to C++, which will enable us to use larger values of N_1, N_2 and N_3. In Table 1.5, the numerical results for different quantities for the laminates with a/h = 5 and 10 are compared with their values reported by Vel and Batra (1999), who used the Eshelby–Stroh formalism. The maximum difference of 4.2% in the two sets of values of the six quantities suggests that the least-squares method provides a reasonably accurate solution for this problem.

	β→	1	10	100	1000	10,000
$\bar{w}$	(0.5, 0.5, -0.5h)	-436.824	-102.422	-32.9858	-21.6946	-10.5531
		(-436.820)	(-102.422)	(-32.9858)	(-21.6942)	(-10.5452)
$\bar{\sigma}_{xx}$	(0.5, 0.5, z)					
	z = 0.5h	-24.6675	-45.1109	-52.1977	-155.576	-694.701
		(-24.6687)	(-45.1110)	(-52.1980)	(-155.574)	(-694.429)
	z = 0.4h+	-19.4585	-32.7715	-11.4111	116.821	676.521
		(-19.4537)	(-32.7713)	(-11.4113)	(116.819)	(676.846)
	z = 0.4h-	-19.4585	-3.29448	-0.13298	0.09864	0.05351
		(-19.4537)	(-3.29446)	(-0.13300)	(0.09864)	(0.05349)
	z = -0.4h+	19.3977	3.26259	0.11506	-0.11458	-0.06898
		(19.3929)	(3.26259)	(0.11507)	(-0.11457)	(-0.06911)
	z = -0.4h-	19.3977	32.6356	11.6322	-112.483	-631.675
		(19.3929)	(32.6355)	(11.6324)	(-112.471)	(-632.023)
	z = -0.5h	24.5962	44.8809	51.4978	150.743	649.367
		(24.5973)	(44.8810)	(51.4980)	(150.733)	(650.206)

	β→	1	10	100	1000	10,000
$\bar{\sigma}_{yy}$	(0.5, 0.5, z)					
	z = 0.5h	-15.6259	-30.7654	-48.9480	-123.891	-442.713
		(-15.6264)	(-30.7654)	(-48.9478)	(-123.889)	(-442.514)
	z = 0.4h+	-12.4002	-23.1401	-23.8576	43.4841	399.777
		(-12.3978)	(-23.1399)	(-23.8573)	(43.4832)	(399.971)
	z = 0.4h-	-12.4002	-2.47206	-0.41064	-0.12230	-0.08903
		(-12.3978)	(-2.47205)	(-0.41064)	(-0.12230)	(-0.08946)
	z = -0.4h+	12.3427	2.33393	0.23404	-0.05935	-0.09017
		(12.3404)	(2.33392)	(0.23404)	(-0.05934)	(-0.09129)
	z = -0.4h-	12.3427	23.4274	24.5512	-40.2356	-371.648
		(12.3404)	(23.4273)	(24.5509)	(-40.2337)	(-374.066)
	z = -0.5h	15.5620	30.9947	49.0753	121.504	415.435
		(15.5625)	(30.9947)	(49.0751)	(121.502)	(417.599)
$\bar{\sigma}_{xz}$	(x=0, y=0.5)					
	z = 0.4h+	0.99214	1.78299	1.64940	1.24653	0.61880
		(0.99210)	(1.78301)	(1.64946)	(1.24664)	(0.59365)
	z = -0.4h+	0.99152	1.78239	1.64878	1.24591	0.61820
		(0.99151)	(1.78241)	(1.64885)	(1.24609)	(0.72700)

Table 1.2. *Normalized results for the sandwich plate of the aspect ratio a/h = 10. Values in the upper (lower) row are from the analytical (least-squares) solution*

	β→	1	10	100	1000	10,000
$\bar{W}$	(0.5, 0.5, -0.5h)	-32.8669	-11.0284	-5.97592	-3.72388	-0.75513
		(-32.8636)	(-11.0283)	(-5.97592)	(-3.72393)	(-0.75586)
$\bar{\sigma}_{xx}$	(0.5, 0.5, z)					
	z = 0.5h	-6.24479	-11.5195	-22.3767	-111.720	-325.689
		(-6.24849)	(-11.5194)	(-22.3764)	(-111.718)	(-325.756)
	$z = 0.4h^+$	-4.71502	-5.97928	10.0974	104.342	323.587
		(-4.69630)	(-5.97819)	(10.0971)	(104.339)	(323.515)

	β→	1	10	100	1000	10,000
	z = 0.4h-	-4.71502	-0.61521	0.08244	0.08837	0.02295
		(-4.69630)	(-0.61510)	(0.08244)	(0.08836)	(0.02297)
	z = -0.4h$^+$	4.63045	0.58473	-0.08814	-0.09061	-0.02446
		(4.61255)	(0.58471)	(-0.08813)	(-0.09061)	(-0.02461)
	z = -0.4h-	4.63045	5.85733	-8.65984	-86.8644	-182.172
		(4.61255)	(5.85718)	(-8.65970)	(-86.8631)	(-183.119)
	z = -0.5h	6.11899	11.0239	20.1365	93.4349	183.620
		(6.12250)	(11.0236)	(20.1366)	(93.4337)	(184.363)
$\bar{\sigma}_{yy}$	(0.5, 0.5, z)					
	z = 0.5h	-4.28086	-9.36617	-20.0972	-74.2391	-201.621
		(-4.28249)	(-9.36611)	(-20.0970)	(-74.2373)	(-201.636)
	z = 0.4h$^+$	-3.31557	-5.92194	-0.10678	58.6449	197.676
		(-3.30643)	(-5.92142)	(-0.10682)	(58.6434)	(197.642)
	z = 0.4h-	-3.31557	-0.74984	-0.17007	-0.08703	-0.06604
		(-3.30643)	(-0.74979)	(-0.17007)	(-0.08704)	(-0.06579)
	z = -0.4h$^+$	3.24276	0.61151	0.00125	-0.08155	-0.06794
		(3.23401)	(0.61150)	(0.00124)	(-0.08154)	(-0.06842)
	z = -0.4h-	3.24276	6.20664	1.52702	-47.3508	-110.275
		(3.23401)	(6.20655)	(1.52704)	(-47.3527)	(-110.197)
	z = -0.5h	4.18246	9.41935	19.2533	63.5372	114.682
		(4.18400)	(9.41922)	(19.2532)	(63.5391)	(114.477)
$\bar{\sigma}_{xz}$	(X=0, Y=0.5)					
	z = 0.4h$^+$	0.49637	0.83806	0.71375	0.47356	0.12573
		(0.49615)	(0.83804)	(0.71377)	(0.47363)	(0.10374)
	z = -0.4h$^+$	0.49166	0.82533	0.69746	0.45722	0.11274
		(0.49144)	(0.82532)	(0.69747)	(0.45722)	(0.12017)

Table 1.3. *Normalized results for the sandwich plate of the aspect ratio a/h = 5. Values in the upper (lower) row are from the analytical (least-squares) solution*

	β→	1	10	100	1000	10,000
$\bar{W}$	(0.5, 0.5, -0.5h)	-1.55045	-0.80673	-0.47167	-0.07767	-0.00174
		(-1.55041)	(-0.80672)	(-0.47167)	(-0.07767)	(-0.00179)
$\bar{\sigma}_{xx}$	(0.5, 0.5, z)					
	z = 0.5h	-1.30381	-3.68270	-17.8693	-63.5162	-92.9483
		(-1.30409)	(-3.68274)	(-17.8692)	(-63.5153)	(-92.9222)
	$z = 0.4h^+$	-0.71151	0.70904	15.5178	62.5785	92.7600
		(-0.70956)	(0.70917)	(15.5175)	(62.5774)	(92.7827)
	z = 0.4h-	-0.71151	0.05412	0.13952	0.05564	0.00816
		(-0.70956)	(0.05413)	(0.13952)	(0.05564)	(0.00814)
	$z = -0.4h^+$	0.53608	-0.03570	-0.06779	-0.01411	-0.00082
		(0.53463)	(-0.03569)	(-0.06780)	(-0.01411)	(-0.00045)
	z = -0.4h-	0.53608	-0.34646	-6.56097	-11.5108	-2.59311
		(0.53463)	(-0.34641)	(-6.56090)	(-11.5113)	(-1.9654)
	z = -0.5h	0.90503	1.92738	7.39749	11.6037	2.58611
		(0.90518)	(1.92735)	(7.39739)	(11.6042)	(2.41811)
$\bar{\sigma}_{yy}$	(0.5, 0.5, z)					
	z = 0.5h	-1.12800	-3.06623	-11.9076	-39.5773	-57.5749
		(-1.12813)	(-3.06624)	(-11.9075)	(-39.5766)	(-57.5841)
	$z = 0.4h^+$	-0.73700	-0.31380	8.78405	38.4871	57.3827
		(-0.73607)	(-0.31374)	(8.78389)	(38.4863)	(57.3665)
	z = 0.4h-	-0.73700	-0.18445	-0.05495	-0.02475	-0.00442
		(-0.73607)	(-0.18444)	(-0.05496)	(-0.02477)	(-0.00463)
	$z = -0.4h^+$	0.58971	0.06560	-0.04952	-0.03041	-0.00524
		(0.58899)	(0.06560)	(-0.04952)	(-0.03040)	(-0.00247)
	z = -0.4h-	0.58971	0.75187	-2.96035	-6.72640	-1.55007
		(0.58899)	(0.75190)	(-2.96026)	(-6.72638)	(-1.46223)
	z = -0.5h	0.83614	2.17320	5.68229	7.57969	1.65537
		(0.83621)	(2.17319)	(5.68223)	(7.57976)	(1.42636)
$\bar{\sigma}_{xz}$	(X=0, Y=0.5)					
	$z = 0.4h^+$	0.22882	0.36501	0.30164	0.11080	0.01593
		(0.22878)	(0.36502)	(0.30166)	(0.11086)	(0.03117)
	$z = -0.4h^+$	0.17656	0.23854	0.16531	0.03726	0.00302
		(0.17652)	(0.23853)	(0.16530)	(0.03723)	(0.00056)

Table 1.4. *Normalized results for the sandwich plate of the aspect ratio a/h = 2. Values in the upper (lower) row are from the analytical (least-squares) solution*

a/h	Variable	Clamped–Clamped		Free–Free	
		Vel and Batra	Present	Vel and Batra	Present
5	$\bar{W}$ (a/2, b/2, h/2)	1.1800	1.1771	1.5250	1.4649
	$\bar{\sigma}_{xx}$ (a/2, b/2, 0)	-4.2350	-4.2757	-6.9870	-7.2402
	$\bar{\sigma}_{xx}$ (a/2, b/2, h)	4.5040	4.5437	7.1800	7.4300
	$\bar{\sigma}_{yy}$ (a/2, b/2, h/3)	-3.7260	-3.7185	-4.784	-4.582
	$\bar{\sigma}_{yy}$ (a/2, b/2, 2h/3)	3.5760	3.5652	4.639	4.4364
	$\bar{\sigma}_{yz}$ (a/2, 0, h/2)	1.4700	1.4711	1.9110	1.8494
10	$\bar{W}$ (a/2, b/2, h/2)	0.4460	0.4457	0.7530	0.72872
	$\bar{\sigma}_{xx}$ (a/2, b/2, 0)	-3.0000	-2.9746	-5.8980	-5.983
	$\bar{\sigma}_{xx}$ (a/2, b/2, h)	3.0320	3.0066	5.9060	5.9901
	$\bar{\sigma}_{yy}$ (a/2, b/2, h/3)	-1.7130	-1.7113	-2.882	-2.7854
	$\bar{\sigma}_{yy}$ (a/2, b/2, 2h/3)	1.6740	1.672	2.845	2.748
	$\bar{\sigma}_{yz}$ (a/2, 0, h/2)	0.7220	0.7243	1.2280	1.1973

Table 1.5. *Normalized results for the [0/90/0] laminated plate with C–C–SS–SS and F–F–SS–SS boundary conditions*

1.4. Remarks

This work differs from the classical Ritz method in that the basis functions are not required to satisfy the essential boundary conditions. In the Ritz method, we can use a penalty parameter to enforce essential boundary conditions in a weak sense and use the present basis functions. In the traditional finite element method (FEM), we discretize each layer into disjoint domains called finite elements (FEs) and the compact support of the FE basis function for a node equals all FEs sharing that node. We could use the FEM by first taking the inner product of the nine governing equations $R_a = 0$ with a nine-dimensional test function and integrating the result over the laminate. The matrix **K** in **K A** = **F** will not be symmetric. We can improve on the accuracy of the numerical solution by either increasing the order of polynomials in the basis functions or reducing the element size or both. Even though we have not tried it, in general, it takes more computational resources than those needed for the present least-squares approach. Of course, only using one FE with the current basis functions is possible. Then, the difference will be in deriving matrices **K** and **F** and satisfying boundary conditions.

1.5. Conclusion

We have numerically solved three-dimensional linear elasticity equations by taking the three transverse stresses, three strain–displacement relations in the xy plane and three displacement components as independent variables and using the least-squares method to minimize the residuals in the expressions for these nine variables. For a simply supported rectangular plate, using complete polynomials of degree 4 in the z direction, and at most, 10 in each of the x and y directions provides, with very few degrees of freedom, an accurate solution relative to the exact solutions with error less than 0.3% in each of the nine variables, even for a very thick plate of aspect ratio 2. However, when edges are either clamped or free, a higher degree polynomial is needed in each direction to achieve the same accuracy. For a sandwich plate with ratio β of the axial modulus of the facesheet to that of the core up to 1000 and aspect ratios varying from 2 to 100, accurate results are obtained for each one of the nine variables. For $\beta = 10{,}000$, the accuracy considerably deteriorates with errors approaching 100%.

All of these computations have been done using a Dell laptop. Limitations of the work include satisfying governing equations and boundary conditions in the least-squares sense rather than pointwise.

1.6. Acknowledgments

The authors gratefully acknowledge the partial support of this work by the Office of Naval Research through grants N00014-18-1-2548 and N00014-20-1-2876 to Virginia Polytechnic Institute and State University, with Dr. Y. D. S. Rajapakse as the Program Manager. The views expressed in this chapter are those of the authors and neither of the funding agency nor of their Institution.

1.7. References

Bahar, L.Y. (1975). A state space approach to elasticity. *Journal of the Franklin Institute*, 299, 33–41.

Elishakoff, I. (2019). *Handbook on Timoshenko–Ehrenfest Beam and Uflyand–Mindlin Plate Theories*. World Scientific, Singapore.

Goldberg, J.E. and Bogdanoff, J.L. (1957). Midwest applied science corporation (MASC). Report, No. 57–7.

Horgan, C.O. and Baxter, S.C. (1998). Saint-Venant's principle for sandwich structures. *Mechanics of Sandwich Structures*, 113–122.

Hyer, M.W. (2009). *Stress Analysis of Fiber-Reinforced Composite Materials*. DEStech Publications, Inc., Lancaster, PA.

Jones, R.M. (1998). *Mechanics of Composite Materials*. CRC Press, Boca Raton.

Kant, T. and Ramesh, C.K. (1981). Numerical integration of linear boundary value problems in solid mechanics by segmentation method. *International Journal for Numerical Methods in Engineering*, 17, 1233–1256.

Moleiro, F., Mota Soares, C.M., Mota Soares, C.A., Reddy, J.N. (2011). A layerwise mixed least-squares finite element model for static analysis of multilayered composite plates. *Computers & Structures*, 89, 1730–1742.

Pagano, N.J. (1969). Exact solutions for composite laminates in cylindrical bending. *Journal of Composite Materials*, 3, 398–411.

Srinivas, S. and Rao, A.K. (1970). Bending, vibration and buckling of simply supported thick orthotropic rectangular plates and laminates. *International Journal of Solids Structures*, 6, 1463–1481.

Toupin, R.A. (1965). Saint-Venant's principle. *Archive for Rational Mechanics and Analysis*, 18, 83–96.

Vel, S.S. and Batra, R.C. (1999). Analytical solutions for rectangular thick laminated plates subjected to arbitrary boundary conditions. *AIAA Journal*, 37, 1464–1473.

Vel, S.S. and Batra, R.C. (2000). The generalized plane strain deformations of thick anisotropic composite laminated plates. *International Journal of Solids Structures*, 37, 715–733.

Vlasov, B.F. (1957). On one case of bending of rectangular thick plates. In *Vestnik Moskovskogo Universiteta. Serieila Matematiki, Mekhaniki, Astronomii, Fiziki, Khimii*, 2(2), 25–34.

2

Stability of Laterally Compressed Elastic Chains

2.1. Introduction

Nominally identical elastic components are frequently arranged periodically in many natural and artificial structures. Examples include multilayer laminated composites (Dodwell and Hunt 2014), lattice structures with regularly placed struts (Cuan-Urquizo and Bhaskar 2018), structures with orderly placed holes and cut-outs in cold-formed rods, beams and plates (Moen and Schafer 2009a) and geological strata of rock and sedimentary soil (Hunt *et al.* 2006; Dodwell *et al.* 2012). While such structures often operate under mechanical compression and may be prone to buckling instability, most studies so far are confined to the apparent stiffness of periodic structures or wave propagation in such systems. This makes the prediction of critical buckling loads and associated mode shape of such periodic structures an important engineering task, which is what we will undertake here.

Perfect spatial periodicity may not be present in many practical structures (natural or artificial). This necessitates the inclusion of *spatial disorder* within the analysis of structural instability. A disorder is any geometrical or physical irregularity that appears in the otherwise identical structural members. Such irregularities may include differences in geometric dimensions of the structural members, or in the material properties, such as elastic modulus, shear modulus, Poisson ratio, etc. Elastically coupled simple structural chains that are periodic in the direction of compression, such as axially compressed multi-span beams, have been studied before (Li *et al.* 1995; Xie 1995). The periodicity there is along the direction of loading. By contrast, here we consider a mechanical problem where nominally identical elastic structures

Chapter written by Andrii IAKOVLIEV, Srinandan DASMAHAPATRA and Atul BHASKAR.

may be coupled across their length, whereas loading is still applied longitudinally. A simple experiment may consist of stacks of sheets that are held together laterally while being compressed longitudinally. While such a simple experiment does not necessarily bring out the behavior of such elastically coupled structures quantitatively, it certainly is illustrative of the qualitative behavior. The essential feature of such elastically connected structures is periodicity *across* the bays and not along them.

Here, we consider a simple chain model as a toy problem, where identical springs couple rigid rods characterized by small random irregularities, represented by small differences in length between the rods. We show that randomness localizes the buckling amplitude, which decays exponentially with distance from the impurity. We associate the localization with the ratio of the strength of the elastic coupling between rods to the magnitude of the variations in rod lengths or a disorder. We also show that the mathematical structure of the localization in the buckling amplitude in the chain and the multilayer structure is similar to wave localization, which was pioneered in the transport properties of conduction electrons in the presence of random impurities.

There exists an extensive literature on wave propagation in periodic elastic media. The solutions to this class of problems are well known and are given by the so-called Floquet–Bloch theorem (Floquet 1883; Bloch 1929). The solutions represent physically extended propagation modes. If disorders exist in periodic systems, modes begin to localize spatially. The spatial localization of the modes is characterized by exponential decay – the rate of decay depends on the ratio of coupling to disorder. While this story is fairly well established in the field of wave propagation, an analogous phenomenon in the context of the stability of periodic elastic structures is less familiar. Moreover, within the few works concerning periodic elastic structures buckling, nearly all previous studies are about periodicity along the loading direction, such as axially compressed multi-span structures (Pierre and Plaut 1989; Li *et al.* 1995; Xie 1995; Xie and Ibrahim 2000). The simple chain considered here as an expository device has the features of being repetitive across bays, while the structure is compressed along the bays.

A substantial body of literature exists on wave propagation in periodic media, such as electromagnetic waves in crystals (Brillouin 2003), waves in periodic elastic structures (Phani *et al.* 2006), the vibration of multi-span beams and chains of oscillators (Hodges and Woodhouse 1983; Castanier and Pierre 1995). In contrast to this, the static instability of structural assemblies of periodically coupled members has been, relatively, less studied. This chapter makes a contribution to this area and reports research into buckling behavior that parallels phenomena frequently encountered in vibration studies (Pierre and Plaut 1989; Elishakoff *et al.* 1995, 1997; Li *et al.* 1995; Xie 1995; Xie and Elishakoff 2000). Analysis of the available literature (Chai *et al.* 1981; Cho and Kim 2001; Rodman *et al.* 2008; Juhász and Szekrényes 2017) also suggests that studies focused on the buckling of layered structures such as laminated composites rarely make a connection with the studies of buckling of periodic structures carried out by researchers from the wave propagation field.

It turns out that there are close mathematical and physical analogies between the instability problem of coupled repetitive structures and those of the propagation of waves in periodic media. Here, we expand on these similarities by drawing upon the loss of wave propagation in lattices induced by random impurities (Anderson 1958) to investigate analogous phenomena in static instability problems. Indeed, we have identified only a few studies of the instability behavior of periodic structures. Pierre and Plaut (1989) studied the effect of irregularity on the instability of a two-span beam, while Xie (1995) studied multi-span disordered continuous beams using Lyapunov exponents as a measure of the localization of the buckling mode. Similarly, Li *et al.* (1995) studied buckling of *axially loaded* multi-span beams with the disorder in a single span. The authors find that the buckling amplitude exhibits weak or strong localization in the vicinity of a disordered segment depending on the strength of coupling between the spans, the magnitude of the disorder and the location of the disordered span. Elishakoff *et al.* (1995), Xie (1997), Li *et al.* (2005) and others studied the instability behavior of nominally periodic plates in the presence of disorder using a model of a long plate with the regularly positioned stiffeners *along its length*. The plate is typically compressed by loading P distributed along the shorter edges that are also simply supported, while the longer edges were considered free. Elishakoff *et al.* (1995) showed that, as the number of stiffeners increases, i.e. as the plate approaches perfect periodicity, it becomes more sensitive to disorder. Furthermore, as stiffeners naturally provide coupling between the neighboring spans of the plate, the critical buckling load and mode shape were shown to be sensitive to a single universal factor, namely, the ratio of the coupling strength to the degree of disorder. A similar plate but with many stiffeners was studied by Xie (1998) where the calculations were simplified by the introduction of a transfer matrix method. Transfer matrices simplify calculations for a large number of periodic structural arrays, and suitable approximations greatly reduce the number of degrees of freedom needed for computation. If the structure contains a large number of spans that are randomly disordered according to a known distribution, we may use Furstenberg's theorem (Furstenberg 1963) in conjunction with the transfer matrix method to estimate the localization factors. Xie (1998) used the numerical approach based on the transfer matrix method to calculate the localization factors for the long but finite plate with $N = 5 \times 10^5$ stiffeners. Xie and Ibrahim (2000) also employed the transfer matrix method to rib-stiffened plates. However, they formulate the transfer matrices using a finite strip method. Xie and Elishakoff (2000) also studied a similar problem of the stiffened plate using the Kantorovich approach.

Nearly all these studies are limited to multi-span structures compressed along their length, the same as the direction of periodicity. By contrast, here we are motivated by the problem of axially loaded *layered structures*. We consider them as structures with periodicity *across bays*, not along them. A close analogy of the class of problem considered here is that of vibration problems of bladed disks coupled by a hub – these blades are like bays, or layers, that can be considered as forming a chain, even though the physical context of such a problem is structural dynamics rather than elastic instability. Moreover, the problems studied in the current work involve relatively simple analytical models. This is to complement and provide insight into studies of industrially relevant structures that involve either extensive computational resources, such as the finite element method (Rikards *et al.* 2001; Moen and Schafer 2009b; Nguyen-Thoi *et al.* 2013), as well as mesh-free and semi-analytical methods (Byklum *et al.* 2004; Peng *et al.* 2006), or a direct experimental characterization (Cheng *et al.* 2013). While the experimental and large-scale numerical studies provide a valuable estimation of the stability of these periodic structures, it is difficult to extract meaningful interpretations on the role structural parameters play. Our study is inspired by the need to gain a simple analytic understanding of the generic instability behavior of this class of problems. In the next section, we describe an elementary experiment of layers of elastic sheets that are compressed while the small lateral coupling is provided by grips.

2.2. Compression of stacked elastic sheets

We examine a structure consisting of several layers of nearly identical elastic sheets coupled at the ends experimentally. These sheets are stacked on top of each other and compressed while being held together. A schematic diagram of the structure and the applied loading is shown in Figure 2.1. We believe that the insights gained from this model will inform the stability of layered composites, strata and similar structures. The samples were made of cardboard sheets, which are much stiffer than paper. A single sheet of the size considered does not bend under its own weight. Such sheets are industrially produced, widely available and have excellent dimensional consistency. It is also easy to cut them into the desired shape, and they require only a few Newtons of loading to trigger buckling. Therefore, complex procedures of milling and then testing using custom-made grips on bulky experimental equipment designed for testing large and stiff samples are not needed. The coupling between the sheets can be simply achieved by gluing their ends to each other. What we seek to achieve with such structures is not their complete experimental characterization but to gain a qualitative idea of the buckling modes involved.

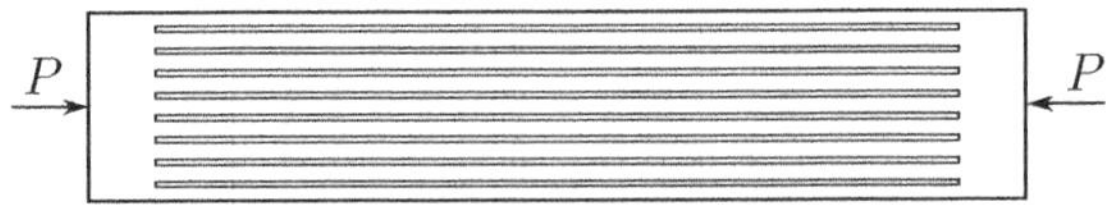

Figure 2.1. *A schematic diagram of a structure consisting of identical layers stacked on top of each other and coupled at the ends. If the number of layers is large, the structure may be considered to approach periodicity in the stacking direction, which is perpendicular to the axis of loading*

We tested several such samples resembling the layered geometry shown in Figure 2.1. The parts of the sheets that were either not glued or attached to the ends were uncoupled and allowed to deform freely. The gaps between the layers were introduced by inserting short strips of cardboard between the layers in the gluing zone. A schematic view of a single layer in the stack with indicated gripping, gluing and glue-free areas is shown in Figure 2.2. The details of the sample geometry and properties are given in Table 2.1. As a consequence of the sample design, the gaps between layers are very small, compared to their thickness, which constrains the deflection of the layers to be directed away from each other, producing a net repulsive effect.

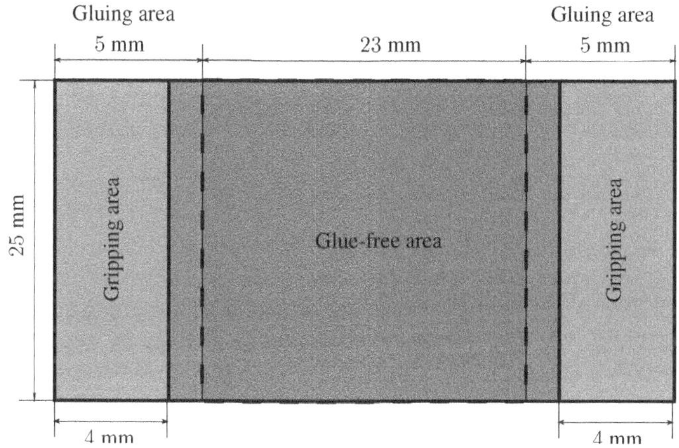

Figure 2.2. *The schematic view of the sample showing gripping area (purple), gluing area (dark purple) and glue-free area (blue). For a color version of this figure, see www.iste.co.uk/challamel/mechanics1.zip*

Parameter	Value
Number of sheets	8
Dimensions of each sheet	25 mm × 25 mm
Modulus of elasticity	2×10^7 MPa (Ayrilmis *et al.* 2008)
Poisson ratio	0.3
Grip size	25 mm × 4 mm

Table 2.1. *Geometric and material properties of the cardboard samples*

The compression tests were performed using a *DEBEN MICROTEST* tensile testing machine. The benefit of this type of machine is its extremely sensitive load cell which allows the testing of materials and structures with little stiffness. The maximum distance between the grips of the machine is 25 mm. The load cell permits a small load resolution of 0.01 N. The samples consisted of eight 25 mm × 23 mm layers of cardboard sheets ≈1 mm thick glued along the strips ≈1 mm wide. The ends were tightly gripped by the two grips of the machine and compressed at a rate of 0.1mm/min to ensure a quasi-static loading process. The testing was carried out in a displacement-controlled mode of the machine and the load–deflection curves were recorded.

(a) (b)

Figure 2.3. *Photographs of the experimentally observed modes. (a) The top and the bottom layers exhibit two opposite deflections, while the middle layers buckle cooperatively in one direction. (b) Rapid decay of the buckling amplitude is observed along the stacking direction. For a color version of this figure, see www.iste.co.uk/challamel/mechanics1.zip*

The buckling mode shapes observed during the experiments are shown in Figure 2.3(a) and 2.3(b). Various modes were observed in the experiment, as the structure turned out to be very sensitive to any small irregularities in the layers and the coupling between them. However, a generic trend appears in most of the test cases. Either the top or the bottom layer exhibits a relatively large buckling amplitude which then reduces quickly along the stacking direction. Often, the layers in the middle would not exhibit any notable deformation, while the layers at the boundaries appear in the post-buckling regime. Even far in the post-buckling regime (as shown in Figure 2.3(a)), the layers in the central part exhibit relatively small amplitude, compared to the boundaries. This behavior resembles the classical wave localization picture (Hodges and Woodhouse 1983; Xie and Wang 1997a) but for a static problem of elastic stability.

It should be noted that stacks of elastic sheets held together pose a kinematic constraint: that sheets cannot pierce each other. Unlike a chain of pendula in oscillatory motion, laterally buckled elastic sheets that are held together at the ends do not allow individual sheets to penetrate each other. This inevitably would bias observed modes to be in favor of those with the highest amplitude at the edges, since the ends are free to move outwards. In a disordered pendulum, or in a model of stacked sheets that mathematically allow sheets to penetrate each other, other possibilities exist that cannot be practically realized.

Motivated by the lateral instability of a stack of nominally identical elastic sheets and similarly layered materials and structures, we propose a toy problem consisting of elastically connected rigid rods that destabilize laterally beyond a critical value when compressed along the length of the rods. The rest of the chapter is concerned with this specific problem mathematically, as it provides analytical opportunities to study the localization behavior of instability modes. The simple experiment presented here is limited only as an illustration of the class of structural problems in layered media that may show similar structural behavior.

2.3. Stability of an elastically coupled cyclic chain

Consider a simple chain of rigid rods coupled elastically by light springs. Although the configuration of coupled rods does not capture the complex elasticity of the layered structure, it allows analytical investigation of the effect of a regular placement of nominally identical members on the character of buckling. The effects of the disorder are treated analytically in the following sections.

A chain of masses coupled with springs and coupled pendula have been traditionally used to study *wave propagation* (Brillouin 2003) and *vibration of periodic structures* in one dimension (Rosenstock and McGill 1968; Hodges and Woodhouse 1983; Castanier and Pierre 1995). Here, we propose a buckling analogy of such chains as a chain of elastically coupled rigid rods, each with one degree of freedom, as shown schematically in Figure 2.4.

Each rigid rod of length L is pinned at the bottom using a hinge with a torsional spring of stiffness k_t. Each rod is subjected to a compressive force P at the top acting along the length of the rods. The rods are coupled by light springs k ($k \ll k_t$), and the rotation of each rod from its equilibrium position is given by angle θ with a subscript denoting the bay number.

Consider, first, the stability of an isolated rod on bearing support. The hinge allows free rotation which is resisted by a torsional spring. As the compressive loading increases from zero, the rod remains stable, while the moments about the hinge support are in equilibrium $PL\sin\theta_i = k_t\theta_i$. In the small-angle approximation $\sin\theta_i \approx \theta_i$, we see that there is a critical value $P_{\text{cr}} = k_t/L$ such that the structure is in stable equilibrium for $P < P_{\text{cr}}$ and destabilizes when $P = P_{\text{cr}}$. Here, a bifurcation occurs, leading to a left or a right buckling mode characterized by the deflection angle θ_b when the bifurcation point is reached.

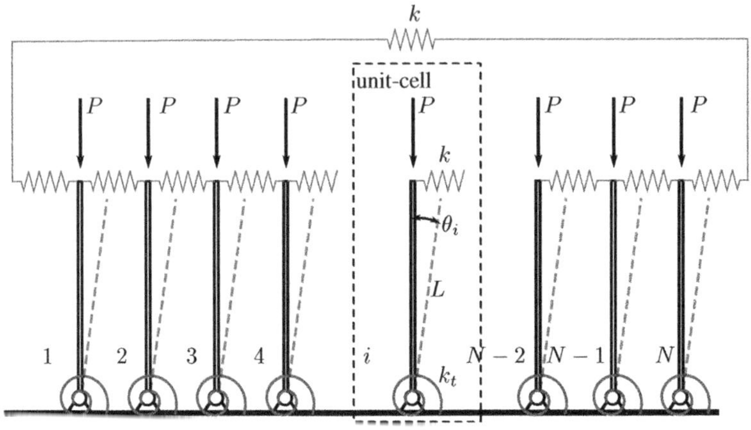

Figure 2.4. *A chain of N identical rigid rods. Each rod has length L and is attached to a hinge bearing and a torsional spring k_t. The rods are coupled with light springs k and each is loaded with a force P. The deflected shape of each rod when buckled is indicated with dashed gray lines and the angle of deflection is labeled by θ. For a color version of this figure, see www.iste.co.uk/challamel/mechanics1.zip*

Now, consider a rigid rod coupled by a light spring k to its neighbors in a cyclic chain. We designate the i-th rod in the chain, as shown within the dashed rectangle in Figure 2.4, as a unit cell within the chain. Applying the moment equilibrium condition for a unit cell leads to a recursive relation between the rotations of the neighboring rods

$$-kL^2\theta_{m-1} + (k_t + 2kL^2)\theta_m - kL^2\theta_{m+1} = PL\theta_m. \qquad [2.1]$$

We divide both sides of equation [2.1] by L and introduce stiffness parameter $\bar{k}_0 = k_t/L + 2kL$ and coupling parameter $\bar{k} = kL$. Then, rearranging terms, we can write

$$-\bar{k}\theta_{m-1} + \bar{k}_0\theta_m - \bar{k}\theta_{m+1} = P\theta_m \qquad [2.2]$$

Now, consider a chain of N rods forming a closed loop. This implies that the rotation of the first rod is equal to that of the last $\theta_1 = \theta_{N+1}$. Equation [2.2], along with the cyclic boundary conditions, can be written in the form of an eigenvalue problem in terms of a circulant matrix

$$\begin{bmatrix} \bar{k}_0 & -\bar{k} & \cdots & -\bar{k} \\ -\bar{k} & \bar{k}_0 & \ddots & \vdots \\ \vdots & \ddots & \ddots & \vdots \\ -\bar{k} & \cdots & -\bar{k} & \bar{k}_0 \end{bmatrix} \begin{Bmatrix} \theta_1 \\ \theta_2 \\ \vdots \\ \theta_N \end{Bmatrix} = P \begin{Bmatrix} \theta_1 \\ \theta_2 \\ \vdots \\ \theta_N \end{Bmatrix}, \qquad [2.3]$$

where the column vector on both sides is the eigenvector and P is the corresponding eigenvalue. More compactly, $\mathbf{C}\boldsymbol{s} = P\boldsymbol{s}$, where C is the circulant matrix of the chain. The components θ_m form a vector of generalized coordinates $\boldsymbol{s}$, and the eigenvalue P is interpreted as the critical buckling load of the chain. The N eigenvectors $\boldsymbol{s}_n$, $n = 1, \ldots, N$ of a circulant matrix are well known (see, for example, Gray (2005) and Olson *et al.* (2014)), and have components $s_{n,m} := (\boldsymbol{s}_n)_m$ that are the N-th roots of unity $(z_N)^N = 1$, so that $s_{n+1,m+1} = z_N^{mn}$:

$$\boldsymbol{s}_n = \frac{1}{\sqrt{N}}\left(1, z_N^{(n-1)}, z_N^{2(n-1)}, \ldots, z_N^{(N-1)(n-1)}\right)^{\mathsf{T}}, z_N := \mathrm{e}^{\frac{2\pi i}{N}}. \quad [2.4]$$

The matrix of eigenvectors of the circulant matrix is a Fourier matrix (Olson *et al.* 2014)

$$\mathbf{F} = \frac{1}{\sqrt{N}}\begin{bmatrix} 1 & 1 & 1 & \cdots & 1 \\ 1 & z_N^1 & z_N^2 & \cdots & z_N^{N-1} \\ 1 & z_N^2 & z_N^4 & \cdots & z_N^{2(N-1)} \\ \vdots & \vdots & \vdots & \ddots & \vdots \\ 1 & z_N^{N-1} & z_N^{2(N-1)} & \cdots & z_N^{(N-1)(N-1)} \end{bmatrix} \quad [2.5]$$

where each entry is the buckling amplitude $s_{n,m} = \theta_m$; here, m indicates the location of the rod and n is an index that labels each eigenvector.

To obtain the eigenvalues, we substitute θ_m back into equation [2.2] and obtain

$$-\bar{k}z_N^{n(m-1)} + \bar{k}_0 z_N^{nm} - \bar{k}z_N^{n(m+1)} = P_n z_N^{nm}.$$

Dividing both sides of the above equation by z_N^{nm} yields

$$-\bar{k}z_N^{-n} + \bar{k}_0 - \bar{k}z_N^{n} = P_n.$$

Further, we scale the critical buckling load (eigenvalue) P_n by the critical load of a single rod $P_{\text{cr}} = k_t/L$ and use $\mathrm{e}^{-i\varphi} + \mathrm{e}^{i\varphi} = 2\cos\varphi$ and $1 - \cos\varphi = 2\sin^2(\varphi/2)$. Thus, the n-th scaled critical buckling load of the chain is

$$\widetilde{P}_n = 1 + 4\widetilde{k}\sin^2\left(\frac{\pi(n-1)}{N}\right) = 1 + 4\widetilde{k}\sin^2(\varphi_n), \quad [2.6]$$

where $\widetilde{k} = kL^2/k_t$ is the ratio of the coupling strength to rod stiffness.

We have introduced $\varphi_n = \pi(n-1)/N$ in equation [2.6], which we call a *general mode number*, analogous to the reciprocal wave numbers (vectors) used in wave propagation literature (Ziman 1972). The periodicity of the eigenvalues $\widetilde{P}_n$ is thus analogous to those in the first Brillouin zone. Each critical buckling load is a function $\widetilde{P}(\phi) = 1 + 4\widetilde{k}\sin^2(\phi)$ defined over the discrete set of N points φ_j, $j = 1, \ldots, N$, so that $\widetilde{P}_n = \widetilde{P}(\varphi_n)$. Since z_N live on a circle, the Brillouin zone is closed under the addition of mode number: for any two modes φ_k, φ_l, $\widetilde{P}(\varphi_k + \varphi_l) = \widetilde{P}(\varphi_{k+l}) = \widetilde{P}_{n+m}$. All the buckling loads that correspond to the first Brillouin zone appear within a single group with the magnitudes ranging between $\widetilde{P}_{min} = 1$ and $\widetilde{P}_{max} = 1 + 4\widetilde{k}$, as shown in Figure 2.5 for $\widetilde{k} = 0.1$. Further, we will refer to such a packed group of buckling loads as an *instability cluster*.

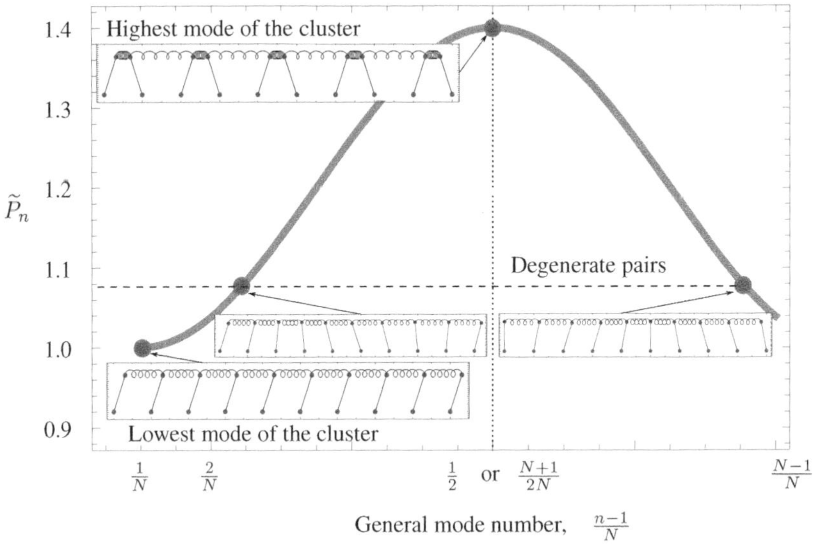

Figure 2.5. *The plot of $\widetilde{P}_{cr}$ versus general mode number $(n-1)/N$ (blue line) for an infinite chain $N \to \infty$. It contains the instability cluster of the arbitrarily long chain. The mode types associated with the chosen general mode numbers are shown with the subplots. For a color version of this figure, see www.iste.co.uk/challamel/mechanics1.zip*

In Figure 2.5, we plot the values of the n-th buckling load of the chain $\widetilde{P}_n$ versus the general mode number $(n-1)/N$, using equation [2.6] for $n = 1, \ldots, N$. The discrete points form a continuous blue line in Figure 2.5 if the matrix size $N \to \infty$ (infinite chain). Indeed, in the limit of an infinite chain, the discrete variable $(n-1)/N$ becomes a continuous one, which is reflected by a continuous blue curve in Figure 2.5. In the same figure, we represent results for a finite chain of size $N = 10$ using red dots on the blue curve. To illustrate the nature of the mode shapes, we present four examples of instability mode shapes for $N = 10$.

In the case of a structure with one degree of freedom, the first instability cluster plotted in Figure 2.5 is the entire spectrum of the eigenvalue problem. The cluster is symmetric with respect to $n = N/2$ for even N N and $n = (N+1)/2$ for odd

N. The exceptions to this rule are the lowest ($n = 1$) and the highest ($n = N/2$ or $n = (N+1)/2$) buckling loads. This implies that the cluster contains pairs of degenerate buckling loads.

Consider the selected mode shapes plotted in Figure 2.5. The smallest buckling load corresponds to the mode with the least energy – all rods deflect in the same direction, and coupling springs do not contribute to the strain energy. This is the buckling mode practically observable in the chain. The highest buckling load of the cluster is associated with the mode in which all the coupling springs contribute equal amounts to the energy. Any modes associated with the degenerate pairs are mere left–right reflections since the chain is invariant under the transformation $(\theta_m) \rightarrow -(\theta_m)$ for all m. The modes suggest that the deflection of the m-th rod in the chain is governed by the spatial modulation of the n-th eigenvector of matrix $\boldsymbol{C}$.

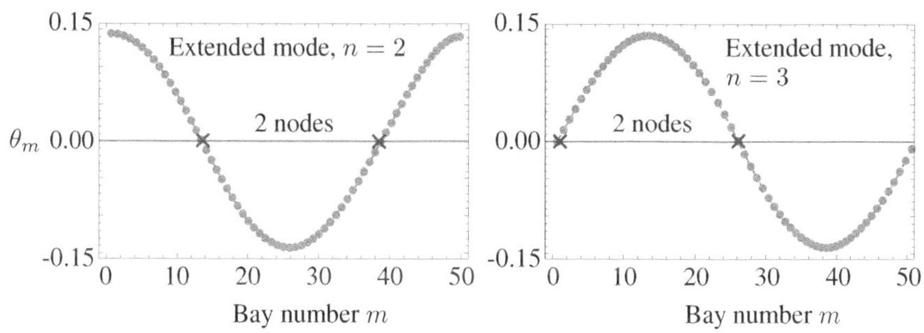

Figure 2.6. *The angular displacement θ_m of the m-th rod plotted versus its number m. Extended buckling mode 2 (left) and extend buckling mode 3 (right) appear as spatially modulated values of θ_m. The spatial coordinate is discrete and corresponds to rod numbers m. For a color version of this figure, see www.iste.co.uk/challamel/mechanics1.zip*

For example, we plot the spatial modes 2 and 3 in Figure 2.6 where each blue circle is the buckling angle of the m-th rod in the chain θ_m, plotted as a function of its position m. Since the chain is discrete, m is a discrete representation of the continuous spatial variable x along the chain. We see from the figure that the rod amplitude is modulated by a periodic function. Thus, we observe here a buckling counterpart of Bloch waves in the one-dimensional periodic chain.

Note the key aspects of the behavior specific to this problem. In contrast to wave propagation and vibration, here we are studying a static problem. Hence, the chain assumes a static configuration such that the buckling amplitudes of the rods are modulated in space due to the periodicity of the structure. Here and later, we call such spatial modulation of buckling amplitudes of discrete members an *extended buckling mode*. This name is chosen by analogy with the wave propagation literature where an extended mode is spatially spread and is associated with the conductivity of matter (Ziman 1972).

Note that only the lowest mode of the chain of rigid rods is achievable, in practice, as the chain fails above the lowest buckling load. To observe the second mode, we need to fix a single rod on the chain to provide an artificial node to the extended buckling mode. Similarly, higher modes are achieved by providing more nodes.

2.4. Elastic stability of two coupled rods with disorder

Until now, we studied a case of perfect periodicity – a cyclic chain of identical rigid rods. However, as any real structure would inevitably possess some degree of disorder, we need to account for it in our modeling.

Coupled pendula serve as excellent models in structural vibration and elastic waves to study the dependence of eigenvalues of system parameters analytically. Here, we analytically consider the stability of a pair of elastically coupled rods under axial loading with the disorder. Such a model allows us to derive the exact dependence of the critical buckling loads on the structural parameters and the irregularity.

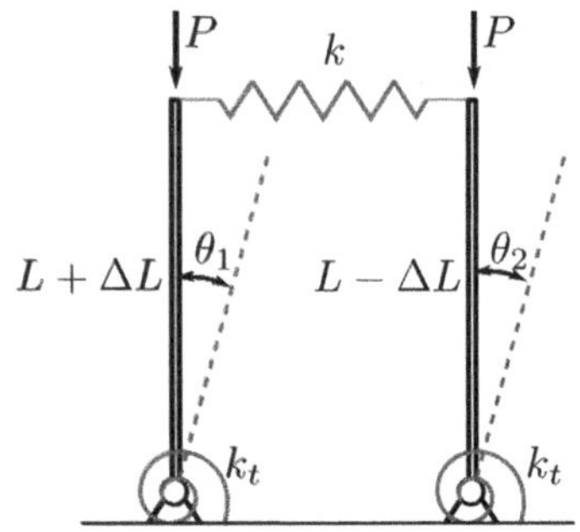

Figure 2.7. *A model of two rigid rods on hinge supports with torsional springs k_t coupled by a weak spring k and loaded with a compressive loading P. The rods have a length imperfection of $\pm\Delta L$. For a color version of this figure, see www.iste.co.uk/challamel/mechanics1.zip*

Here, we start by considering two rigid rods coupled by a spring and assume that they possess a small irregularity. For example, let their lengths be slightly different, as shown in the schematic diagram in Figure 2.7. Moment equilibrium about hinges under the small-angle approximation results in a system of two linear equations

$$\begin{cases} PL_1\theta_1 = k_t\theta_1 + k\left[L_1\theta_1 - L_2\theta_2\right]L_1 \\ PL_2\theta_2 = k_t\theta_2 + k\left[L_2\theta_2 - L_1\theta_1\right]L_2, \end{cases} \qquad [2.7]$$

where $L_1 = L + \Delta L$ and $L_2 = L - \Delta L$. We divide the first equation by $L + \Delta L$ and the second equation by $L - \Delta L$ and write

$$\begin{cases} -P\theta_1 + \dfrac{k_t}{L+\Delta L}\theta_1 + k\left[(L+\Delta L)\theta_1 - (L-\Delta L)\theta_2\right] = 0, \\ -P\theta_2 + \dfrac{k_t}{L-\Delta L}\theta_2 + k\left[(L-\Delta L)\theta_2 - (L+\Delta L)\theta_1\right] = 0, \end{cases} \quad [2.8]$$

which can be arranged formally as the following generalized eigenvalue problem:

$$\mathbf{K}\boldsymbol{\theta} = P\mathbf{K}_g\boldsymbol{\theta}. \quad [2.9]$$

Here, $\mathbf{K}$ is the stiffness matrix, $\mathbf{K}_g$ is the geometric stiffness matrix and $\boldsymbol{\theta} = [\theta_1, \theta_2]$ contain the buckling angle amplitudes of each rod. Note that due to normalization by $L - \Delta L$ and $L + \Delta L$, the geometric stiffness matrix is the identity matrix.

The expressions for the two eigenvalues $P_{1,2}$ are easy to calculate analytically as

$$\begin{aligned} P_{1,2} = &\frac{k_t L^2 + kL^2(L^2 - \Delta L^2)}{k_t(L^2 - \Delta L^2)} \\ &\mp \frac{L\sqrt{k_t^2\Delta L^2 + k(L^2 - \Delta L^2)\left(kL^4 - \Delta L^2\left(kL^2 + 2k_t\right)\right)}}{k_t(L^2 - \Delta L^2)}. \end{aligned} \quad [2.10]$$

Let us introduce a non-dimensional measure of disorder ε as the ratio between the perturbation in rod length ΔL and the nominal rod length L, i.e. $\varepsilon = \Delta L/L$ and the dimensionless coupling $\widetilde{k} = \bar{k}L/k_t = kL^2/k_t$ as a ratio of the coupling stiffness k and the torsional stiffness of each rod k_t. We then define a dimensionless buckling load as $\widetilde{P} = P/P_{cr}$, where $P_{cr} = k_t/L$ is the buckling load of an individual rod. Therefore, the expressions in equation [2.10] can be written in the dimensionless form

$$\widetilde{P}_{1,2} = \widetilde{k} + \frac{1}{1-\varepsilon^2} \mp \sqrt{\widetilde{k}^2 + \frac{\varepsilon^2\left(1 - 2\widetilde{k}(1-\varepsilon^2)\right)}{(1-\varepsilon^2)^2}}. \quad [2.11]$$

Note that $\widetilde{P}_{1,2}$ depend only on the coupling $\widetilde{k}$ and disorder ε. We assume disorder to be small, i.e. $\varepsilon \ll 1$, and use a power series expansion for $\widetilde{P}_{1,2}$

$$\begin{cases} \widetilde{P}_1 = 1 + \left(2 - \frac{1}{2\widetilde{k}}\right)\varepsilon^2 + O\left(\varepsilon^3\right) \\ \widetilde{P}_2 = (1 + 2\widetilde{k}) + \frac{1}{2\widetilde{k}}\varepsilon^2 + O\left(\varepsilon^3\right). \end{cases} \quad [2.12]$$

Equation [2.12] provides a leading order approximation of the exact functional forms in equation [2.11] for small disorder.

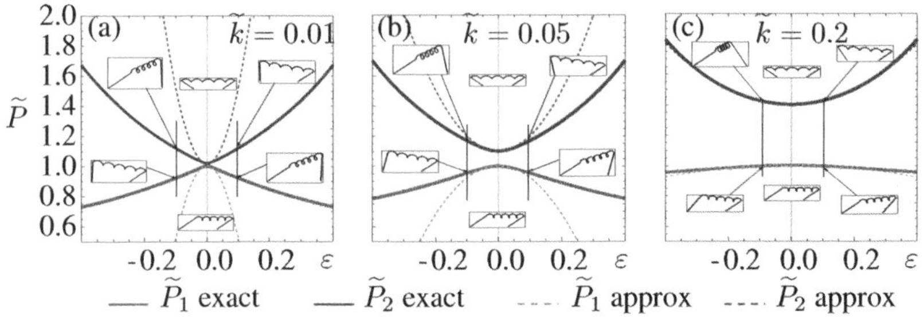

Figure 2.8. *Plots of the first and the second critical buckling loads of two coupled rods as functions of disorder* ε *for non-dimensional coupling (a)* $\widetilde{k} = 0.01$*, (b)* $\widetilde{k} = 0.05$ *and (c)* $\widetilde{k} = 0.2$*. The solid lines represent the exact solution from equation* [2.11]*, while the dashed lines correspond to the second-order series expansion from equation* [2.12]*. For a color version of this figure, see www.iste.co.uk/challamel/mechanics1.zip*

It is apparent that when $\varepsilon = 0$, we have a perfectly ordered structure with critical loads $\widetilde{P}_1 = 1$ and $\widetilde{P}_2 = 1 + 2\widetilde{k}$. These are the same as the lowest and the highest buckling loads of an ordered chain given by equation [2.6]. This is consistent with the Floquet–Bloch theorem as the solutions of the unit cell should specify the fundamental states which are then extended periodically for an entire medium. When ε is close to zero, the critical buckling loads have a functional dependence on disorder, given by equation [2.12]. As the disorder increases further, the approximate solution is expected to diverge from the exact solution in equation [2.11]. The change of the solution with the disorder ε for various values of coupling is demonstrated in Figure 2.8(a) for $\widetilde{k} = 0.01$, Figure 2.8(b) for $\widetilde{k} = 0.05$ and Figure 2.8(c) for $\widetilde{k} = 0.2$. The buckling load trajectories, as functions of the disorder parameter ε, show differences for different coupling strengths. This dependence is either magnified by small coupling or reduced by a strong coupling.

The discussed model is particularly useful to demonstrate how buckling of a structure made of identical members is affected by coupling and by the disorder. We generalize it as follows. The system matrix $\mathbf{A} = \mathbf{K}_g^{-1}\mathbf{K}$ of the two identical uncoupled structural members is

$$\mathbf{A} = \begin{bmatrix} K & 0 \\ 0 & K \end{bmatrix}, \qquad [2.13]$$

where K is some stiffness function of structural parameters. The eigenvalues of $\mathbf{A}$ are identical for all values of structural parameters and constitute the buckling loads of uncoupled members. We now introduce coupling between the members via a coupling matrix

$$\mathbf{A} = \begin{bmatrix} K & 0 \\ 0 & K \end{bmatrix} + \begin{bmatrix} k & -k \\ -k & k \end{bmatrix}. \qquad [2.14]$$

Here, k is the coupling coefficient that may also depend on several structural parameters. Then, the eigenvalues of the system matrix are $\lambda_1 = K$, $\lambda_2 = K + 2k$. When plotted as functions of one or two parameters of interest, the eigenvalues appear as trajectories or surfaces in the parameter space. The eigenvalue trajectories (surfaces) coincide in the absence of coupling. The introduction of coupling then essentially breaks the degeneracy, and the trajectories (surfaces) separate. However, if $k = 0$ in a point (or on a line) in a parameter plane (or space), this causes the trajectories (surfaces) to cross over that point (line).

Finally, let us introduce small irregularities, via a disorder matrix, to the sum of the previous matrix A as

$$\mathbf{A} = \begin{bmatrix} K & 0 \\ 0 & K \end{bmatrix} + \begin{bmatrix} k & -k \\ -k & k \end{bmatrix} + \begin{bmatrix} \epsilon_1 & 0 \\ 0 & \epsilon_2 \end{bmatrix}. \qquad [2.15]$$

In this case, the crossing in the crossing point (crossing line) is removed and replaced by the repulsion of the respective trajectories (surfaces), as can be seen from Figure 2.8. Note that the trajectories as functions of the disorder are studied.

By writing a generic parameter-dependent system matrix as a sum of member stiffnesses, coupling and disorder, we present a comprehensive matrix description of all structures considered in the present study. All specific problems are partial examples of this representation. The matrix $\mathbf{A}$ can be large and can have circulant or tridiagonal structure. The current discussion will still be valid.

2.5. Spatial localization of lateral buckling in a disordered chain of elastically coupled rigid rods

Now, we extend the analysis for two rods to a long chain of coupled rods. For now, consider all rods to be identical (no disorder). We also abandon cyclic periodicity $\theta_1 \neq \theta_{N+1}$. Therefore, the system matrix $\mathbf{C}$ in equation [2.16] is tridiagonal

$$C = \begin{bmatrix} \bar{k}_0 & -\bar{k} & \cdots & 0 \\ -\bar{k} & \bar{k}_0 & \ddots & \vdots \\ \vdots & \ddots & \ddots & \vdots \\ 0 & \cdots & -\bar{k} & \bar{k}_0 \end{bmatrix}. \qquad [2.16]$$

Note that $C_{1,N} = C_{N,1} = 0$, unlike in equation [2.3]. Here, as before, $\bar{k} = kL$ and $\bar{k}_0 = k_t/L + 2kL$, where k is the strength of the coupling spring and k_t is the strength of the torsional spring in each rod. The exact spectrum of the tridiagonal matrix that describes the Coxeter graph $\mathbf{A}_l$ for $l \geq 2$ is known (Goodman *et al.* 2012). It is easy to show that the matrix $\mathbf{C}$ can be transformed into a form consisting of $\mathbf{A}_l$ and the identity matrix. Denoting $\beta = \bar{k}_0/\bar{k}$, we can rewrite the matrix as

$$\mathbf{C} = \beta\mathbf{I} - \bar{k}\mathbf{A}_l, \qquad [2.17]$$

where $\mathbf{I}$ is the identity matrix and $\mathbf{A}_l$ is the adjacency matrix of the A_l-Coxeter graph (a linear chain). The eigenvalues of $\mathbf{C}$ are then β, plus the eigenvalues of $\mathbf{A}_l$ times $\widetilde{k}$. Hence, after rescaling the eigenvalues P by the critical load of a single rod $P/P_{cr} = k_t/L$ and considering that $\widetilde{k} = \bar{k}L/k_t = kL^2/k_t$, the scaled buckling loads of the chain are

$$\widetilde{P}_n = 1 + 2\widetilde{k} - 2\widetilde{k}\cos\varphi_n. \qquad [2.18]$$

Here, we see that the exact calculation of the critical loads of an infinite but ordered chain is still possible.

We now abandon the assumption that all rods are identical and proceed with introducing disorder. Note that there are no known methods to calculate the exact spectrum of matrices, such as the one in equation [2.16] but with random elements. Therefore, iterative methods are often employed. The transfer matrix method is extensively used in the literature on waves in periodic media and has long become a standard tool (Benaroya 1997). It is particularly efficient in treating perfectly ordered structures with small random irregularities (Xie and Wang 1997a, b). Here, we start with the definition of the transfer matrix for all identical rods and then introduce disorder into it. We use equation [2.2] to formulate a transfer matrix $\mathbf{T}_m$, which controls how the state vector associated with the m-th rod $\mathbf{s}_m = \{\theta_m, \theta_{m-1}\}^\top$ is changed to that corresponding to the $m+1$-th rod: $\mathbf{s}_{m+1} = \{\theta_{m+1}, \theta_m\}^\top$

$$\begin{Bmatrix} \theta_{m+1} \\ \theta_m \end{Bmatrix} = \begin{bmatrix} (2 + \frac{1}{\widetilde{k}} - \frac{\widetilde{P}}{\widetilde{k}}) & -1 \\ 1 & 0 \end{bmatrix} \begin{Bmatrix} \theta_m \\ \theta_{m-1} \end{Bmatrix}. \qquad [2.19]$$

Here, $\widetilde{k} = k\frac{L^2}{k_t}$ and $\widetilde{P} = P\frac{L}{k_t}$. In the vector–matrix notation

$$\boldsymbol{s}_{m+1} = \boldsymbol{T}_m \boldsymbol{s}_m, \quad m = 1, \ldots, N, \qquad [2.20]$$

where $\boldsymbol{T}_m$ is the transfer matrix between the state of the unit cell $m+1$ described by the state vector $\boldsymbol{s}_{m+1}$ and that of the unit cell m described by the state vector $\boldsymbol{s}_m$ in a chain with N unit cells (rods). The whole chain or rigid rods may thus be described by a cumulative product of transfer matrices

$$\boldsymbol{s}_N = \boldsymbol{T}_{N-1}\boldsymbol{T}_{N-2}\cdots\boldsymbol{T}_m\cdots\boldsymbol{T}_2\boldsymbol{s}_1 \triangleq \boldsymbol{B}_N\boldsymbol{s}_1. \qquad [2.21]$$

Given a small disorder in each rod, either to the length of each rod L, to the spring coupling k or the torsion coefficient k_t, the transfer matrices $\boldsymbol{T}_m$ become unequal, and the system of N coupled rods is no longer perfectly ordered. In what follows, we will treat the disorder as random.

Anderson (1958) studied the asymptotic behavior of disordered lattices to show that the exponential decay of wave-function amplitude due to disorder leads to the development of an insulating state of a conductor, as all electrons appear in highly localized states. Later, the ergodic multiplicative theorem by Oseledets (1968) and Furstenberg's theorem on non-commuting random products (Furstenberg 1963) were developed to characterize the asymptotic behavior of products of random matrices as Lyapunov exponents. The first adaptations of these two theorems to the cases of vibration of multi-span beams were made by Ariaratnam and Xie (1995) and Castanier and Pierre (1995). At the same time, Xie (1995) introduced the idea of products of random matrices to studying multi-span beams in buckling. Here, we apply the analytical framework proposed by Xie (1995) to a different buckling problem. Note that, in the present study, the direction of periodicity is lateral to that of loading, in contrast to multi-span beams compressed axially.

Furstenberg's theorem allows the asymptotic behavior of the products in equation [2.21] to be calculated, quantifying the average change of norms under successive transformations $\boldsymbol{T}_m : \boldsymbol{s}_m \to \boldsymbol{s}_{m+1}$. The symbol $\|\cdot\|$ below refers to a choice of norms, for both vectors, the L_2-norm and, for matrices, the operator norm, L_∞. For the case at hand, the determinants of all $\boldsymbol{T}_m$ are 1, ensuring that the transformations are volume-preserving. Under these conditions, for $\boldsymbol{T}_m$, $m = 1, \ldots, N$, non-singular, independent, identically distributed $d \times d$ random matrices, such that no two matrices share common eigenvectors, the following two limits exist:

$$\lim_{N\to\infty} \frac{1}{N} \ln \|\boldsymbol{B}_N \boldsymbol{s}_1\| = \lim_{N\to\infty} \frac{1}{N} \ln \|\boldsymbol{T}_{N-1} \cdots \boldsymbol{T}_m \cdots \boldsymbol{T}_2 \boldsymbol{s}_1\| = \lambda \qquad [2.22]$$

and

$$\lim_{N\to\infty} \frac{1}{N} \ln \|\boldsymbol{B}_N\| = \lambda, \text{ where} \|\boldsymbol{B}_N\| = \max_{\boldsymbol{x}} \frac{\|\boldsymbol{B}_N \boldsymbol{x}\|}{\|\boldsymbol{x}\|}. \qquad [2.23]$$

The application of Furstenberg's theorem to the chain of rigid rods described by equation [2.21] leads to

$$\lim_{N\to\infty} \frac{\|\boldsymbol{B}_N \boldsymbol{s}_1\|}{\|\boldsymbol{B}_{N-1} \boldsymbol{s}_1\|} = e^\lambda \Rightarrow \lim_{N\to\infty} \frac{\psi_N + 1}{1 + \frac{1}{\psi_{N-1}}} = e^\lambda \qquad [2.24]$$

where $\psi_j = (\theta_j/\theta_{j-1})^2$. In the $N \to \infty$ limit, $\psi_N = \psi$ independent of N, and the solution to the resulting quadratic equation for ψ yields

$$\lim_{N\to\infty} (\theta_N/\theta_{N-1}) = \exp(\lambda), \sqrt{-1}. \qquad [2.25]$$

It expresses how the angles in the configuration change from rod to rod and characterizes the exponential localization of the angles of rotation. For the entire chain, this relation can be rewritten as

$$|\theta_{N+1}| = e^{\lambda N} |\theta_1|. \qquad [2.26]$$

If $\lambda > 0$, the angles of rotation will grow exponentially along the chain. At the same time, the extended modes of the structure are characterized by the Fourier matrix or complex exponents. If real Lyapunov exponents are non-zero, they govern how quickly buckling amplitudes of each rod deviate from the extended modes. The λ factors are called *localization factors* and are generic descriptors of the behavior in the disordered periodic structure.

To understand the difference between the ordered system and the disordered one, we first calculate Lyapunov exponents for the case of identical rods and springs. For this ordered case ($\boldsymbol{T}_m = \boldsymbol{T}$ for all m), $\|\boldsymbol{B}_N\|$ is the largest singular value of $\boldsymbol{B}_N$ which is the $(N-1)$-th power of the largest singular values of $\boldsymbol{T}$, since $\boldsymbol{T}_{N-1}, \boldsymbol{T}_{N-2}, \cdots, \boldsymbol{T}_m, \cdots, \boldsymbol{T}_2$ are all identical and equation [2.21] is simplified to

$$\begin{Bmatrix} \theta_{N+1} \\ \theta_N \end{Bmatrix} = \begin{bmatrix} \left(2 + \frac{1}{\widetilde{k}} - \frac{\widetilde{P}}{\widetilde{k}}\right) & -1 \\ 1 & 0 \end{bmatrix}^{N-1} \begin{Bmatrix} \theta_2 \\ \theta_1 \end{Bmatrix}. \qquad [2.27]$$

The eigenvalues of $\boldsymbol{T}$ are

$$\Lambda_{1,2} = \frac{1 + 2\widetilde{k} - \widetilde{P} \pm \sqrt{(1-\widetilde{P})(4\widetilde{k} - \widetilde{P} + 1)}}{2\widetilde{k}}. \qquad [2.28]$$

Xie (1995) uses Rayleigh's quotient and Oseledets' theorem (1968) to relate the eigenvalues $e_1 = e_{max}$ and $e_2 = e_{min}$ of the matrix $\boldsymbol{B}_N^\top \boldsymbol{B}_N$ to Lyapunov exponents λ_1 and λ_2 as

$$\lambda_1 = \lambda_{max} = \lim_{N\to\infty} \frac{1}{N-1} \ln e_{max}, \qquad [2.29]$$

$$\lambda_2 = \lambda_{min} = \lim_{N\to\infty} \frac{1}{N-1} \ln e_{min}. \qquad [2.30]$$

Substituting eigenvalues of $\boldsymbol{T}$ given by equation [2.28] into the expressions for the Lyapunov exponents given by equations [2.29] and [2.30], we obtain:

$$\lambda_1 = \max_{1,2} \lambda_{1,2} = \max_{1,2} \operatorname{sgn}(\Lambda_{1,2}) \ln |\Lambda_{1,2}|. \qquad [2.31]$$

The functional dependencies of the largest factor λ_1 on compressive loading $\widetilde{P} = P/P_{cr} = PL/k_t$ for various values of coupling $\widetilde{k} = kL^2/k_t$ and for the perfectly ordered case are plotted in Figure 2.9. To understand the plot, consider that a transfer matrix in equation [2.27] is a function of the loading parameter $\widetilde{P}$. If the loading is continuously increased from zero, the plot of λ_1 as a function of $\widetilde{P}$ has three distinctive regions (as seen in Figure 2.9), each of which corresponds to a different physical response of the structure. These regions correspond to the structure being stable, buckling (instability cluster) and the failed structure for loads higher than those inside the instability cluster. The chain of rigid rods is stable until the first buckling load of the cluster is reached ($\widetilde{P} < 1$). In this region of $\widetilde{P}$, λ_1 is of no practical interest as buckling does not occur, whereas, within the instability cluster that corresponds to a set of values of $\widetilde{P} \in [1, 1 + 4\widetilde{k}]$, λ_1 would reflect localization. Naturally, the real part of λ_1 happens to be zero in this region of $\widetilde{P}$, which corresponds to the absence of localization in a perfectly ordered structure. Finally, for the loads beyond the instability cluster, i.e. $\widetilde{P} > 1 + 4\widetilde{k}$, localization is of no interest, since the structure has already failed.

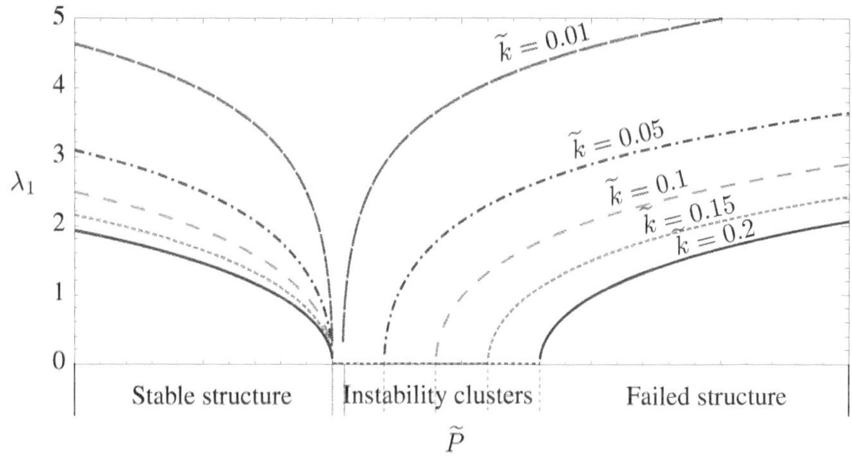

Figure 2.9. *Plots of λ_1 factors as functions of loading parameter $\widetilde{P} = P/P_{cr}$ for various coupling strengths $\widetilde{k}$. For a color version of this figure, see www.iste.co.uk/challamel/mechanics1.zip*

The magnitudes of λ_1 for each of the three regions in Figure 2.9 become apparent when we study the eigenvalues of $\boldsymbol{T}$ given by equation [2.28] and λ_1 given by equation [2.31]. Lyapunov exponents λ_1 are real and positive for $\widetilde{P} < 1$ and complex for $1 < \widetilde{P} < 1 + 4\widetilde{k}$ due to the negative radical in equation [2.28]. Finally, λ_1 is real and positive again for $\widetilde{P} > 1 + 4\widetilde{k}$. The specific points that correspond to the buckling loads within the instability cluster such as $\widetilde{P} = 1$, $\widetilde{P} = 1 + 4\widetilde{k}$ and others in-between correspond to $\Lambda_1 = 0$ and $\lambda_1 = 0$.

We now introduce disorder into each rigid rod. Furstenberg's theorem only tells us that such an asymptotic value λ exists to which the product of random matrices should converge. It does not specify how to calculate it. The previously used equations [2.28] and [2.31] are not applicable here, since equation [2.27] does not hold as all the transfer matrices are now different. Therefore, we have to use numerical algorithms to calculate localization factors. One such algorithm was employed by Xie (1995), which we will make use of as well.

Consider the disorder to be introduced into each rigid rod by changing its restoring torsional spring stiffness k_t. The new torsional spring of each disordered rod has a stiffness $k_t = \bar{k}_t + k_\varepsilon$, where $\bar{k}_t$ is some mean value of torsional spring stiffness and k_ε is a small parameter which is a measure of disorder. Values of k_ε are taken from a uniform distribution. Note that in contrast to the previous case, where the irregularity was introduced via length imperfection ΔL, here, k_ε is a small parameter that has a dimension of torsional stiffness, i.e. Newton-meters per radian, and $k_\varepsilon \ll \bar{k}_t$. Recall that $\widetilde{P} = PL/k_t$ and $\widetilde{k} = kL^2/k_t$. Therefore, we can rewrite the first entry of the transfer matrix in equation [2.27] introducing the disorder in torsional stiffness as

$$2 + \frac{k_t}{kL^2} - \frac{P}{kL} = 2 + \frac{\bar{k}_t + k_\varepsilon}{kL^2} - \frac{P}{kL} = 2 + \frac{1+\varepsilon}{\widetilde{k}} - \frac{\widetilde{P}}{\widetilde{k}}, \qquad [2.32]$$

defining the *dimensionless torsional stiffness perturbation* $\varepsilon = k_\varepsilon/k_t$. The m-th random transfer matrix obtained from equation [2.27] becomes

$$\boldsymbol{T} = \begin{bmatrix} \left(2 + \frac{1+\varepsilon}{\widetilde{k}} - \frac{\widetilde{P}}{\widetilde{k}}\right) & -1 \\ 1 & 0 \end{bmatrix}. \qquad [2.33]$$

Now, to obtain the value for the localization factor λ, we use a numerical procedure to evaluate the limit in equation [2.22]. We calculate the norm of the product of transfer matrices by the initial state vector $\|\boldsymbol{B}_N\|$ numerically and then calculate the logarithm and take the limit. We perform such calculations for $N = 10, 20, 50, 100$ and 1000 rods, as we want to replace the theoretical limit of $N \to \infty$ by a large enough number and be sure that the numerical calculation has converged. Practically, it turned out that the values of the localization factor do not change by more than $10\% - 15\%$, but we plot λ as a function of $\widetilde{P}$ for $N = 1000$ to have smooth curves that are shown in Figure 2.10. Here, the localization factor λ is plotted for the same disorder magnitude $\varepsilon = 0.1$ but for various couplings.

The plots in Figure 2.10 give valuable insight into the physics of localization of buckling modes. We observe that the strength of localization increases as coupling to disorder ratio $\widetilde{k}/\varepsilon \to 1$. If the coupling to disorder ratio $\widetilde{k}/\varepsilon \gg 1$, the chain exhibits weak localization and the respective λ curves approach those from the perfectly ordered case, as in Figure 2.9. The values of λ are no longer zero in the instability cluster. Instead, they also depend on $\widetilde{k}/\varepsilon$. In such a way, we observe that disorder affects the width of the instability cluster and the ordering of the buckling loads in it.

In Figure 2.11, we plot the extended mode shapes of the chain with $N = 100$ rigid rods, numerically calculated from the eigenvalue problem $\boldsymbol{C}_\varepsilon \boldsymbol{s} = P\boldsymbol{s}$ that describes the chain. Here, $\boldsymbol{C}_\varepsilon$ is 100×100 matrix, similar to the one in equation [2.16] but with a small random disorder introduced in each element in the main diagonal. These modes are associated with the four lowest buckling loads $\widetilde{P}$ within an instability cluster of the chain. The individual angles of rotations θ_m, $m = 0, \ldots, 99$ constitute the entries of the respective eigenvectors of $\boldsymbol{C}_\varepsilon$. In the case of no disorder ($\varepsilon = 0$), we observe periodically extended modes, as expected from a perfectly ordered structure, as seen in Figure 2.11, the left column. However, even for a small disorder, the extended modes deviate from those given by the Fourier matrix, as seen in Figure 2.11, the center column. This case corresponds to weak localization that occurs for a large coupling-to-disorder ratio. However, as we increase the disorder until the coupling-to-disorder ratio approaches one, we observe a strong or Anderson-type localization of the modes, as shown in Figure 2.11, the right column. Note that the location of the rod with maximum localized amplitude is random, but the exponential decay of the amplitude from this rod is symmetric, as indicated by both localization factors in Figure 2.10.

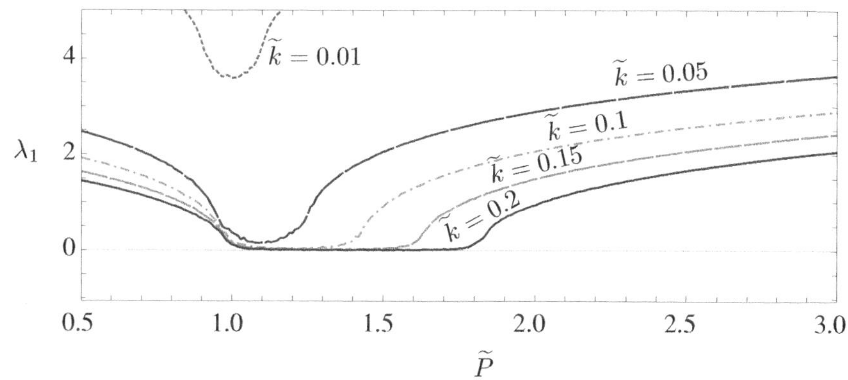

Figure 2.10. *The localization factor λ plotted as a function of $\widetilde{P}$ for disorder magnitude $\varepsilon = 0.1$ and for various values of coupling $\widetilde{k}$. For a color version of this figure, see www.iste.co.uk/challamel/mechanics1.zip*

To better demonstrate how the buckling amplitude localizes, we choose several eigenvectors of $\boldsymbol{C}_\varepsilon$ and, in Figure 2.12, plot the corresponding buckling modes of the chain for the case of the strong localization, i.e. $\widetilde{k}/\varepsilon \to 1$, on the (natural) logarithmic scale $\ln\theta_m$, $m = 0, \ldots, 99$. The linear dependence, with slope λ in the logarithmic plot, agrees with the limiting behavior derived in equation [2.26], that the magnitudes of θ_m and θ_{m+1} are related by a multiplicative factor e^λ. The lines in Figure 2.12 correspond to the Lyapunov factor λ plotted in Figure 2.10.

For example, the linear regression model of the angles of rotation $\ln\theta_m$ for mode 1 (green line) gives slope $\lambda = 0.900$ in the range of $N \in [0, 20)$ and $\lambda = -0.987$ in the range of $N \in (20, 99]$. At the same time, the numerically computed critical buckling load for mode 1 is $\widetilde{P} = 0.8879$. The localization factor predicted by Furstenberg's theorem for this value of buckling load and for $\widetilde{k} = 0.1$, $\varepsilon = 0.1$ is obtained from the green line in Figure 2.10 and is equal to $\lambda = 0.968$, which matches the magnitude of λ computed numerically extremely well.

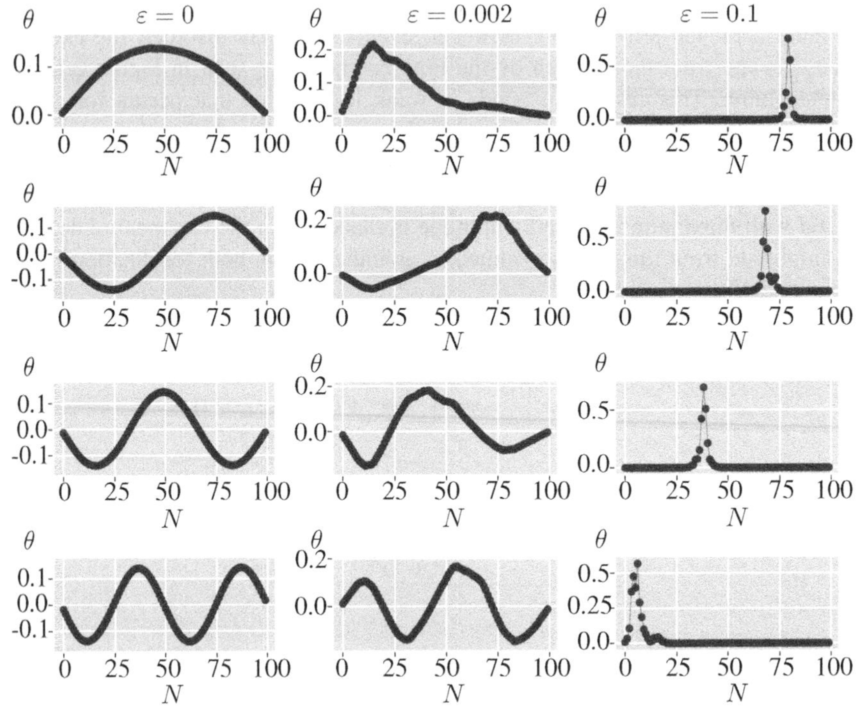

Figure 2.11. *Four lowest extended buckling modes of a chain with $N = 100$ rigid rods for various magnitudes of disorder. Left column: four extended modes of the chain with no disorder, center column: four extended modes of the chain with small disorder and right column: four extended modes of the chain with moderate disorder. For a color version of this figure, see www.iste.co.uk/challamel/mechanics1.zip*

Finally, we observe that the slopes of the modes 1 and 99 are the largest and the slopes of the modes in-between, such as mode 76, are much smaller. This is a reflection of the dependence of λ on the compressive loading $\widetilde{P}$ shown in Figure 2.10. Modes 1 and 99 correspond to the values of $\widetilde{P}$ at the boundaries of the instability cluster for which λ is relatively large, compared to mode 76 which corresponds to $\widetilde{P}$ inside the instability cluster. Therefore, we conclude that the localization factors accurately describe the behavior of the randomly disordered long chain of rigid rods and are a good measure for the localization of the bucking amplitude. Furthermore, we conclude that the symmetric exponential decay of the buckling amplitude from a random location on the chain is a generic phenomenon that depends only on the geometry of the structure, coupling between the members and the magnitude of the disorder.

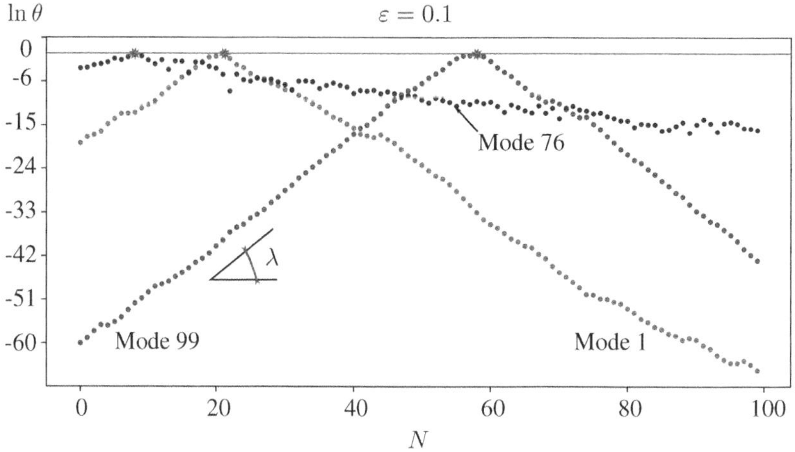

Figure 2.12. *Plots of the selected buckling modes of the disordered chain in a natural logarithm scale. The modes are plotted for the chain with the coupling strength* $k = 0.1$ *and disorder magnitude* $\varepsilon = 0.1$ *which constitutes the case of strong localization. For a color version of this figure, see www.iste.co.uk/challamel/mechanics1.zip*

2.6. Conclusion

In this chapter, we considered the problem of elastic stability of a multilayer structure and that of an expository model, i.e. a chain of elastically coupled rigid rods. The latter is an abstraction that we hope will provide analytical insight into a host of instability problems concerning periodic elastic structures. While wave propagation and normal mode vibration of the structures with translation symmetry are extensively studied in the literature, only a few papers are devoted to the buckling of such structures. We provided a brief summary of the literature, drawing attention to the relevance of periodic structures buckling, which has been less extensively studied. It is noted that the buckling of layered structures is particularly interesting, which may possess translation symmetry along the line perpendicular to the application of load.

We performed an illustrative experiment on a stack of elastic sheets glued at the ends to show that the buckling amplitude of each sheet localizes along the stacking direction, which is perpendicular to the axis of loading. We proposed a minimal model comprising of a chain of nominally identical elastically coupled rigid rods. We introduced the disorder in this model to analytically investigate the localization of buckling modes.

The modeling in the absence of disorder showed that the magnitudes of the buckling loads of the chain are densely packed within a range that we termed an *instability cluster*. The mode numbers associated with the buckling loads within the instability cluster appear inside the buckling analog of the first Brillouin zone. The buckling amplitudes of each rod are spatially modulated by the periodic functions along the chain and form the so-called extended buckling modes. This behavior is characterized by the eigenvectors of the chain, which are columns of the Fourier matrix.

We modeled the effect of the disorder on buckling of two elastically coupled rigid rods next. We reported that the buckling mode of the structure is highly sensitive to the disorder for small coupling and relatively insensitive to it in the case of strong coupling. Therefore, the coupling to disorder ratio is the parameter that drives the sensitivity to the irregularities.

Finally, we modeled a long chain of elastically coupled rigid rods with small random disorder introduced in each rod. We reported that, in contrast to the extended Fourier modes of the perfectly periodic case, the modes of the disordered chain exhibit localization. The exponential decay of the amplitude of the localized modes is determined by the Lyapunov exponent, a property introduced from the literature on products of random matrices, which we also call the localization factor. We used Furstenberg's theorem of products of random matrices and numerical methods to calculate the localization factors for the chain and reported their values for systems of large numbers of rods. Similar to the case of two coupled rods, the chain exhibits a stronger localization as the coupling to disorder ratio reduces. For the coupling to disorder ratio close to one, we observe a strong or Anderson-type localization of buckling amplitudes along the chain. In this case, the buckling mode shows that only a few neighboring rods exhibit large amplitudes, the deviations of the remaining rods being exponentially suppressed.

2.7. References

Anderson, P.W. (1958). Absence of diffusion in certain random lattices. *Physical Review*, 109(5), 1492–1505.

Ariaratnam, S. and Xie, W.-C. (1995). Wave localization in randomly disordered nearly periodic long continuous beams. *Journal of Sound and Vibration*, 181, 7–22.

Ayrilmis, N., Candan, Z., Hiziroglu, S. (2008). Physical and mechanical properties of cardboard panels made from used beverage carton with veneer overlay. *Materials & Design*, 29(10), 1897–1903.

Benaroya, H. (1997). Waves in periodic structures with imperfections. *Composites Part B: Engineering*, 28(1–2), 143–152.

Bloch, F. (1929). Über die Quantenmechanik der Elektronen in Kristallgittern. *Zeitschrift für Physik*, 52(7), 555–600.

Brillouin, L. (2003). *Wave Propagation in Periodic Structures*. Courier Corporation, North Chelmsford.

Byklum, E., Steen, E., Amdahl, J. (2004). A semi-analytical model for global buckling and postbuckling analysis of stiffened panels. *Thin-Walled Structures*, 42(5), 701–717.

Castanier, M. and Pierre, C. (1995). Lyapunov exponents and localization phenomena in multi-coupled nearly periodic systems. *Journal of Sound and Vibration*, 183(3), 493–515.

Chai, H., Babcock, C., Knauss, W. (1981). One dimensional modelling of failure in laminated plates by delamination buckling. *International Journal of Solids and Structures*, 17(11), 1069–1083.

Cheng, B., Wang, J., Li, C. (2013). Compression tests and numerical analysis of perforated plates containing slotted holes in steel pylons. *Thin-Walled Structures*, 67, 129–143.

Cho, M. and Kim, J.-S. (2001). Higher-order zig-zag theory for laminated composites with multiple delaminations. *Journal of Applied Mechanics*, 68(6), 869.

Cuan-Urquizo, E. and Bhaskar, A. (2018). Flexural elasticity of woodpile lattice beams. *European Journal of Mechanics – A/Solids*, 67, 187–199.

Dodwell, T.J. and Hunt, G.W. (2014). Periodic void formation in chevron folds. *Mathematical Geosciences*, 46(8), 1011–1028.

Dodwell, T.J., Peletier, M.A., Budd, C.J., Hunt, G.W. (2012). Self-similar voiding solutions of a single layered model of folding rocks. *Journal of Applied Mathematics*, 72(1), 444–463.

Elishakoff, I., Li, Y., Starnes Jr., J.H. (1995). Buckling mode localization in elastic plates due to misplacement in the stiffener location. *Chaos, Solitons & Fractals*, 5(8), 1517–1531.

Elishakoff, I., Li, Y., Starnes Jr., J.H. (1997). Passive control of buckling deformation via Anderson localization phenomenon. *Chaos, Solitons and Fractals*, 8(1), 59–75.

Floquet, G. (1883). Sur les équations différentielles linéaires à coefficients périodiques. *Annales scientifiques de l'École normale supérieure*, 12, 47–88.

Furstenberg, H. (1963). Noncommuting random products. *Transactions of the American Mathematical Society*, 108(3), 377–428.

Goodman, F.M., de la Harpe, P., Jones, V.F.R. (2012). *Coxeter Graphs and Towers of Algebras*. Springer Science & Business Media, Berlin.

Gray, R.M. (2005). Toeplitz and circulant matrices: A review. *Foundations and Trends® in Communications and Information Theory*, 2(3), 155–239.

Hodges, C.H. and Woodhouse, J. (1983). Vibration isolation from irregularity in a nearly periodic structure: Theory and measurements. *The Journal of the Acoustical Society of America*, 74(3), 894–905.

Hunt, G.W., Edmunds, R., Budd, C.J., Cosgrove, J.W. (2006). Serial parallel folding with friction: A primitive model using cubic B-splines. *Journal of Structural Geology*, 28(3), 444–455.

Juhász, Z. and Szekrényes, A. (2017). The effect of delamination on the critical buckling force of composite plates: Experiment and simulation. *Composite Structures*, 168, 456–464.

Li, Y., Elishakoff, I., Starnes Jr., J.H. (1995). Buckling mode localization in a multi-span periodic structure with a disorder in a single span. *Chaos, Solitons & Fractals*, 5(6), 955–969.

Li, F.M., Wang, Y.S., Hu, C., Huang, W.H. (2005). Localization of elastic waves in periodic rib-stiffened rectangular plates under axial compressive load. *Journal of Sound and Vibration*, 281(1–2), 261–273.

Moen, C.D. and Schafer, B.W. (2009a). Elastic buckling of cold-formed steel columns and beams with holes. *Engineering Structures*, 31(12), 2812–2824.

Moen, C.D. and Schafer, B.W. (2009b). Elastic buckling of thin plates with holes in compression or bending. *Thin-Walled Structures*, 47(12), 1597–1607.

Nguyen-Thoi, T., Bui-Xuan, T., Phung-Van, P., Nguyen-Xuan, H., Ngo-Thanh, P. (2013). Static, free vibration and buckling analyses of stiffened plates by CS-FEM-DSG3 using triangular elements. *Computers & Structures*, 125, 100–113.

Olson, B.J., Shaw, S.W., Shi, C., Pierre, C., Parker, R.G. (2014). Circulant matrices and their application to vibration analysis. *Applied Mechanics Reviews*, 66(4), 040803.

Oseledets, V.I. (1968). A multiplicative ergodic theorem. Characteristic Ljapunov, exponents of dynamical systems. *Transactions of the Moscow Mathematical Society*, 19, 33.

Peng, L., Liew, K., Kitipornchai, S. (2006). Buckling and free vibration analyses of stiffened plates using the FSDT mesh-free method. *Journal of Sound and Vibration*, 289(3), 421–449.

Phani, A.S., Woodhouse, J., Fleck, N.A. (2006). Wave propagation in two-dimensional periodic lattices. *The Journal of the Acoustical Society of America*, 119(4), 1995–2005.

Pierre, C. and Plaut, R.H. (1989). Curve veering and mode localization in a buckling problem. *ZAMP Journal of Applied Mathematics and Physics*, 40(5), 758–761.

Rikards, R., Chate, A., Ozolinsh, O. (2001). Analysis for buckling and vibrations of composite stiffened shells and plates. *Composite Structures*, 51(4), 361–370.

Rodman, U., Saje, M., Planinc, I., Zupan, D. (2008). Exact buckling analysis of composite elastic columns including multiple delamination and transverse shear. *Engineering Structures*, 30(6), 1500–1514.

Rosenstock, H.B. and McGill, R.E. (1968). Vibrations of disordered solids. *Physical Review*, 176(3), 1004–1014.

Xie, W.-C. (1995). Buckling mode localization in randomly disordered multispan continuous beams. *AIAA Journal*, 33(6), 1142–1149.

Xie, W.-C. (1997). Buckling mode localization in nonhomogeneous beams on elastic foundations. *Chaos, Solitons and Fractals*, 8(3), 411–431.

Xie, W.-C. (1998). Buckling mode localization in rib-stiffened plates with randomly misplaced stiffeners. *Computers & Structures*, 67(1–3), 175–189.

Xie, W.-C. and Elishakoff, I. (2000). Buckling mode localization in rib-stiffened plates with misplaced stiffeners – Kantorovich approach. *Chaos, Solitons & Fractals*, 11(10), 1559–1574.

Xie, W.-C. and Ibrahim, A. (2000). Buckling mode localization in rib-stiffened plates with misplaced stiffeners – a finite strip approach. *Chaos, Solitons & Fractals*, 11(10), 1543–1558.

Xie, W.-C. and Wang, X. (1997a). Vibration mode localization in one-dimensional systems. *AIAA Journal*, 35(10), 1645–1652.

Xie, W.-C. and Wang, X. (1997b). Vibration mode localization in two-dimensional systems. *AIAA Journal*, 35(10), 1653–1659.

Ziman, L. (1972). *Principles of the Theory of Solids*, 2nd edition, Cambridge University Press, Cambridge.

3

Analysis of a Beck's Column over Fractional-Order Restraints via Extended Routh–Hurwitz Theorem

3.1. Introduction

Column stability resting on elastic and viscoelastic foundations, in the presence of follower agencies, has been studied extensively (Beck 1952; Ziegler 1952; Plaut and Infante 1970; Elishakoff 2001, 2005). The fascinating dynamic stability of a column with inherent paradoxes (Smith and Herrmann 1972) depending on the kind of external restraint (Hermann–Schmidt and/or Ziegler) has attracted several scientists to help represent the mechanics of rods and columns in different engineering fields.

Indeed, problems dealing with rocket thrusts, wind-exposed structures, biomechanics of the spine, as well as polymer-reinforced stratified structural glass are often modeled as structural elements subjected to follower forces.

In the presence of external restraints with energy dissipating properties, different kinds of mathematical descriptions of the external restraints have been proposed.

In recent years, the behavior of dissipative materials has been studied by means of fractional differential calculus in terms of the so-called Caputo fractional derivative and the Riemann–Liouville fractional integral. Indeed, the experimental data on hereditary materials show that power laws may be efficiently used to capture the material behavior (Bologna *et al.* 2019, 2020) and, in a rheological context (Mainardi 2010), it leads to the introduction of the springpot element. Other studies to capture the material behavior in stochastic dynamics have been proposed recently (Spanos

Chapter written by Emanuela BOLOGNA, Mario DI PAOLA and Massimiliano ZINGALES.

and Miller 1994; Pinnola 2016). Springpot is a one-dimensional element defined in terms of two parameters, i.e. the order of the springpot β, $0 \leq \beta \leq 1$ and the anomalous modulus $C_\beta > 0$ with physical dimensions $[C_\beta] = \frac{FT^\beta}{L^2}$, which may be related to the elastic properties introducing the relaxation time τ_0 as $C_\beta = E_0 (\tau_0)^\beta$. Such an element with intermediate behavior among elastic springs and viscous dashpot is widely used nowadays to define several types of materials including, as limiting cases, elastic $(\beta = 0)$ and viscous elements $(\beta = 1)$. Moreover, a simple spring corresponds to $\beta = 0$ and $\frac{d^\beta f}{dt^\beta} = \frac{d^0 f}{dt^0} = f$, while the case of $\beta = 1$ corresponds to a first-order derivative, i.e. $\frac{d^\beta f}{dt^\beta} = \frac{df}{dt} = \dot{f}$, which is a Newtonian dashpot.

The presence of springpot as external restraints yields, on one hand, a more reliable description of the external material foundation. On the other hand, the mathematical prediction of the dynamic stability of Beck's column became more challenging and several studies have been devoted to this topic in recent years.

A different approach of the extension of the Routh–Hurwitz criterion has been recently reported for a bed of independent springpot (Bologna *et al.* 2017).

In this study for the fractional-order dynamical systems, the authors introduce a more general representation of the external restraint by means of the fractional Zener model. This specific restraint is a generalization of the well-known Zener models that involves a fractional Kelvin–Voigt model, complete with an elastic spring forming a series element. The presence of the Zener element represents a specific generalization of the Routh–Hurwitz theorem since, for fractional-order dynamical systems, the conventional condition involving the sign of the real part of the problem eigenvalues no longer holds. The well-known Routh–Hurwitz criterion used to handle the sign of the eigenvalues can no longer be applied. In this regard, the authors introduced a novel approach that involves a conformal mapping of the complex plane yielding complex problem eigenvalues that can be solved by the extended Routh–Hurwitz method (Bologna *et al.* 2017; Bologna and Zingales 2018). Some numerical examples reporting the critical loads applied to the column have been reported in the paper.

3.2. Material hereditariness

Conventional approaches to material behavior has led to two limit cases: i) in the presence of a complete recovery in cycle tensile tests, the concept of material elasticity is introduced, and ii) in the presence of mechanical behavior, so that complete energy dissipation (no recovery) is observed, the concept of viscous material is reported. Real types of materials show intermediate behavior and the term material hereditariness is reported. The term hereditariness indicates the behavior of an intermediate material between elastic solid and viscous liquid; this behavior is typical of polymers, human bones, various mortars and resins used in construction, some families of rocks and other materials. The hereditariness material is therefore characterized by two asymptotic behaviors, that of the solid elastic and that of the viscous liquid.

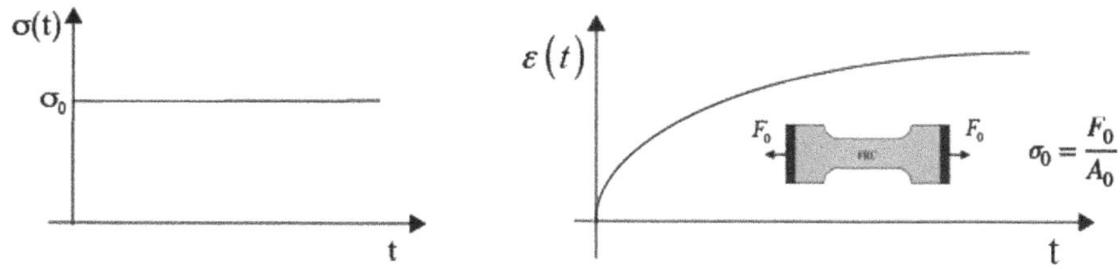

Figure 3.1. *Creep test*

Two main mechanical tests allow us to construct the hereditary behavior of materials, namely the creep and relaxation tests.

In a creep test, the specimen of the material undergoes a step stress history (Figure 3.1), in which the stress is instantaneously increased to some value σ_0 at $t = 0$ and is then fixed. The typical strain response consists of i) an instantaneous increase in strain at $t = 0$, followed by ii) continued straining in time at a non-constant rate and (iii) an asymptotic approach to some limit value as time increases. Let $J(t, \sigma_0)$ denote the strain at time t when the value of the stress is σ_0. Then, $J(t, \sigma_0) = 0$ when $t < 0$; ii) jumps to value $J(0, \sigma_0)$ at $t = 0$ and iii) $J(t, \sigma_0)$ monotonically increases to the limit value denoted by $J(\infty, \sigma_0)$ as $t \to \infty$. The jump in strain $J(0, \sigma_0)$ at $t = 0$ indicates instantaneous elasticity. The relations σ_0 versus $J(0, \sigma_0)$ and σ_0 versus $J(\infty, \sigma_0)$ describe, respectively, instantaneous elastic response and the long-time or equilibrium elastic response. The function $J(t, \sigma_0)$ has a different dependence on time t and stress σ_0 for each material and is therefore considered to be material dependent. As we assume that a linear dependence of the step stress is involved, the $J(t, \sigma_0) = \sigma_0 J(t)$ and the time-dependent function $J(t)$ is known as the material creep function.

In the relaxation test, the specimen is subjected to a step strain history (Figure 3.2), in which the strain is instantaneously increased to some value ε_0 at $t = 0$ and is then fixed. The typical stress history required to produce this strain history consists of i) an instantaneous increase in stress at $t = 0$, followed by ii) a gradual monotonic decrease of stress at a non-constant rate and iii) an asymptotic approach to some non-zero limit value as time increases. $G(t, \varepsilon_0)$ denotes the stress at time t, when the value of the strain is fixed at ε_0. Then, i) $G(t, \varepsilon_0) = 0$ when $t < 0$, ii) $G(t, \varepsilon_0)$ jumps to the value $G(0, \varepsilon_0)$ at $t = 0$ and iii) $G(t, \varepsilon_0)$ monotonically decreases to the non-zero limit value denoted by $G(\infty, \varepsilon_0)$ as $t \to \infty$. The jump in stress $G(t, \varepsilon_0)$ at $t = 0$ is another indication of instantaneous elasticity. The relations $G(0, \varepsilon_0)$ versus ε_0 and $G(\infty, \varepsilon_0)$ versus ε_0 also describe, respectively, instantaneous elastic response and the long-time or equilibrium elastic response. $G(t, \varepsilon_0)$ has a different dependence on time t and strain ε_0 for each material and is therefore considered a material property. As the linearity condition is fulfilled, $G(t, \varepsilon_0) = \varepsilon_0 G(t)$ and the time-dependent material function $G(t)$ is known as a relaxation function.

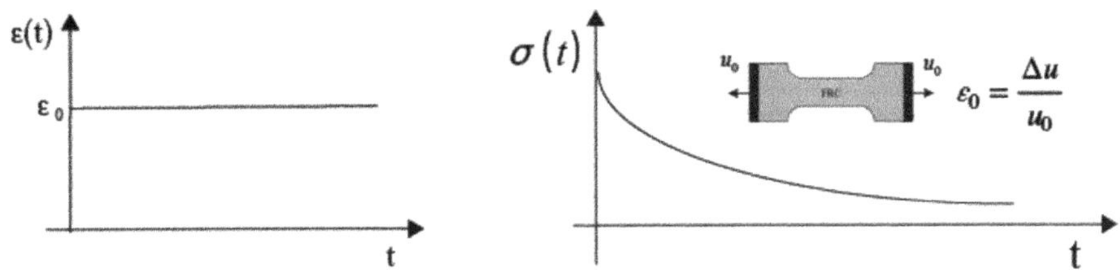

Figure 3.2. *Relaxation test*

A generic stress/strain history, namely $\sigma(\tau)$ and $\varepsilon(\tau)$ with $\tau \leq t$, yields:

$$\sigma(t) = \int_0^t G(t-\tau)d\varepsilon(\tau) + \varepsilon_0 G(t) = \int_0^t G(t-\tau)\dot{\varepsilon}(\tau)d\tau + \varepsilon_0 G(t) \quad [3.1a]$$

$$\varepsilon(t) = \int_0^t J(t-\tau)d\sigma(\tau) + \sigma_0 J(t) = \int_0^t J(t-\tau)\dot{\sigma}(\tau)d\tau + \sigma_0 J(t) \quad [3.1b]$$

Equations [3.5a] and [3.5b] are defined in terms of the Boltzmann superposition with $d\sigma = \dot{\sigma}dt$ and $d\varepsilon = \dot{\varepsilon}dt$ increments, where $[\dot{\cdot}] = \frac{d}{dt}$. The Laplace transform of equations [3.1a] and [3.1b] with $\sigma_0 = 0$ and $\varepsilon_0 = 0$ yields the fundamental relation between the creep and relaxation of linear hereditariness in the Laplace domain as:

$$\hat{J}(s)\hat{G}(s) = \frac{1}{s^2} \quad [3.2]$$

where $\hat{G}(s)$ and $\hat{J}(s)$ are the Laplace transform of relaxation and creep functions, respectively. This fundamental relationship expresses the circumstance that $G(t)$ and $J(t)$ are functions related to each other in the domain of Laplace. It follows that if it is determined through experimental tests $G(t)$, the function $J(t)$ is determined accordingly and vice versa.

Another approach used to develop constitutive equations for linear hereditary response involves mechanical analogs. These are mechanical devices formed by combining linear elastic springs and linear viscous dampers in series or parallel. The devices can be shown to exhibit a time-dependent response that is similar to that observed in hereditary materials, namely creep under a constant load and force relaxation under a constant deformation. For this reason, these devices are treated as mechanical analogs of hereditariness response. Since the springs and dampers are described by linear equations, as are the equations for the kinematics of deformation and force transmission, there is a linear relation between the overall force and deformation. These approaches are known as rheological models, which will be discussed in the following.

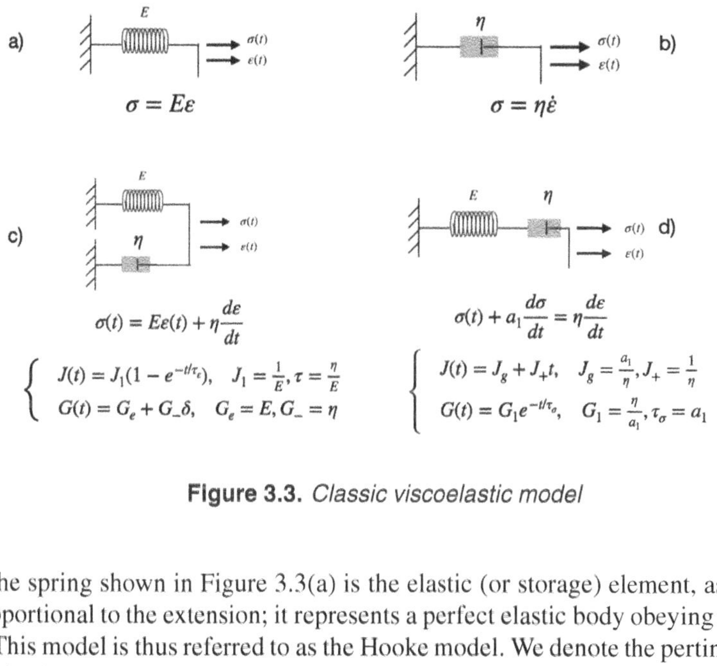

Figure 3.3. *Classic viscoelastic model*

The spring shown in Figure 3.3(a) is the elastic (or storage) element, as the force is proportional to the extension; it represents a perfect elastic body obeying the Hooke law. This model is thus referred to as the Hooke model. We denote the pertinent elastic modulus by m. In this case, there is no creep or relaxation, so the creep compliance and the relaxation modulus are constant functions $J(t) \equiv 1/E$, $G(t) \equiv E$.

The dashpot in Figure 3.3(b) is the viscous (or dissipative) element, with the force being proportional to the rate of extension; it represents a perfectly viscous body obeying the Newton law. By denoting the pertinent viscosity coefficient by ν, we observe that the Hooke and Newton models represent the limiting cases of viscoelastic bodies.

A branch constituted by a spring parallel to a dashpot is known as the Voigt model (see Figure 3.3(c)), where τ_ε is the retardation time.

A branch constituted by a spring in series with a dashpot is known as the Maxwell model (see Figure 3.3(d)), where τ_σ is the relaxation time. The Voigt model exhibits an exponential (reversible) strain creep but no stress relaxation, which is also referred to as the retardation element. The Maxwell model exhibits an exponential (reversible) stress relaxation and a linear (non-reversible) strain creep, which is also referred to as the relaxation element.

By increasing the number of simple elements to the Kelvin–Voigt model, other more accurate models are obtained in the simulation of viscoelastic behavior. Such models are called the SLS (Standard Linear Solid) or Zener model.

The simplest viscoelastic model is obtained by adding a spring, either in series to a Voigt model, or parallel to a Maxwell model. In this way, according to the combination rule, we add a positive constant, both to the Voigt-like creep compliance and to the Maxwell-like relaxation modulus, so that $J_g > 0$ and $G_e > 0$. Such a model was considered by Zener with the denomination SLS, which will also be referred to here as the Zener model. In Figure 3.4(a), we have a Zener model with a spring in series to a Kelvin Voigt model.

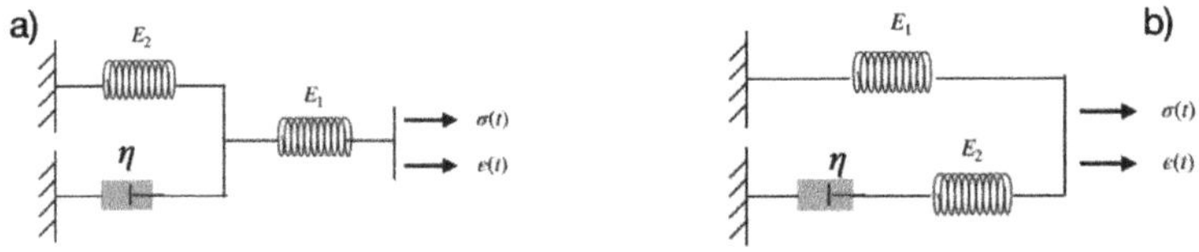

$$\dot{\sigma}(t) + \alpha\sigma(t) = E_1\left[\dot{\varepsilon}(t) + \beta\varepsilon(t)\right] \qquad \dot{\sigma}(t) + \beta\sigma(t) = \alpha\eta\dot{\varepsilon}(t) + \beta E_1\varepsilon(t)$$

$$\alpha = \frac{E_1 + E_2}{\eta} \qquad \beta = \frac{E_2}{\eta}$$

Figure 3.4. *a) Zener model: a spring in series to a Voigt model and b) Zener model: a spring in parallel to a Maxwell model*

The model shown in Figure 3.4(b), consists of a Hooke model in parallel to a Maxwell model, with α and β the same for both.

Similarly, the functional stress–strain relations reported have an equivalent differential formulation in terms of elastic (Hookean) and viscous (Newtonian) elements.

3.2.1. *Linear hereditariness: fractional-order models*

Linear hereditariness is certainly the field of the most extensive applications of fractional calculus, in view of its ability to model hereditary phenomena with long memory. The analysis starts from the power-law creep to justify the introduction of the operators of fractional calculus into the stress–strain relationship.

Let us consider the hereditariness of a material with creep compliance,

$$J(t) = \frac{1}{G_0 \Gamma(1+\beta)} \left(\frac{t}{\tau_s} \right)^{\beta} \tag{3.3}$$

where $\Gamma(\cdot)$ is the Euler–Gamma function, β, and $[\tau_s] = T$ are the material parameters that may be estimated through a best-fitting procedure of experimental data and G_0 is the elastic modulus of the material. Such creep behavior is found to be of great interest in a number of creep experiments; it is usually referred to as the power-law creep.

Using the reciprocity relationship [3.2] in the Laplace domain, we can find for such a hereditariness solid, its relaxation modulus and then the corresponding relaxation spectrum. After simple manipulations, we obtain

$$G(t) = \frac{G_0}{\Gamma(1-\beta)} \left(\frac{t}{\tau_s} \right)^{-\beta} \tag{3.4}$$

For our hereditariness solid exhibiting power-law creep, the stress/strain relationship in the creep representation can be easily obtained by inserting the creep law equation [3.3] into the integral equation [3.2]. Straightforward manipulations show that the power-law functional class in equations [3.3] and [3.4] satisfies the conjugation relation, which yields, upon substitution into equations [3.1a] and [3.1b], the following constitutive relations:

$$\sigma(t) = \frac{G_0 \tau_s^{\beta}}{\Gamma(1-\beta)} \int_0^t (t-\tau)^{-\beta} \dot{\varepsilon}(\tau) d\tau = G_0 \tau_s^{\beta} \left(D_{0+}^{\beta} \varepsilon \right)(t) \tag{3.5a}$$

$$\varepsilon(t) = \frac{1}{G_0 \tau_s^{\beta} \Gamma(\beta+1)} \int_0^t (t-\tau)^{\beta} \dot{\sigma}(\tau) d\tau = \frac{1}{\tau_s^{\beta} G_0} \left(I_{0+}^{\beta} \sigma \right)(t) \tag{3.5b}$$

where $\left[{}_C D_0^{+\beta} \right]$ is the Fractional Caputo derivative and $[I_{o+}^{\beta}]$ is the Riemann–Liouville fractional integral. The constitutive equations [3.5a] and [3.5b] have been modeled with the introduction of a new rheological element, the springpot (Figure 3.5), after Scott-Blair. Springpot is a mechanical element with mechanical properties that are intermediate between those of the Hooke model and a pure viscous fluid (Newton model). The use of fractional calculus in linear hereditariness leads us to generalize the classical mechanical models, in that the basic Newton element (dashpot) is substituted by springpot. The springpot (Figure 3.5) is defined in terms of two parameters, i.e. $C_{\beta} = G_0 \tau_s^{\beta} \geq 0$ and β, with $\beta \in [0, 1]$, whose constitutive relation is given in equations [3.5a] and [3.5b]. Such an element is widely used nowadays to define several types of materials including as limiting cases, elastic ($\beta = 0$) and viscous elements ($\beta = 1$). More precisely, a simple spring corresponds to $\beta = 0$ and $\frac{d^{\beta} f}{dt^{\beta}} = \frac{d^0 f}{dt^0} = f$, while the case of $\beta = 1$ corresponds to a first-order derivative, i.e. $\frac{d^{\beta} f}{dt^{\beta}} = \frac{df}{dt} = \dot{f}$, which is a Newtonian dashpot.

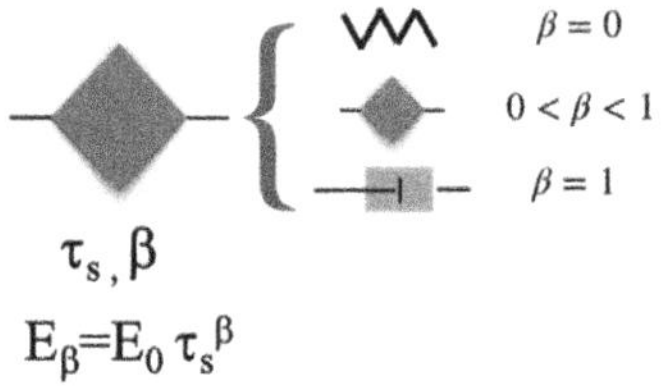

Figure 3.5. *Springpot element*

As the springpot element is connected with springs in series and parallel, the generalization of the rheological models, as shown in section 3.2.1, is obtained (see Figure 3.6).

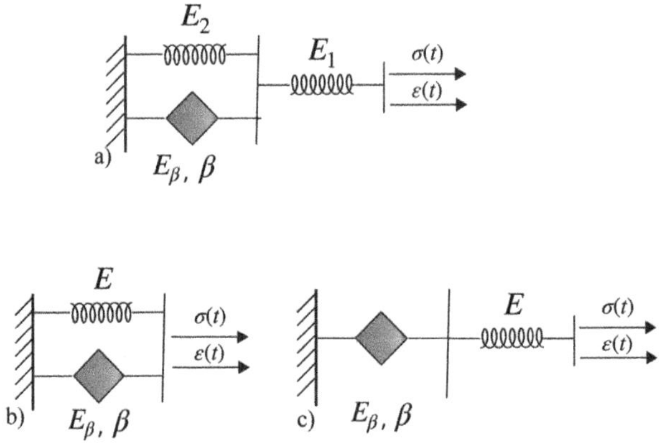

Figure 3.6. *Fractional models*

Moreover, the elastic spring and the springpot in series in Figure 3.6(c) are described by a constitutive equation

$$\sigma(t) = E\varepsilon(t) + E_{\beta}(D^{\beta}\varepsilon)(t) \qquad [3.6]$$

that is the generalization of the Kelvin–Voigt element. In the presence of spring in parallel, the constitutive equation can be written as

$$\tau_s(D^{\beta}\sigma)(t) + \sigma(t) = E_{\beta}(D^{\beta}\varepsilon)(t) \qquad [3.7]$$

with $E_\beta = E\tau_s^\beta$ and the combination corresponds to the Zener element, as shown in Figure 3.6(a), with the governing equation in the form

$$E_{\beta}(D^{\beta}\varepsilon)(t) + E_2\varepsilon = \frac{E_\beta}{E_1}(D^{\beta}\sigma)(t) + (1+\alpha)\sigma(t) \qquad [3.8]$$

and as $E_\beta = E_2\tau_s^\beta$ we get the constitutive equation of the fractional Zener model, in the form:

$$\tau_s^{\beta}(D^{\beta}\varepsilon)(t) + \varepsilon = \alpha\tau_s^{\beta}(D^{\beta}\sigma)(t) + \gamma \qquad [3.9]$$

with $\gamma = \dfrac{E_1 + E_2}{E_1 E_2}$ and $\alpha = \dfrac{K_2}{K_1}$ that is formally analogous to the classical Zener model, as we replace fractional derivative with integer-order derivative and viscosity $\eta = E_2\tau_0$.

In the next section, the stability of the column over generalized foundation is considered.

3.3. Dynamic equilibrium of an elastic cantilever over a fractional-order foundation

Let us consider a Bernoulli–Euler elastic column with cross-section A, length L and rotary inertial properties across the centroidal axis denoted as J_1. The Young modulus of elasticity has been denoted as E and the material density as ρ, both of which are assumed as homogeneous along the column axis, denoted as x_3 in Figure 3.7. The column is clamped at $x_3 = 0$, which is supported by a bed of vertical and independent springpots, connected in series and parallels with two springs to compose the fractional Zener model described in the previous section, assumed to be homogeneous along the x_3-axis. The column is subjected to a follower axial load, denoted as P, at the free end (Figure 3.7). In the following, we restrict our analysis to the onset of bifurcations from the undeformed (straight) configuration (Figure 3.7) in the $(x_2 - x_3)$ plane, defined by the vertical displacement field $w(x_3)$ of the column axis.

The governing equation of the column may be obtained in a classical fashion with the equilibrium of a column element (Figure 3.8), where we denote $\ddot{[\bullet]} = \dfrac{\partial^2}{\partial t^2}$ and F_f the reaction of the fractional-order, hereditary support of the column foundation.

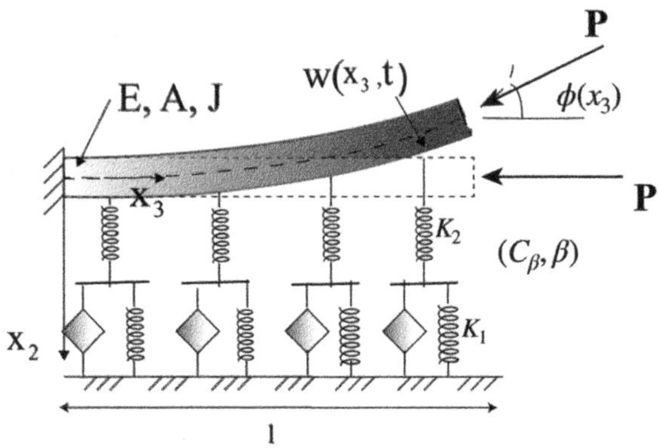

Figure 3.7. *Onset of the column bifurcations and its generic section*

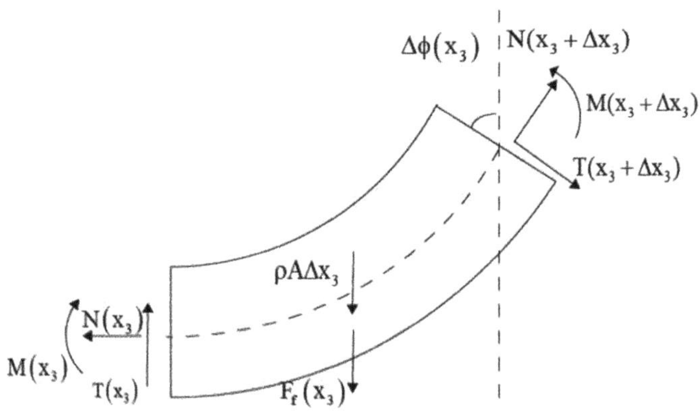

Figure 3.8. *Onset of the column bifurcation*

The equilibrium equation of the column can be written as (omitting time argument):

$$N(x_3+\Delta x_3)cos(\phi(x_3)+\Delta\phi(x_3))-N(x_3)\cos(\phi(x_3))+ \\ +T(x_3)\sin(\phi(x_3))-T(x_3+\Delta x_3)\sin(\phi(x_3)+\Delta\phi(x_3))=0 \quad [3.10a]$$

$$T(x_3)\sin(\phi(x_3))+N(x_3)\sin(\phi(x_3))-T(x_3+\Delta x_3)\cos(\phi(x_3)+\Delta\phi(x_3))+ \\ -N(x_3+\Delta x_3)\sin(\phi(x_3)+\Delta\phi(x_3))+\rho A\frac{\partial^2 w}{\partial t^2}\Delta x_3+F_f\,(x_3)=0 \quad [3.10b]$$

$$T(x_3+\Delta x_3)\cos(\phi(x_3)+\Delta\phi(x_3))\Delta x_3+T(x_3+\Delta x_3)\sin(\phi(x_3)+ \\ +\Delta\phi(x_3))\Delta w(x_3)+M(x_3)-M(x_3+\Delta x_3)=0 \quad [3.10c]$$

where, in the latter equation, we neglect the contribution of the axial stress moment that is a higher-order infinitesimal. Since the onset of bifurcation is analyzed, the following approximations hold time $\sin(x) \cong x$; $\cos(x) \cong 1$, yielding the equilibrium equations in linearized formulation:

$$N(x_3 + \Delta x_3) - N(x_3) + T(x_3)\phi(x_3) + \\ - T(x_3 + \Delta x_3)(\phi(x_3) + \Delta\phi(x_3)) = 0 \quad [3.11a]$$

$$T(x_3)\phi(x_3) + N(x_3)\phi(x_3) - T(x_3 + \Delta x_3) + \\ - N(x_3 + \Delta x_3)(\phi(x_3) + \Delta\phi(x_3)) + \rho A \frac{\partial^2 w}{\partial t^2}\Delta x_3 + F_f(x_3) = 0 \quad [3.11b]$$

$$T(x_3 + \Delta x_3)\Delta x_3 + T(x_3 + \Delta x_3)(\phi(x_3) + \\ + \Delta\phi(x_3))\Delta w(x_3) + M(x_3) - M(x_3 + \Delta x_3) = 0 \quad [3.11c]$$

After some straightforward manipulations, and neglecting higher-order contributions, the governing equations of the onset of bifurcation can be written as:

$$\frac{\partial N(x_3, t)}{\partial x_3} = 0 \quad [3.12a]$$

$$\rho A \frac{\partial^2 w(x_3, t)}{\partial t^2} + F_f(x_3, t) - \frac{\partial T(x_3, t)}{\partial x_3} - N(x_3, t)\frac{\partial \phi(x_3, t)}{\partial x_3} = 0 \quad [3.12b]$$

$$\frac{\partial M(x_3, t)}{\partial x_3} = T(x_3, t) \quad [3.12c]$$

yielding, after substitutions $\phi(x_3, t) = -\frac{\partial w(x_3, t)}{\partial x_3}$; $N(x_3, t) = -P$; $M(x_3, t) = -EJ\frac{\partial^2 w(x_3, t)}{\partial {x_3}^2}$ the governing equation of the transverse displacement of the column resting over a generalized foundation:

$$\rho A \frac{\partial^2 w(x_3, t)}{\partial t^2} + EJ\frac{\partial^4 w(x_3, t)}{\partial {x_3}^4} + P\frac{\partial^2 w(x_3, t)}{\partial {x_3}^2} + s(x_3, t) = 0 \quad [3.13]$$

where the reaction support $[s(x_3, t)] = F/L$, which is governed by the fractional differential equation for the fractional Zener model introduced in the previous section:

$$C_\beta D^\beta w + K_2 w = \frac{C_\beta}{K_1} D^\beta s + (1 + \alpha)s \quad [3.14]$$

where $[K_1] = [K_2] = K/L^2$ are the spring stiffness of the restraint and $[C_\beta] = F/L^2$ is the anomalous viscosity of the restraint that depends on the differentiation-order,

β, and as far as $\beta = 1$ the Newtonian viscosity η is recovered. In order to achieve a more convenient formulation for the external restraint constitutive equation, we set $C_\beta = K_2 \tau_s^\beta$ and, in such a context, equation [3.14] becomes

$$\tau_s^\beta D^\beta w + w = \alpha \tau_s^\beta D^\beta s + \gamma \qquad [3.15]$$

The boundary value problem associated with equation [3.13] is defined by using the initial and boundary conditions:

$$w\left(0,t\right) = \left.\frac{dw}{dx_3}\right|_0 = 0; \qquad EJ \left.\frac{\partial^2 w}{\partial {x_3}^2}\right|_L = EJ \left.\frac{\partial^3 w}{\partial {x_3}^3}\right|_L = 0 \qquad [3.16]$$

$$w\left(x_3, 0\right) = 0 \qquad ; \dot{w}\left(x_3, 0\right) = 0 \qquad [3.17]$$

It can be observed that [3.10a], [3.10b] and [3.10c] rule the mechanics of the column resting on a fractional-order foundation, which is formally analogous to the well-known column on an elastic (Beck 1952) and viscous (Plaut and Infante 1970) foundation.

The onset of stability of Beck's column resting on the fractional-order foundation is analyzed hereinafter in the non-dimensional form obtained by casting equation [3.13] in the form:

$$\frac{\partial^2 \overline{w}}{\partial \bar{t}^2} + \frac{\partial^4 \overline{w}}{\partial \zeta^4} + \lambda^2 \frac{\partial^2 \overline{w}}{\partial \zeta^2} + \tau_0^2 \bar{s}\left(\zeta, t\right) = 0 \qquad [3.18a]$$

$$\alpha \tau_s^\beta D_{0+}^\beta \bar{s} + \gamma \bar{s} = \tau_s^\beta D_{0+}^\beta \bar{w} + \bar{w} \qquad [3.18b]$$

where $\zeta = \dfrac{x_3}{L}$; $\bar{t} = \dfrac{t}{\tau_0}$; $\tau_0 = \dfrac{PAL^4}{EJ}$; $\bar{w} = \dfrac{w}{L}$; $\lambda^2 = \dfrac{PL^2}{EJ}$ and $\bar{s}\left(\zeta, \bar{t}\right) = \dfrac{s\left(\zeta, \bar{t}\right)}{k_1 L}$ with the associate boundary conditions:

$$\bar{w}\left(0,t\right) = \left.\frac{\partial \bar{w}}{\partial \zeta}\left(\zeta, t\right)\right|_0 = \left.\frac{\partial^2 \bar{w}}{\partial \zeta^2}\left(\zeta, t\right)\right|_1 = \left.\frac{\partial^3 \bar{w}}{\partial \zeta^3}\left(\zeta, t\right)\right|_1 = 0 \qquad [3.19]$$

Despite these formal analogies, the stability analysis of the column in the presence of fractional-order time-derivatives needs a specific core, which will be discussed in the next section.

3.4. Stability analysis of Beck's column over fractional-order hereditary foundation

In this section, the evaluation of the critical instability load of Beck's column is outlined. Stability analysis is conducted, as usual, by introducing the variable separation $\bar{w}(\zeta, \bar{t}) = v(\zeta)y(\bar{t})$ and in order to satisfy equation [3.18b], analogous factorization of the load term $\bar{s}(\zeta, \bar{t}) = v(\zeta)r(\bar{t})$, it is given by:

$$\frac{d^4 v(\zeta)}{d\zeta^4} + \lambda^2 \frac{d^2 v(\zeta)}{d\zeta^2} - \omega^2 v(\zeta) = 0 \quad [3.20a]$$

$$\frac{d^2 y(\bar{t})}{d\bar{t}^2} + \tau_0^2 r(\bar{t}) + \omega^2 y(\bar{t}) = 0 \quad [3.20b]$$

with function ω^2 being an unknown separation constant. The time-dependent function $r(\bar{t})$ related to the Zener foundation reaction is provided by the solution of the fractional differential equation:

$$\alpha \tau_s{}^\beta \left(D_{0+}^\beta r\right)(\bar{t}) + \gamma r(\bar{t}) = \tau_s{}^\beta \left(D_{0+}^\beta y\right)(\bar{t}) - y(\bar{t}) \quad [3.21]$$

which is coupled with the governing equations in equation [3.20a] to provide the transverse displacement of the column.

The boundary conditions for the shape function $v(\zeta)$ can be written as:

$$v(0) = \left.\frac{dv}{d\zeta}\right|_0 = \left.\frac{d^2 v}{d\zeta^2}\right|_1 = \left.\frac{d^3 v}{d\zeta^3}\right|_1 = 0 \quad [3.22]$$

In addition, we observe that the governing equation for the shape function involves the parameter ω^2 that has the role of unknown dynamical frequency, as shown in equation [3.20b]. The separation constant is $\omega^2 \in \mathbb{C}$, with $\mathbb{C}$ as the complex field, so that $\omega^2 = \omega_r + i\omega_i$, with $\omega_r, \omega_i \in \mathbb{R}$ and $i = \sqrt{-1}$ as the imaginary unit.

The generalized Zener model of external restraints reduces, for limiting values of the material parameters (i.e. $\alpha \to 0$, with $k_2 \neq 0$) to the fractional Kelvin–Voigt model. In this case, use $\alpha \to 0$, with $k_2 \to 0$, so that only the springpot is retained in the external restraints. The fractional-order Maxwell model is instead obtained as $\alpha \to \infty$.

3.4.1. *The characteristic polynomial*

The exact solution of equation [3.20a] involves linear combinations of transcendental and trigonometric functions that, as we introduce the boundary condition, yields the eigenvalue equation among the axial load λ and the dynamic frequency ω. Such a relation, yielding a condition $\omega = \omega(\lambda)$, corresponds to a different time evolution of the column displacement for each choice of λ. The explicit relation $\omega = \omega(\lambda)$ is very cumbersome, which is beyond the scope of this chapter. In the following, we resort to an approximate representation of the shape function $v(\zeta)$ by assuming a two-term linear combination as:

$$v(\zeta) = A_1\phi_1(\zeta) + A_2\phi_2(\zeta) = \phi_j(\zeta) A_j \quad j = 1, 2 \quad [3.23]$$

where in the latter equation [3.23], we use the sum on the repeated index convention. Shape functions $\phi_j(\zeta)$ in equation [3.23] belong to a set of functions that satisfy the boundary conditions $\phi_j(\zeta) = \left.\dfrac{d\phi_j(\zeta)}{d\zeta}\right|_{\zeta=0} = 0$ and A_j are the unknown coefficients. The use of the two-term approximation of the space-dependent solution $v(\zeta)$ is a well-established procedure in the context of elastic stability of non-conservative systems. The reduction method introduced in equation [3.23] yields, in the case of the purely viscous foundation obtained for $\beta = 1$ in the proposed analysis, an overestimation of the critical load of almost 4%, with respect to the exact solution of equation [3.20a]. A similar consideration holds true for the case of elastic foundation where the overestimation of the critical load obtained with the two-term expansion with respect to the exact solution is almost 1% (Plaut and Infante 1970). Introducing equation [3.23] into equation [3.20a] yields:

$$\varepsilon(\zeta) = A_j\left[\frac{\partial^4\phi_j(\zeta)}{\partial\zeta^4} + \lambda^2\frac{\partial^2\phi_j(\zeta)}{\partial\zeta^2} - \omega^2 A_j\phi_j(\zeta)\right] \neq 0 \qquad [3.24]$$

with $j = 1, 2$ and the sum convention being used. The unbalance function, $\varepsilon(\zeta)$, projected onto the manifold defined by two functions $\phi_j(\zeta)$ yields two algebraic equations in terms of constants A_j that can be written as:

$$\int_0^1 \varepsilon(\zeta)\,\phi_1(\zeta)\,d\zeta = \varepsilon_{11}\left(\omega^2,\lambda^2\right)A_1 + \varepsilon_{21}\left(\omega^2,\lambda^2\right)A_2 = 0 \qquad [3.25a]$$

$$\int_0^1 \varepsilon(\zeta)\,\phi_2(\zeta)\,d\zeta = \varepsilon_{12}\left(\omega^2,\lambda^2\right)A_1 + \varepsilon_{22}\left(\omega^2,\lambda^2\right)A_2 = 0 \qquad [3.25b]$$

where ε_{jk} are the coefficients. The system of algebraic equation in equations [3.25a] and [3.25b] can be solved by an inhomogeneous set of constants A_j, only if the determinant:

$$r\left(\omega^2,\lambda^2\right) = \varepsilon_{11}\left(\omega^2,\lambda^2\right)\varepsilon_{22}\left(\omega^2,\lambda^2\right) - \varepsilon_{21}\left(\omega^2,\lambda^2\right)\varepsilon_{12}\left(\omega^2,\lambda^2\right) = 0 \qquad [3.26]$$

that corresponds to a specific condition among the separation constant ω^2 and the square of the non-dimensional load λ. For any specified value of the axial load, a value of dynamic frequency $\omega^2(\lambda^2)$ is obtained by [3.26]. In the case of the springpot external restraint (namely $\alpha \to \infty$), the solution of equation [3.20b] is obtained as a series of two-parameter Mittag–Leffler functions as:

$$y(t) = \sum_{k=0}^{\infty}\frac{(-1)^k}{k!}\left(\frac{\omega^2}{\tau_0^{1-\beta}\tau_\beta}\right)^k t^{2(k+1)-1}E^{(k)}_{2-\beta,2+\beta k}\left(-\frac{\omega^2}{\tau_0^{1-\beta}\tau_\beta}t^{2-\beta}\right) \qquad [3.27]$$

where $E_{\lambda,\mu}^{(k)}(\cdot)$ is the two-parameter Mittag–Leffler function, with $k = (0, 1, 2, \ldots)$ (Podlubny 1998). The stability of the solution in equation [3.27] cannot be inferred by direct inspection and, in such a case, several strategies to assess the stability of dynamical systems governed by fractional differential equations (FDE) have been proposed. Moreover, the Laplace transform of equation [3.20b] yields conditions on stability by checking the poles of the Laplace transform $\mathcal{L}\left[y\left(\bar{t}\right)\right] = \hat{y}\left(s\right)$ (Atanackovic and Stankovic 2004; Atanackovic *et al.* 2015).

Similar considerations hold true for the case of generalized Zener external restraints and, in those cases, a modal space approach leading to decoupled fractional differential equations involving the separation constant ω is obtained.

In the following, a different approach to stability of FDE is proposed under the assumption of rational values of the derivation order.

3.4.2. *State-space representation of the dynamic equilibrium equation*

The transverse displacement $w(\zeta, t)$ of cantilever columns, introduced in the previous section, is governed by the two differential equations [3.20a] and [3.20b]. Separated solutions involve the Mittag–Leffler series, which requires the knowledge of the function $\omega^2 = \omega^2(\lambda)$ by equation [3.20b], but no considerations on the stability of the solution may be inferred. In this section, we aim to provide a solution to the stability problem of the fractional-order Beck's column with an original approach. Let us assume, in the following, that the differentiation order $\beta = \dfrac{p}{q}$, with $p, q \in \mathbb{N}$ and $p \leq q$ since $0 \leq \beta \leq 1$, may be expanded. Under these circumstances, the time variations of the transverse displacement $y(\bar{t})$ are governed by a coupled system of FDE that can be written as (Rossikhin and Shitikova 2001):

$$\frac{d^2 y}{d\bar{t}^2}\left(\bar{t}\right) + r\left(\bar{t}\right) + \omega^2\left(\lambda\right) y\left(\bar{t}\right) = 0 \qquad [3.28]$$

The composition rule of the fractional-order operators namely $D_{0+}^{\beta_1}(D_{0+}^{\beta_2} f) = D_{0+}^{\beta_1+\beta_2}$ allows us to recast equation [3.28] in the form:

$$\sum_{l=1}^{2q} c_l \left(D_{0+}^{l\bar{\beta}} y\right)\left(\bar{t}\right) + \omega^2\left(\lambda\right) y\left(\bar{t}\right) + r\left(\bar{t}\right) = 0 \qquad [3.29a]$$

$$\sum_{l=1}^{p} a_l \left(D_{0+}^{l\bar{\beta}} r\right)\left(\bar{t}\right) + \gamma r\left(\bar{t}\right) - \sum_{l=1}^{p} b_l \left(D_{0+}^{l\bar{\beta}} y\right)\left(\bar{t}\right) + y\left(\bar{t}\right) = 0 \qquad [3.29b]$$

with the order $\bar{\beta} = 1/q$ and the coefficients $c_l = 0\, \forall l \neq 2q$, $c_{2q} = 1$, $a_l = b_l = 0 \forall l \neq p$ $a_l = \tau_s^{\beta}$ and $b_l = \tau_\varepsilon^{\beta}$ for $l = p$.

Under these circumstances, equation [3.29a] can be expanded in terms of a state variable vector $\mathbf{x}(\bar{t})$ with order mxp, with $m = 2q$:

$$\mathbf{x}(\bar{t}) = \begin{bmatrix} y_1(\bar{t}) \\ y_2(\bar{t}) \\ y_3(\bar{t}) \\ \vdots \\ y_{p+1}(\bar{t}) \\ \vdots \\ y_m(\bar{t}) \\ r_1(\bar{t}) \\ r_2(\bar{t}) \\ \vdots \\ r_p(\bar{t}) \end{bmatrix} = \begin{bmatrix} y(\bar{t}) \\ \left(D_{0+}^{\bar{\beta}} y\right) \\ \left(D_{0+}^{2\bar{\beta}} y\right) \\ \vdots \\ \left(D_{0+}^{r\bar{\beta}} y\right) \\ \vdots \\ \left(D_{0+}^{(m-1)\bar{\beta}} y\right) \\ p((\bar{t}) \\ \left(D_{0+}^{\bar{\beta}} r\right) \\ \vdots \\ \left(D_{0+}^{(p-1)\bar{\beta}} r\right) \end{bmatrix} \qquad [3.30]$$

with the supplementary $2q-1$ and $p-1$ identities:

$$\left(D_{0+}^{\bar{\beta}} y_l\right)(\bar{t}) - y_{l+1}(\bar{t}) = 0 \qquad l = 1, 2, ..., 2q-1 \qquad [3.31a]$$

$$\left(D_{0+}^{\bar{\beta}} r_l\right)(\bar{t}) - r_{l+1}(\bar{t}) = 0 \qquad l = 1, 2, ..., p-1 \qquad [3.31b]$$

The introduction of the state variable vector and the state identities into equations [3.30] and [3.31a] allows us to express equation [3.29a] as a system of $\bar{\beta}$-order FDE as:

$$\begin{aligned} &c_1 1\left(D_{0+}^{\bar{\beta}} y_1\right)(\bar{t}) + h_{12} y_2 = 0 \\ &\vdots \\ &c_{mm}\left(D_{0+}^{\bar{\beta}} y_{m-1}\right)(\bar{t}) + r(\bar{t}) + h_{m1} y_1(\bar{t}) = 0 \end{aligned} \qquad [3.32]$$

with $h_{i,i+1} = 1$ for $i = 1, 2, ..., 2q-1$ and , $h_{m1} = \omega^2$, $h_{12} = h_{23} = h_{34} = \cdots = 1$. The supplementary FDE for the state variable of the Zener foundation can be written as:

$$\begin{aligned} &c_{m+j,m+j}\left(D_{0+}^{\bar{\beta}} r_j\right)(\bar{t}) + h_{m+j,m+j+1} r_{j+1} = 0 \\ &\vdots \\ &c_{m+p,m+p}\left(D_{0+}^{\bar{\beta}} r_{p-1}\right)(\bar{t}) - h_{m+p,m+1} r_1(\bar{t}) + c_{m+p,m}\left(D_{0+}^{\bar{\beta}} y_{p-1}\right)(\bar{t}) \\ &\qquad + h_{m+p,1} y_1(\bar{t}) = 0 \end{aligned} \qquad [3.33]$$

with $c_{ij} = \delta_{ij}$ and δ_{ij} as the Kronecker delta for $i \neq m$ and $i \neq m + p$ and $j \neq p$. $c_{mm} = \tau_0^2$ and $c_{m+p,1} = \alpha\tau_s^\beta$ and $c_m + p, p = \tau_s^\beta$.

The system of equation [3.32] can be written in the matrix form as:

$$\mathbf{C}\left(D_{0+}^{\bar{\beta}}\mathbf{x}\right)(\bar{t}) + \mathbf{H}\left(\omega^2\right)\mathbf{x}(\bar{t}) = 0 \qquad [3.34]$$

and in the matrix form:

$$\mathbf{C} = \begin{bmatrix} 1 & 0 & 0 & \cdots & 0 & 0 \\ 0 & 1 & \vdots & \ddots & & 0 \\ \vdots & 0 & 1 & \ddots & \ddots & \vdots \\ \vdots & \vdots & \ddots & 1 & \ddots & \vdots \\ 0 & 0 & 0 & \cdots & \ddots & 0 \\ 0 & 0 & \cdots & \tau_s^\beta & \cdots & \alpha\tau_s^\beta \end{bmatrix} \qquad [3.35]$$

and the frequency-dependent matrix $\mathbf{H}\left(\omega^2\right)$ is given by:

$$\mathbf{H}(\omega^2) = \begin{bmatrix} 0 & 1 & 0 & \cdots\cdots\cdots & & & 0 \\ 0 & 0 & 1 & \cdots\cdots\cdots & & & 0 \\ \vdots & \vdots & \ddots & \ddots & \cdots & \cdots & \cdots \\ \vdots & \vdots & \cdots & \ddots & \cdots & \cdots & \cdots \\ 0 & 0 & \cdots & \omega^2 & 1 & \cdots & 0 \\ 0 & \vdots & \vdots & \cdots & \ddots & \cdots & 1 \\ 1 & 0 & \cdots\cdots & & 0 & \alpha\gamma & 0 \end{bmatrix} \qquad [3.36]$$

Equation [3.34] can be rewritten in the form:

$$\left(D_{0+}^{\bar{\beta}}\mathbf{x}\right)(\bar{t}) + \mathbf{V}(\omega^2)\mathbf{x}(\bar{t}) = \mathbf{0} \qquad [3.37]$$

with matrix $\mathbf{V}\left(\omega^2\right)$ the dynamic matrix of the system related to the frequency-dependent matrix $\mathbf{H}\left(\omega^2\right)$ as:

$$\mathbf{V}(\omega^2) = \mathbf{C}^{-1}\mathbf{H}(\omega^2) \qquad [3.38]$$

The system of fractional differential equations given in equation [3.37] is stable as the eigenvalues of the dynamic matrix lie in a specific sector of the complex plane. In order to investigate system stability, we find a solution in the form:

$$\mathbf{x}(\bar{t}) = \mathbf{\Phi} E_{\bar{\beta}}\left(-\bar{t}^{-\bar{\beta}}\chi\right) \qquad [3.39]$$

with $\mathbf{\Phi} \in \mathbb{C}^m$ and $\chi \in \mathbb{C}$ are the eigenvectors and eigenvalues of the matrix $\mathbf{C}(\omega^2)$. The substitution of equation [3.39] into equation [3.34] yields the algebraic system:

$$(\mathbf{I}\chi - \mathbf{C}(\omega^2))\mathbf{\Phi} = \mathbf{0} \qquad [3.40]$$

where $\mathbf{I}$ is the identity matrix to the eigenvalue equation:

$$\det[\mathbf{C}\left(\omega^2\right) - \mathbf{I}\chi] = 0 \qquad [3.41]$$

Equation [3.40] is an algebraic equation for the problem eigenvalues $\chi(\omega^2)$, which can always be expressed as:

$$\chi^{m+1} + \chi^{m-1}C_\beta + A_m(\omega^2) = 0 \qquad [3.42]$$

where A_m is a β-dependent coefficient that will be specified as the order of fractional derivative is prescribed. The explicit values of the roots $\chi_1(\omega), \cdots, \chi_m(\omega)$ of the algebraic secular equation in equation [3.42] cannot be obtained for any values of the order β, and some explicit values have been obtained in previous papers (Bologna *et al.* 2017). In the following, we introduce a different approach to the stability of fractional-order Beck's column based on the extension of the Routh–Hurwitz criterion.

As far as the assumption of rational order derivative is removed, then the real (or irrational) order of derivation may be decomposed, as an example, in terms of the series of partial fractions and an unbounded number of state functions will be involved in the analysis.

3.4.3. *Stability analysis of fractional-order Beck's column via the extended Routh–Hurwitz criterion*

The direct solution of the secular equation in equation [3.42] provides the values $\chi_j(\omega^2) = \chi_j(\omega^2(\lambda))$ $j = 1, 2..., m$ of the system eigenvalues that can be $\chi_j \in \mathbb{R}$ or $\chi_j, \chi_j^* \in \mathbb{C}$, where $*$ denotes the complex conjugate. The dynamic stability of the fractional-order set of differential equations in equation [3.32] is provided as (Marden 1949; Matignon 1996; Barnett 1970):

$$\arg\left(\chi_j\right) \leq \left|\frac{\pi\bar{\beta}}{2}\right| \quad \forall j = 1, 2, \ldots, m \qquad [3.43]$$

where we denote $arg(\chi_j)$ as the argument of the complex number χ_j evaluated in the principal Riemannian manifold that is $arg(\chi_j) = tg^{-1}\left(\frac{\chi_{j,I}}{\chi_{j,R}}\right)$, with $\chi_{j,I}$ and $\chi_{j,R}$ as the imaginary and real components of the complex number χ_j, respectively.

Stability analysis conducted by the direct evaluation of the roots of the characteristic equation [3.42] is a formidable task since no closed-form expression of the roots $\chi(\omega^2)$ is available under the general differential order β. The search for eigenvalues $\chi_j(\omega^2)$ that fall into the instability region (Figure 3.9) can be observed under cases $\beta = 1$, resorting to the well-established Routh–Hurwitz criterion $arg(\chi_j) \leq \left|\frac{\pi}{2}\right| \forall j$, which corresponds to $\chi_{j.R} \leq 0 \; \forall j$ (Marden 1949). Similar considerations also hold for the case $\beta = 0$, yielding dynamic stability as $\chi_j \in \mathbb{R}^- \forall j$. In addition, we observe that under such conditions, the Hermann–Smith paradox is involved, meaning that the stability load λ is not affected by the external elastic restraints ($\beta = 0$) (Beck 1952).

Indeed, it is well known that the evaluation of the number of roots χ_j $j = 1, 2 \dots m$ with positive real parts is provided by the evaluation of the number of change of signs of the determinant sequence (Marden 1949) $\Delta_1, \Delta_2, \dots \Delta_n$, namely $V(1, \Delta_1, \Delta_2, \dots, \Delta_n)$, with V being the number of signs changed. The Routh–Hurwitz theorem in (Marden 1949) can be obtained by the Sturm sequence applying the characteristic equation, followed by an appropriate rotation of the complex plane counterclockwise by the phase $\theta = i\pi/2$.

A different case is involved as fractional-order hereditary restraints are considered with $0 \leq \bar{\beta} \leq 1$, as shown in equation [3.29a]. Indeed, under such circumstances, a vector of the complex phase with origin in $x = 0$ corresponds to dynamic instability as $-\dfrac{\bar{\beta}\pi}{2} \leq arg(\chi_j) \leq \dfrac{\bar{\beta}\pi}{2}$, as shown in Figure 3.9.

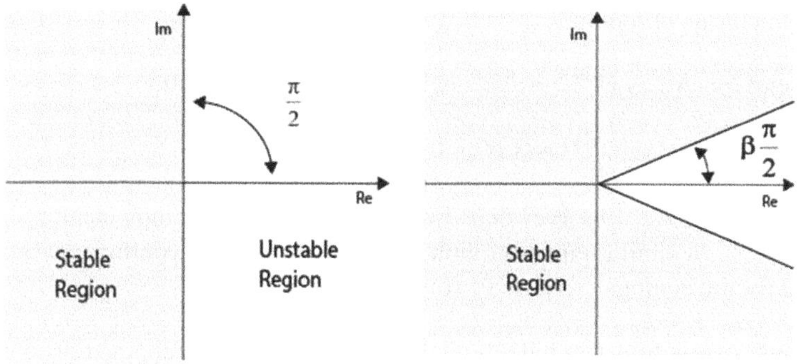

Figure 3.9. *Stability region on the complex plane*

In order to extend the Routh–Hurwitz stability theorem to this latter case, we consider the complex of conjugated χ_j roots on the borders of the sector with arguments:

$$arg(\chi_j) = \frac{\bar{\beta}\pi}{2} \qquad arg(\chi_j^*) = -\frac{\bar{\beta}\pi}{2} \qquad [3.44]$$

The introduction of phase shifts

$$\varphi_j = (1-\beta)\frac{\pi}{2} \quad \varphi_j^* = (1+\beta)\frac{\pi}{2} \qquad [3.45]$$

as:

$$\bar{\chi}_j = \chi_j e^{i\varphi_j}; \qquad \bar{\chi}_j^* = \chi_j^* e^{i\varphi_j{}^*} \qquad [3.46]$$

yields arguments of the couples $\bar{\chi}_j$ as:

$$arg(\bar{\chi}_j) = \frac{\pi}{2}; \qquad arg(\bar{\chi}_j^*) = -\frac{\pi}{2} \qquad [3.47]$$

which corresponds to complex roots on the imaginary axis. Complex conjugated roots in the instability region with $arg(\chi_j) \leq \left|\frac{\beta\pi}{2}\right|$ yield after the phase shifts, $Re(\bar{\chi}_j) > 0$; $Re(\bar{\chi}_j^*) > 0$. Complex conjugated roots within the stable region with $arg(\chi_j) \geq \frac{\beta\pi}{2}$ and $arg(\chi_j^*) \geq \frac{\beta\pi}{2}$ yield, after the phase shifts: $Re(\bar{\chi}_j) \leq 0$; $Re(\bar{\chi}_j^*) \leq 0$. Under these circumstances, as we introduce the phase shifts in equation [3.45] as:

$$\chi_j = \bar{\chi}_j e^{-i\varphi_j} \qquad [3.48]$$

and we replace this in equation [3.42], we obtain a complex coefficient eigenvalue equation in terms of roots $\bar{\chi}_j$:

$$s(\omega^2) = N_{m+1}\bar{\chi}^{m+1} + N_{m-1}\bar{\chi}^{m-1}C_\beta + A_m(\omega^2) = 0 \qquad [3.49]$$

where $N_{m+1} = e^{-i\varphi_j(m+1)}$ and $N_{m-1} = e^{-i\varphi_j(m-1)}$.

In addition, we observe that by introducing the complementary rotation $\chi_j^* = \bar{\chi}_j^* e^{i\varphi_j^*}$, a similar equation with complex conjugated coefficients, $s^*(\omega^2)$, is obtained for the conjugated roots $\bar{\chi}_j^*$ (Figure 3.10).

Indeed, a straightforward algebraic manipulation yields a characteristic equation coalescing equation [3.49]. The sign of the real parts $\bar{\chi}_{jR}$ of the roots of equation [3.49], corresponds henceforth to the roots χ_j of the characteristic equation in equation [3.42] that may be: i) $Re(\bar{\chi}_j) < 0 \; \forall j \Rightarrow arg(\chi_j) \geq \frac{|\beta\pi|}{2} \; \forall j$; ii) $Re(\chi_j) = 0 \Rightarrow arg(\chi_j) = \frac{|\beta\pi|}{2} \; \forall j$. Previous consideration yields that phase shift of the stability sector in the complex plane represents a conformal mapping of the $\mathbb{C}$ plane with one to one correspondence. The aforementioned considerations suggest that the stability of fractional-order dynamical systems may be studied via the Routh–Hurwitz criterion applied to the complex characteristic equation in equation [3.49]. The Routh–Hurwitz criterion will be applied over the real coefficient characteristic equation $S^2(\omega^2(\lambda)) = 0$ of order 2m, obtained by the product

$S^2(\omega^2(\lambda)) = s(\omega^2)s^*(\omega^2)$. Signs of the real part of the roots of the equation $S^2(\omega^2(\lambda)) = 0$ can be obtained by the sign of the sequence of determinants.

$$p = V(1, \Delta_1, \ldots, \Delta_n) \qquad [3.50a]$$

$$q = V(1, -\Delta_1, \ldots, (-1)^n \Delta_n) \qquad [3.50b]$$

where Δ_k (Marden 1949) has the structure of:

$$\Delta_k = \begin{vmatrix} m_1 & m_3 & \cdots & m_{2k-1} & -n_2 & -n_4 & \cdots & -n_{2k-2} \\ 1 & m_2 & \cdots & m_{2k-2} & -n_1 & -n_3 & \cdots & -n_{2k-3} \\ 0 & m_1 & \cdots & m_{2k-3} & 0 & -n_2 & \cdots & -n_{2k-4} \\ 0 & 1 & \cdots & \vdots & \vdots & 0 & \vdots & \vdots \\ 0 & 0 & \cdots & m_{2k} & 0 & 0 & \cdots & -n_{k-1} \\ 0 & n_2 & \cdots & n_{2k-2} & m_1 & m_3 & \cdots & -m_{k-3} \\ \vdots & n_1 & \cdots & n_{2k-3} & 1 & m_2 & \cdots & \vdots \\ 0 & 0 & \cdots & n_k & 0 & 0 & \cdots & -m_{k-1} \end{vmatrix} \qquad [3.51]$$

for $k = 1, \cdot, n$, with m_j and n_j, with $j = 1, \cdot, n$ are the coefficients of the characteristic polynomial, m_j are the coefficients of real part and n_j are the coefficients of the imaginary part.

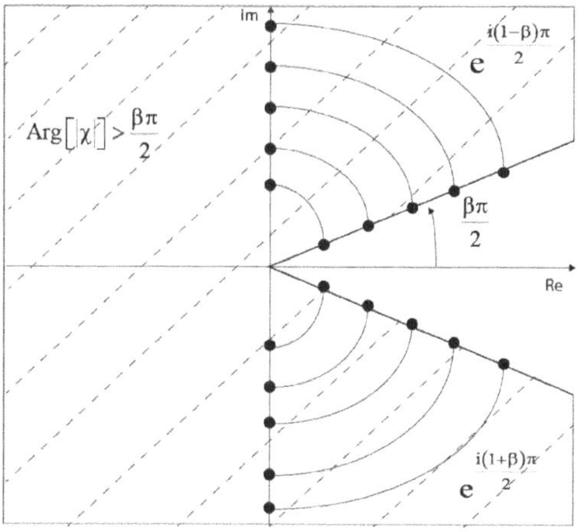

Figure 3.10. *Stability region and complex map*

3.5. Numerical application

We report our findings in Figures 3.11–3.13 for the cases that we analyzed, and we can conclude that the presence of a fractional Kelvin–Voigt model, a Fractional

Zener, in which the spring with stiffness $k_1 \to 0$, keeping $C_\beta \neq 0$, does not remove the Hermann–Smith and Ziegler paradox as already observed in the classical approach to the problem.

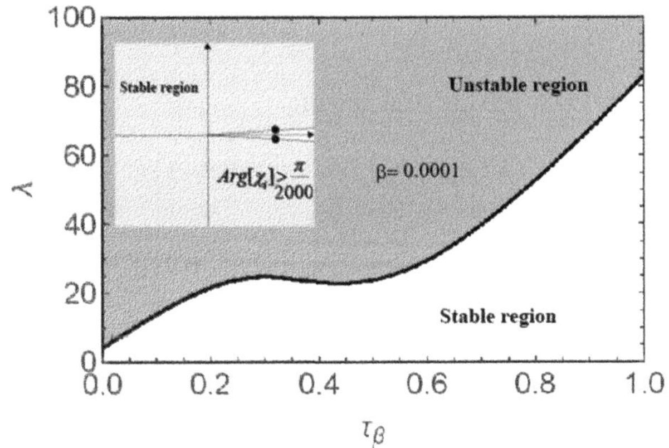

Figure 3.11. *Plot of the critical load for* $\beta = 0.0001$

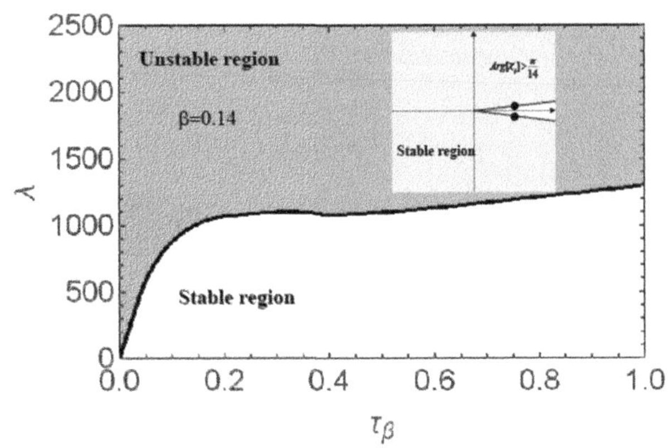

Figure 3.12. *Plot of the critical load for* $\beta = 0.14$

Parametric analysis shows the influence of the parameter τ_s and the effect of the order of derivation on the stability value of the applied load. It can be observed from the reported parametric analysis that as $\beta = 10^{-4}$ ($\beta \to 0$ external elastic restraints), the load λ is almost insensitive to the presence of the external restraints, as predicted by the Hermann–Smith paradox. The analysis has been carried out introducing the anomalous time scaling $\tau_\beta = (\tau_s)^\beta$ for the abscissa and material parameter.

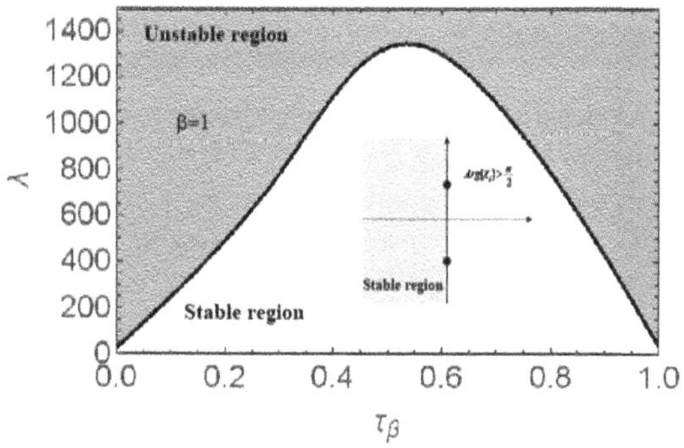

Figure 3.13. *Plot of the critical load for* $\beta = 1$

In the case of the viscous restraint, the common value $\lambda = 4.13$ is obtained as $\tau_s \to 0$ for any value of the differentiation order β, as predicted by the Ziegler paradox (Ziegler 1952).

3.6. Conclusion

The analysis of Beck's column resting on an external foundation is a challenging and still unsolved problem in the field of mechanics. Indeed, the presence of well-established paradoxes arising from stability analysis and involving the evaluation of critical loads yields unreliable solutions in terms of stability. In this study, we aimed to replace the external restraints with a generalized version of material behavior involving fractional-order calculus that proved to be very efficient in representing real material behavior. In the presence of fractional-order constitutive equations, the usual version of the Routh–Hurwitz criterion, based on the sign of the real part of the eigenvalues of the dynamic matrix of the problem, cannot be used since the stability region is not the entire half-positive complex plane. In this study, we also showed that in the presence of more general foundations, besides the classical fractional type, the Routh–Hurwitz theorem may be generalized by introducing a conformal mapping of the boundaries of the stability region that moves to the complex-half plane. Some numerical results assessing the reliability of the proposed approach have been shown in the study.

3.7. References

Atanackovic, T.M. and Stankovic, B. (2004). On a system of differential equations with fractional derivatives arising in rod theory. *Journal of Physics A: Mathematical and General*, 37(4), 1241.

Atanackovic, T.M., Janev, M., Konjik, S., Pilipovic, S., Zorica, D. (2015). Vibrations of an elastic rod on a viscoelastic foundation of complex fractional Kelvin–Voigt type. *Meccanica*, 50(7), 1679–1692.

Barnett, S. (1970). Greatest common divisor of two polynomials. *Linear Algebra and its Applications*, 3(1), 7–9.

Beck, M. (1952). Die knicklast des einseitig eingespannten, tangential gedrückten stabes. *Zeitschrift für angewandte Mathematik und Physik ZAMP*, 3(3), 225–228.

Bologna, E. and Zingales, M. (2018). Stability analysis of Beck's column over a fractional-order hereditary foundation. *Proceedings of the Royal Society A: Mathematical, Physical and Engineering Sciences*, 474(2218), 20180315.

Bologna, E., Deseri, L., Zingales, M. (2017). A state-space approach to dynamic stability of fractional-order systems: The extended Routh-Hurwitz theorem. *AIMETA2017*, Salerno, 4–7 September.

Bologna, E., Graziano, F., Deseri, L., Zingales, M. (2019). Power-laws hereditariness of biomimetic ceramics for cranioplasty neurosurgery. *International Journal of Non-Linear Mechanics*, 115, 61–67.

Bologna, E., Di Paola, M., Dayal, K., Deseri, L., Zingales, M. (2020). Fractional-order nonlinear hereditariness of tendons and ligaments of the human knee. *Philosophical Transactions of the Royal Society A*, 378(2172), 20190294.

Elishakoff, I. (2001). Euler's problem revisited: 222 years later. *Meccanica*, 36(3), 265–272.

Elishakoff, I. (2005). Controversy associated with the so-called "follower forces": Critical overview. *Applied Mechanics Reviews*, 58(2), 117–142.

Mainardi, F. (2010). *Fractional Calculus and Waves in Linear Viscoelasticity: An Introduction to Mathematical Models*. Imperial College Press, London.

Marden, M. (1949). *Geometry of Polynomials*. American Mathematical Society, Providence, RI.

Matignon, D. (1996). Stability results for fractional differential equations with applications to control processing. *Multiconference on Computational Engineering in Systems Applications*, vol. 2, Lille, France, 963–968.

Pinnola, F.P. (2016). Statistical correlation of fractional oscillator response by complex spectral moments and state variable expansion. *Communications in Nonlinear Science and Numerical Simulation*, 39, 343–359.

Plaut, R.H. and Infante, E. (1970). The effect of external damping on the stability of Beck's column. *International Journal of Solids and Structures*, 6(5), 491–496.

Podlubny, I. (1998). *Fractional Differential Equations: An Introduction to Fractional Derivatives, Fractional Differential Equations, to Methods of their Solution and Some of their Applications*. Academic Press, San Diego, CA.

Rossikhin, Y.A. and Shitikova, M. (2001). A new method for solving dynamic problems of fractional derivative viscoelasticity. *International Journal of Engineering Science*, 39(2), 149–176.

Smith, T. and Herrmann, G. (1972). Stability of a beam on an elastic foundation subjected to a follower force. *Journal of Applied Mechanics*, 39(2), 628–629.

Spanos, P.D. and Miller, S.M. (1994). Hilbert transform generalization of a classical random vibration integral. *Journal of Applied Mechanics*, 61, 575–581.

Ziegler, H. (1952). Die stabilitätskriterien der elastomechanik. *Ingenieur-Archiv*, 20(1), 49–56.

4

Localization in the Static Response of Higher-Order Lattices with Long-Range Interactions

The static behavior of a generalized axial lattice with direct and indirect neighboring interactions is studied under uniform tensile loading. The two-neighbor interaction lattice is composed of different elastic springs connected to adjacent nodes and to next adjacent nodes, with possible different stiffness values for each interaction. The boundary nodes of this one-dimensional generalized lattice are assumed to be fixed for one end and loaded by a prescribed elongation at the free other end. The static response of this generalized axial lattice under uniform tensile load is analytically studied, from the resolution of a linear fourth-order difference equation. It is shown that the so-called higher-order boundary conditions have a crucial influence on the possible localization process of the displacement in the vicinity of the clamped end. The referenced homogeneous response may be perturbed when considering non-symmetric higher-order boundary conditions at the fixed nodes (or truncated higher-order boundary conditions). This non-symmetric higher-order boundary condition is also interpreted in terms of local weakening of the stiffness in the vicinity of the clamped section. A non-monotonic boundary layer is exhibited for the discrete displacement field of this generalized lattice. This discrete boundary layer is analyzed from an asymptotic approach, and is approximated through a simplified formulae based on the exponential law, with a change of sign from one particle to the next.

Chapter written by Noël CHALLAMEL and Vincent PICANDET.

4.1. Introduction

Periodic subsystems, which possess periodic or almost periodic microstructures, are basic structural elements that may be introduced in a refined representation of many physical and mechanical problems (see Banakh and Kempner 2010 or Wang *et al.* 2020). The microstructure is controlled by some characteristic sizes with respect to the structural scale. Microstructure effects may play a decisive role in the appearance of some localization phenomena. A possible approach to describing these periodic microstructures is based on lattice mechanics, formulated in terms of discrete cells, connected by some short- or long-range interactions. The wave dispersive properties of these periodic lattices are well studied, and are available in standard textbooks devoted to the wave propagation in heterogeneous or microstructured media (Brillouin 1946; Born and Huang 1954; Aşkar 1986; Maugin 1999; Craster and Kaplunov 2013). In this chapter, we will focus on linear elastic interactions, for each particle, with restoring force based on the interaction with the adjacent particle, as well as with its second neighboring. This lattice may be labeled as a generalized lattice with a general interaction dependence between particles next to the considered adjacent particle. Such a long-range lattice is the generalization of simple lattices considered by Lagrange (1759, 1788) (one-dimensional string or axial lattices) or by Born and von Kármán (1912) (including three-dimensional elastic lattices) based on direct neighboring interaction. The wave dispersive properties in a generalized infinite lattice (with short- and long-range interactions) have been analytically studied by Brillouin (1946), Eaton and Peddieson (1973) and Roseneau (1987). As described in the seminal book by Maugin (1999), lattices with multiple short- and long-range interactions may play a key role in the understanding of non-local and nonlinear generalized wave propagation, whose stability property may be sensitive to the interaction complexity (see more recently Tarasov (2010) or Michelitsch *et al.* (2019)). Generalized lattices may also be strongly related to integral-based non-local elastic media (see, for instance, Eringen and Kim (1977)).

The vibration properties of such generalized elastic lattices of finite size have been extensively studied during the last decades, with closed-form solutions for linear elastic interactions. Pipes (1966), Chen (1970, 1971), Eaton and Peddieson (1973) or more recently Challamel *et al.* (2018, 2019) calculated the natural frequencies of a finite generalized lattice (with direct and indirect neighboring interactions). Pipes (1966) and Chen (1970, 1971) considered a specific finite lattice with N-neighbor interaction characterized by equal stiffness for each interaction. Eaton and Peddieson (1973) obtained the eigenfrequencies of a fixed–fixed, fixed–free and free–free generalized lattice with N-neighbor interaction, with possible differentiation between short- and long-range elastic interactions. Eaton and Peddieson (1973) also approximated the dynamic equation of the generalized lattice (with short- and long-range interactions) by a continuous rod model with higher-order derivatives (higher-order gradient elasticity model). Carcaterra *et al.* (2015) related the topology of the microstructure connections of the generalized lattice (with short- and long-range interactions) to some asymptotic properties of a gradient continuous medium (with convergence properties for sufficiently small lattice spacing). Challamel *et al.* (2018) recently reconsidered the vibration behavior of this finite generalized lattice with relevant scaling parameters, based on an asymptotic behavior of the generalized lattice which converges towards the continuous local rod model for sufficiently small spacing parameters. Challamel *et al.* (2018), following the works of Rosenau (1987), also built an approximated continuous non-local rod model without additional higher-order spatial derivatives for capturing the length scale dependence of the generalized lattice. This non-local rod model is shown to be equivalent to a stress gradient Eringen model (in the sense of the differential model of Eringen (1983). The eigenfrequencies of this non-local rod model fit the exact ones of the generalized lattice with strong accuracy. The scale dependence of the eigenfrequencies of this generalized lattice is found to be controlled by the lattice spacing and the stiffness ratio of the direct and indirect neighboring interactions. Challamel *et al.* (2018) (see also Challamel *et al.* 2019) showed that the calculation of the eigenfrequency of the generalized lattice needs to solve a higher-order difference boundary value problem. The eigenfrequency solutions of Eaton and Peddieson (1973) have been confirmed again in the presence of symmetrical or antisymmetrical higher-order boundary conditions. However, for some unsymmetrical higher-order boundary conditions, the scale dependence of the eigenfrequencies may be drastically affected, thus showing the strong dependence of the mechanical problem to the choice of higher-order boundary conditions. This strong dependence of the mechanical response to the choice of higher-order boundary conditions is revisited in this chapter, and restricted to a static analysis for a generalized lattice under uniform tensile loading.

Exact resolution of linear lattice problems can be attained from the mathematical properties of linear difference equations (see the monographs of Goldberg 1958 or Elaydi 2005). Exact solutions of axial lattice with direct and indirect neighboring interactions have been obtained in statics by Gazis and Wallis (1964), Toupin and Gazis (1964) and Mindlin (1965). Mindlin (1965) also included the influence of a third neighboring interaction. Toupin and Gazis (1964) considered the static behavior of a lattice chain with direct and next-to-direct elastic interaction, loaded by a system of axial tensile loads. They pointed out the existence of a boundary layer for this problem, and the possible coupling exponential regime for the displacement with a change of sign from one particle to the next. Triantafyllidis and Bardenhagen (1993) obtained numerical solutions for a nonlinear axial lattice under uniform axial load and with possible interaction with direct or other adjacent elements. Such a nonlinear discrete problem is the generalization of FPU nonlinear lattices (Fermi *et al.* 1955) with both direct and next-to-direct interactions. Very few analytical solutions are available for the static response of nonlinear lattices (see

Hérisson *et al.* (2016) for the derivation of exact analytical solutions of the static response of FPU systems under non-uniform distributed axial load). Triantafyllidis and Bardenhagen (1993) and Hérisson *et al.* (2016) characterized the localization phenomenon due to the nonlinear restoring force between adjacent particles (these last problems can be viewed as the discrete investigation of the nonlinear Ericksen (1975) continuous bar). This localization phenomenon is different from the one considered in this chapter where only linear elastic interactions are considered. In this chapter, it will be shown that even in the presence of pure linear interactions, the specific effect of additional long-range interaction may be responsible for some localization phenomena. Charlotte and Truskinovsky (2002, 2008) investigated the static behavior of a lattice composed of direct and next-to-direct linear elastic interaction and showed the strong influence of discrete-based boundary conditions, through a soft and a hard loading device (displacement- or force-based boundary conditions). Charlotte and Truskinovsky (2002) also performed a generic study of the mathematical property of the associated linear difference equation of the generalized lattice, with possible negative sign for the generalized stiffness.

In this chapter, the static response of a generalized lattice with direct and next-to-direct elastic interactions is studied. An exact solution is developed for this generalized lattice under uniform tensile load with symmetrical and unsymmetrical higher-order boundary conditions at the clamped end. This solution is obtained from the resolution of a linear fourth-order difference equation. For unsymmetrical higher-order boundary conditions, an oscillatory boundary layer is exhibited in the vicinity of the clamped end. This boundary layer, specific from the discreteness of this lattice problem, is analytically characterized from an asymptotic approach, with oscillating terms from one node to another.

4.2. Two-neighbor interaction – general formulation – homogeneous solution

The generalized axial lattice with two-neighbor interactions is considered, as shown in Figure 4.1. This lattice is composed of $(n+1)$ concentrated masses (or particles). The axial displacement of a mass at node i is denoted by u_i. The spacing between each mass is assumed to be uniform and is denoted by a.

The axial spring stiffness of the direct neighboring interaction is denoted by k_1, whereas the axial spring stiffness of the second neighboring interaction is denoted by k_2. The lattice is composed of equal concentrated massed m attached at each node, except at the end nodes of the lattice. The fixed boundary conditions is ensured by fixing the first node of the lattice and by assuming that the second neighboring is attached to the fixed node with a spring stiffness $2k_2$ that is twice the stiffness value k_2 inside the lattice. Such a stiffness calibration is similar to that considered for calibrating the rotational stiffness at the lattice clamped boundary condition of the Hencky system (see Hencky (1920) who already mentioned that the calibrated stiffness of the clamped section is twice the one inside the beam domain, or see Challamel *et al.* (2014) and Wang *et al.* (2017) for a recent discussion of this problem in conjunction with a non-local beam mechanics insight).

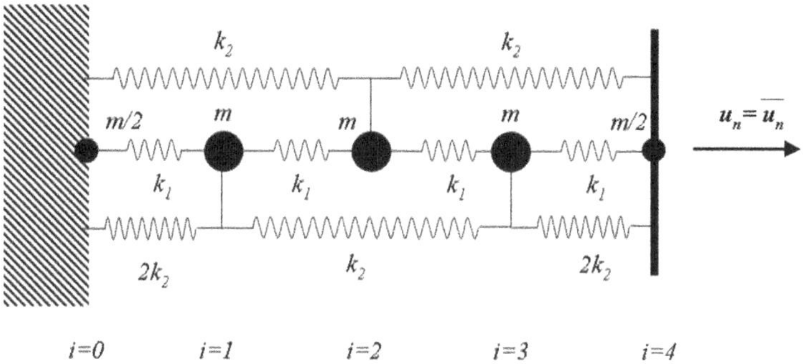

Figure 4.1. *Extension of a generalized lattice with direct and indirect interactions under displacement controlled test* $u_n = \overline{u}_n$ *; symmetrical higher-order boundary conditions at the clamped section* $u_{-1} = -u_1$ *; n=4; uniform loading*

The balance equation can be obtained from the following difference equation:

$$N_{i+1/2} - N_{i-1/2} = 0$$

with

$$N_{i+1/2} = k_1 \left(u_{i+1} - u_i\right) + k_2 \left(u_{i+2} - u_i\right) + k_2 \left(u_{i+1} - u_{i-1}\right) \quad [4.1]$$

It is possible to introduce the following scaling law for the stiffness:

$$k_1 = \beta_1 \frac{EA}{a}, \; k_2 = \beta_2 \frac{EA}{4a} \quad [4.2]$$

with the following parameter constraints $\beta_1 \geq 0$, $\beta_2 \geq 0$. EA is the axial rigidity of the continuous axial bar, asymptotically obtained for a number n of nodes tending to infinite. We also have $\beta_1 + \beta_2 = 1$ so that the parameters β_1 and can be considered as weighting coefficients for the short- and long-range interactions. Moreover, it is physically reasonable to assume that the weighting positive coefficients β_i decrease for the long-range interaction, which means that:

$$\beta_1 \geq \beta_2 \geq 0 \quad \Rightarrow \quad \frac{1}{2} \leq \beta_1 \leq 1 \text{ and } 0 \leq \beta_2 \leq \frac{1}{2} \qquad [4.3]$$

The difference equations can be reformulated as:

$$\frac{N_{i+1/2} - N_{i+1/2}}{a} = 0$$

with

$$N_{i+1/2} = EA\left[\beta_1 \frac{u_{i+1} - u_i}{a} + \beta_2 \frac{u_{i+2} - u_i}{4a} + \beta_2 \frac{u_{i+1} - u_{i-1}}{4a}\right] \qquad [4.4]$$

The mixed differential–difference equation is obtained by coupling both equations:

$$EA\left[\beta_1 \frac{u_{i+1} - 2u_i + u_{i-1}}{a^2} + \beta_2 \frac{u_{i+2} - 2u_i + u_{i-2}}{4a^2}\right] = 0 \qquad [4.5]$$

In this chapter, the case $\beta_2 \geq 0$ will be investigated, which is physically motivated by the positivity of the associated lattice potential energy. Gazis and Wallis (1964) or Toupin and Gazis (1964) remark that in this case, the amplitude of the displacement varies exponentially with the position of the particle, but the sign changes from one particle to the next (in a boundary layer). Gazis and Wallis (1964) and Toupin and Gazis (1964) also considered the case $\beta_2 \leq 0$ with exponential decaying in the boundary zone (but this case is associated with a non-definite positive energy function of the generalized lattice).

This difference equation (equation [4.5]) may have been obtained equivalently from energy arguments. For instance, starting from the fixed–fixed lattice system, the potential energy of the linear elastic lattice with two closest neighbors may be written as:

$$W = \sum_{i=0}^{n-1} \frac{k_1}{2}\left(u_{i+1} - u_i\right)^2 + \sum_{i=0}^{n-2} \frac{k_2}{2}\left(u_{i+2} - u_i\right)^2 + k_2\left(u_1 - u_0\right)^2 + k_2\left(u_n - u_{n-1}\right)^2 \qquad [4.6]$$

with the following parameter constraints $k_1 \geq 0$, $k_2 \geq 0$ so as to ensure the definite positivity of the potential energy. For a fixed boundary condition at the clamped section, and prescribed displacement at the other end $u_n = \overline{u_n}$ (uniaxial tensile loading test), the kinematic constraints are imposed such that:

$$u_0 = 0 \text{ and } u_n = \overline{u_n} \qquad [4.7]$$

Equation [4.6] can also be equivalently written in terms of fictitious nodes outside the domain:

$$W = \sum_{i=0}^{n-1} \frac{k_1}{2}\left(u_{i+1} - u_i\right)^2 + \sum_{i=0}^{n-2} \frac{k_2}{2}\left(u_{i+2} - u_i\right)^2 + \frac{k_2}{4}\left(u_1 - u_{-1}\right)^2 + \frac{k_2}{4}\left(u_{n+1} - u_{n-1}\right)^2 \qquad [4.8]$$

with the antisymmetrical higher-order additional boundary conditions:

$$u_{-1} = -u_1 \text{ and } u_{n+1} = 2\overline{u_n} - u_{n-1} \qquad [4.9]$$

The potential energy of the linear elastic lattice with two closest neighbors is then written as:

$$\begin{aligned} W = & \sum_{i=0}^{n-1} \beta_1 \frac{EA}{2}\left(\frac{u_{i+1} - u_i}{a}\right)^2 \times a + \sum_{i=0}^{n-2} \beta_2 \frac{EA}{2}\left(\frac{u_{i+2} - u_i}{2a}\right)^2 \times a \\ & + \beta_2 EA\left(\frac{u_1 - u_0}{2a}\right)^2 \times a + \beta_2 EA\left(\frac{u_n - u_{n-1}}{2a}\right)^2 \times a \end{aligned} \qquad [4.10]$$

For the prescribed displacement-controlled test, the tensile load is calculated from the discrete displacement field solution:

$$F = k_1\left(\overline{u_n} - u_{n-1}\right) + 2k_2\left(\overline{u_n} - u_{n-1}\right) + k_2\left(\overline{u_n} - u_{n-2}\right) \qquad [4.11]$$

The dimensionless parameters of the problem can be introduced as:

$$F^* = \frac{F}{EA}, \quad u_i^* = \frac{u_i}{L} \qquad [4.12]$$

so that the dimensionless load–displacement relationship at the end of the generalized lattice can be written as:

$$F^* = n\left(\beta_1 + \frac{\beta_2}{2}\right)\left(\overline{u_n^*} - u_{n-1}^*\right) + n\frac{\beta_2}{4}\left(\overline{u_n^*} - u_{n-2}^*\right) \qquad [4.13]$$

This linear fourth-order difference equation [4.5] can be equivalently rewritten as:

$$u_{i+1} - 2u_i + u_{i-1} + \frac{\beta_2}{4}(u_{i+2} - 4u_{i+1} + 6u_i - 4u_{i-1} + u_{i-2}) = 0 \quad [4.14]$$

which can be presented in a condensed formula:

$$\Delta\left(1 + \frac{a^2\beta_2}{4}\Delta\right)u_i = 0 \quad [4.15]$$

where Δ is the discrete Laplacian. Equation [4.15] directly shows that the solution of this fourth-order difference equation can be decomposed into a long wave and a boundary layer contribution:

$$u_i = \tilde{u}_i + \hat{u}_i \quad [4.16]$$

where the long-wave solution is governed by a discrete Laplace equation:

$$\Delta \tilde{u}_i = 0 \quad [4.17]$$

whose solution is linear

$$\tilde{u}_i = A_1 + A_2 i \quad [4.18]$$

whereas the second solution is obtained from a second-order difference equation:

$$\left(1 + \frac{a^2\beta_2}{4}\Delta\right)\hat{u}_i = 0 \quad [4.19]$$

The solution of this linear second-order difference equation is sought in the form (see Goldberg 1958; Elaydi 2005)

$$\hat{u}_i = \sum_{p\in\{1,2\}} C_p f_p^i \quad [4.20]$$

where the two solutions of the characteristic equation f_p are detailed below

$$f^2 + \left(\frac{4}{\beta_2} - 2\right)f + 1 = 0 \quad [4.21]$$

whose solutions may be expressed by:

$$f_{1,2}-1-\frac{2}{\beta_2}\pm\sqrt{\left(1-\frac{2}{\beta_2}\right)^2-1} \qquad [4.22]$$

Recall that $\beta_2\in[0;1]$ (and in fact for physical reasons, it is reasonable to assume that $\beta_2\le\beta_1$), so that the solutions may be equivalently reexpressed by:

$$f_{1,2}=-\cosh\theta\pm\sinh\theta \text{ with } \theta=a\cosh\left(1+2\frac{\beta_1}{\beta_2}\right) \qquad [4.23]$$

The solution to equation [4.14] can be finally written as the sum of the two solutions:

$$u_i=A_1+A_2 i+A_3(-1)^i\cosh(\theta i)+A_4(-1)^i\sinh(\theta i) \qquad [4.24]$$

This is in accordance with the solution derived by Challamel *et al.* (2019) for dynamic problems, when the inertia effects are neglected. This is also the solution obtained by Toupin and Gazis (1964) with symmetry considerations $A_1=A_3=0$. Toupin and Gazis (1964) also investigated the case $\beta_2\le 0$ with a purely exponential long-range solution, but without physical justification of such a sign (in this chapter, we only treat the case $\beta_2\ge 0$). Charlotte and Truskinovsky (2002) also presented a parametric study of the generic solutions of the fourth-order difference equation [4.5], and also obtained a similar solution expressed by equation [4.24] for positive stiffness parameters (which is the physically based constrained parameters of the problem).

The four boundary conditions associated with the case presented in Figure 4.1 are:

$$u_0=0\ ,\ u_{-1}=-u_1\ ,\ u_n=\overline{u_n} \text{ and } u_{n+1}-\overline{u_n}=\overline{u_n}-u_{n-1} \qquad [4.25]$$

Inserting the solution presented in equation [4.24] in the four boundary conditions (equation [4.25]) gives the four constants of the solution:

$$A_1=A_3=A_4=0 \text{ and } A_2=\frac{\overline{u_n}}{n} \qquad [4.26]$$

which means that the discrete displacement field is given for this problem as:

$$u_i = \frac{i}{n}\overline{u_n} \qquad [4.27]$$

The unperturbed homogeneous solution is obtained in this case. Using equation [4.27], the load–displacement relationship is obtained

$$F = \frac{k_1 + 4k_2}{n}\overline{u_n} = \frac{EA}{L}\overline{u_n} \quad \Rightarrow \quad F^* = \overline{u_n^*} \qquad [4.28]$$

This solution can also be developed from the difference equations:

$$i = 0 : u_0 = 0$$

$$i = 1 : \quad \beta_1\left(u_2 - 2u_1\right) + \frac{\beta_2}{4}\left(u_3 - 3u_1\right) = 0$$

$$i \in \{2, \ldots, n-2\} : \quad \beta_1\left(u_{i+1} - 2u_i + u_{i-1}\right) + \frac{\beta_2}{4}\left(u_{i+2} - 2u_i + u_{i-2}\right) = 0$$

$$i = n-1 : \quad \beta_1\left(u_n - 2u_{n-1} + u_{n-2}\right) + \frac{\beta_2}{4}\left(2u_n - 3u_{n-1} + u_{n-3}\right) = 0$$

$$i = n : u_n = \overline{u_n} \qquad [4.29]$$

4.3. Two-neighbor interaction – localization in a weakened problem

For symmetrical higher-order boundary conditions $u_{-1} = -u_1$, the spring associated with the indirect interaction in the vicinity of the fixed point has a stiffness $2k_2$. For the displacement controlled test, this configuration leads to the homogeneous solution with a linearly varying discrete displacement field. $u_{-1} = 0$ for the so-called truncated (or unsymmetrical) higher-order boundary conditions (see Figure 4.2). In this case, the spring associated with the indirect interaction in the vicinity of the fixed point has a stiffness k_2 (instead of $2k_2$ considered in the previous case). This case can also be viewed as a perturbation of the homogeneous one with a local weakening in the vicinity of the clamped section.

The difference equation [4.5] is still valid and may have been obtained equivalently from energy arguments, with a similar potential energy of the linear elastic lattice with two closest neighbors corrected at the boundaries at:

$$W=\sum_{i=0}^{n-1}\frac{k_1}{2}\left(u_{i+1}-u_i\right)^2+\sum_{i=0}^{n-2}\frac{k_2}{2}\left(u_{i+2}-u_i\right)^2+\frac{k_2}{2}\left(u_1-u_0\right)^2+k_2\left(u_n-u_{n-1}\right)^2 \quad [4.30]$$

with the prescribed displacements at the end nodes given by equation [4.7].

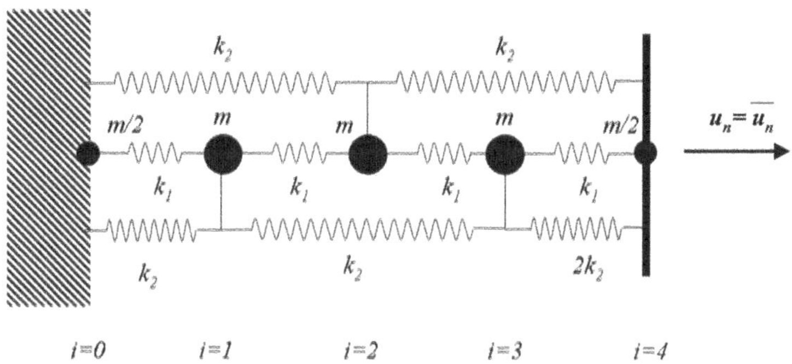

Figure 4.2. *Extension of a generalized lattice with direct and indirect interactions under displacement controlled test* $u_n=\overline{u}_n$ *; truncated higher-order boundary conditions at the clamped section; n=4;* $u_{-1}=0$ *; uniform loading*

Equation [4.30] can also be equivalently written in terms of fictitious nodes outside the domain:

$$W=\sum_{i=0}^{n-1}\frac{k_1}{2}\left(u_{i+1}-u_i\right)^2+\sum_{i=0}^{n-2}\frac{k_2}{2}\left(u_{i+2}-u_i\right)^2+\frac{k_2}{2}\left(u_1-u_{-1}\right)^2+\frac{k_2}{4}\left(u_{n+1}-u_{n-1}\right)^2 \quad [4.31]$$

with the truncated (or unsymmetrical) higher-order additional boundary conditions:

$$u_{-1}=0 \text{ and } u_{n+1}=2\overline{u_n}-u_{n-1} \quad [4.32]$$

The solution can also be presented by the following difference equations:

$$i=0: u_0=0$$

$$i=1: \quad \beta_1\left(u_2-2u_1\right)+\frac{\beta_2}{4}\left(u_3-2u_1\right)=0$$

$$i\in\{2,...,n-2\}: \quad \beta_1\left(u_{i+1}-2u_i+u_{i-1}\right)+\frac{\beta_2}{4}\left(u_{i+2}-2u_i+u_{i-2}\right)=0$$

$$i=n-1: \quad \beta_1\left(u_n-2u_{n-1}+u_{n-2}\right)+\frac{\beta_2}{4}\left(2u_n-3u_{n-1}+u_{n-3}\right)=0$$

$$i=n: u_n=\overline{u_n} \qquad [4.33]$$

The fourth-order difference equation [4.14] is still valid (only the higher-order boundary conditions differ between the two cases represented in Figures 4.1 and 4.2). The solution is expressed by equation [4.24], whose four constants can be identified from the four boundary conditions given by equation [4.34] for the problem with an unsymmetrical higher-order boundary condition at the clamped section (which can be viewed as a perturbed case of the homogeneous one):

$$u_0=0\ ,\ u_{-1}=0\ ,\ u_n=\overline{u_n}\ \text{and}\ u_{n+1}-\overline{u_n}=\overline{u_n}-u_{n-1} \qquad [4.34]$$

The four constants $\left(A_1, A_2, A_3, A_4\right)$ can be calculated as:

$$\begin{cases} \dfrac{A_1}{\overline{u_n}}=\dfrac{\sinh(\theta n)}{\sinh(\theta n)+n\left[\sinh(\theta n)+\sinh(\theta n+\theta)\right]} \\ \dfrac{A_2}{\overline{u_n}}=\dfrac{\sinh(\theta n)+\sinh(\theta n+\theta)}{\sinh(\theta n)+n\left[\sinh(\theta n)+\sinh(\theta n+\theta)\right]} \\ \dfrac{A_3}{\overline{u_n}}=\dfrac{-\sinh(\theta n)}{\sinh(\theta n)+n\left[\sinh(\theta n)+\sinh(\theta n+\theta)\right]} \\ \dfrac{A_4}{\overline{u_n}}=\dfrac{\cosh(\theta n)}{\sinh(\theta n)+n\left[\sinh(\theta n)+\sinh(\theta n+\theta)\right]} \end{cases} \qquad [4.35]$$

We have the remarkable property:

$$A_3=-A_1\ \text{and}\ A_3\cosh(\theta n)+A_4\sinh(\theta n)=0 \qquad [4.36]$$

The load–displacement relationship can be calculated in this case from equation [4.13], which may highlight the reduced stiffness of this weakened system:

$$\frac{F^*}{u_n^*} = n\frac{A_2}{u_n} = \frac{1}{1+\dfrac{\sinh(\theta n)}{n\left[\sinh(\theta n)+\sinh(\theta n+\theta)\right]}} \leq 1 \tag{4.37}$$

The reduced stiffness in equation [4.37] can be approximated with a strong accuracy using the asymptotic property:

$$\lim_{n\to\infty}\frac{\sinh(\theta n)}{\sinh(\theta n)+\sinh(\theta n+\theta)} = \frac{1}{1+e^{\theta}} = \frac{e^{-\theta}}{1+e^{-\theta}} \tag{4.38}$$

Furthermore, the remarkable property can be used:

$$\frac{e^{-\theta}}{1+e^{-\theta}} = \frac{1}{1+e^{\theta}} = \frac{1}{1+\cosh\theta+\sqrt{\cosh^2\theta-1}} = \frac{1-\sqrt{\beta_1}}{2}$$

with

$$\cosh\theta = \frac{1+\beta_1}{1-\beta_1} \tag{4.39}$$

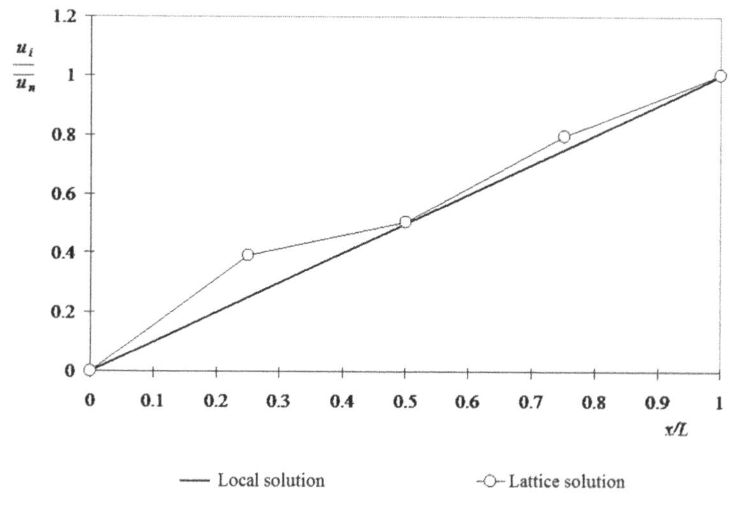

Figure 4.3. *Displacement field in the generalized lattice under uniform loading – n=4 – uniform loading – truncated higher-order boundary conditions; β_1=0.01; β_2=0.99*

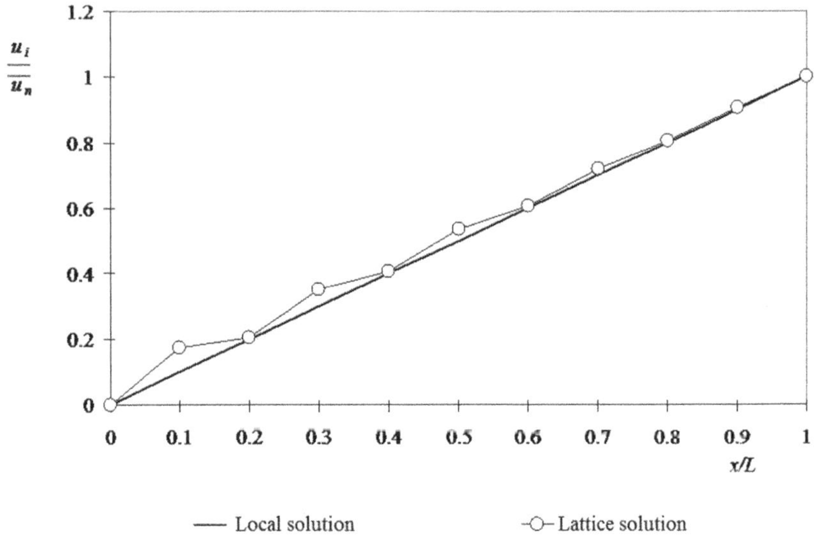

Figure 4.4. *Displacement field in the generalized lattice under uniform loading – n=10 – uniform loading – truncated higher-order boundary conditions; β_1=0.01; β_2=0.99*

It can be concluded that the reduced stiffness can be evaluated with a simplified formulae:

$$\lim_{n\to\infty}\frac{FL}{EAu_n}=\frac{1}{1+\frac{1-\sqrt{\beta_1}}{2n}}\leq 1 \qquad [4.40]$$

Based on $\beta_1=0.01$; $\beta_2=0.99$ (asymptotic case valid for a lattice with predominant indirect interactions), Figure 4.3 shows, for a small number of elements $n=4$, the displacement field in the lattice with the so-called truncated higher-order boundary conditions (or unsymmetrical higher-order boundary conditions), when compared to symmetrical higher-order boundary conditions.

The inhomogeneous displacement field is clearly shown with such higher-order boundary conditions, due to the second part of the solution in exponential form (see also Figure 4.4 with a more pronounced oscillatory behavior in the vicinity of the clamped section).

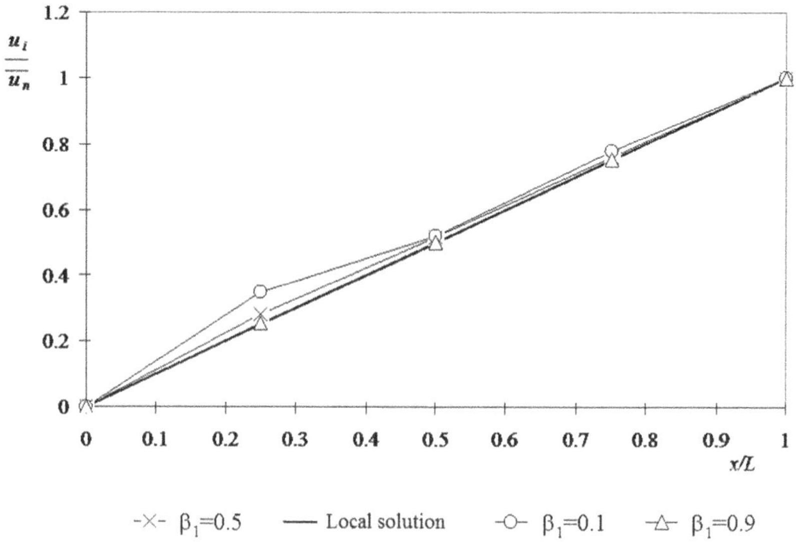

Figure 4.5. *Displacement field in the generalized lattice under uniform loading – n=4 – uniform loading – truncated higher-order boundary conditions; influence of the dimensionless coefficient* $\beta_1 \in \{0.1;0.5;0.9\}$

Note that the values $\beta_1=0.01$ and $\beta_2=0.99$ are quite extremal, because such higher-order lattice is mainly governed by next-to-direct interactions instead of direct interactions. The localization phenomenon is characterized by some oscillation terms in the vicinity of the clamped section (or reduced section). Gazis and Wallis (1964) and Toupin and Gazis (1964) already remarked that in a similar problem, the amplitude of the displacement varies exponentially with the position of the particle, but the sign changes from one particle to the next (in a boundary layer). It is worth mentioning that Charlotte and Truskinovsky (2002) also concluded that for such a physically based problem (with positive short- and long-range stiffness parameters) that the boundary layer contains some oscillations at the scale of the lattice modulated by an exponential envelope.

The localization phenomenon in the vicinity of the fixed point is even more pronounced for a larger number of elements where an oscillatory behavior is observed (see Figure 4.4). In Figure 4.4, n has been chosen equal to $n=10$, when compared to $n=4$ in Figure 4.4 with the same weighting coefficients $\beta_1=0.01$ and $\beta_2=0.99$.

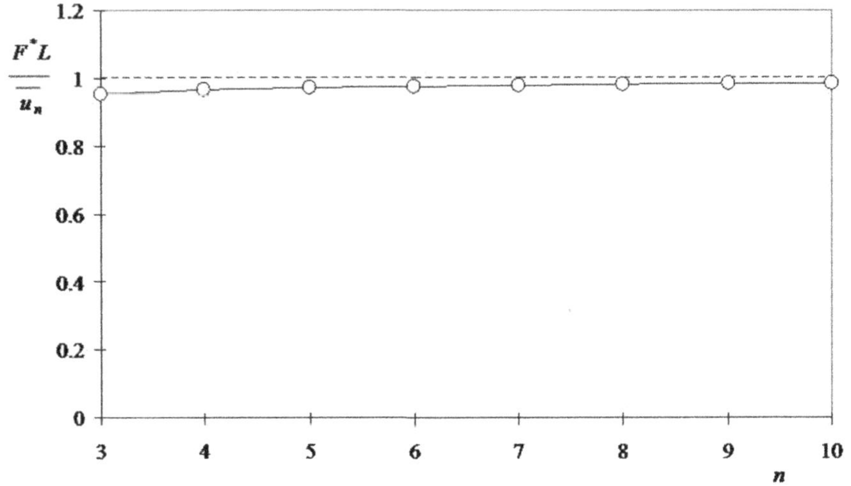

Figure 4.6. *Equivalent stiffness versus n; uniform loading – truncated higher-order boundary conditions;* $\beta_1 = 0.5$

Figure 4.5 shows that the localization phenomenon is more pronounced for predominant next-to-direct interactions (small values of β_1 and large values of β_2 with $0 \leq \beta_1 \leq \beta_2 \leq 1$). When β_1 approaches unity, the solution in the case of unsymmetrical boundary conditions asymptotically converges towards the homogeneous solution valid for the symmetrical higher-order boundary condition. The inhomogeneous solution tends towards the homogeneous one for a sufficiently large number n of elements, as shown in Figure 4.6.

In this part, an approximated solution will be investigated for a better understanding of the localization phenomenon controlled by the boundary layer in the vicinity of the fixed point. We here employed a methodology followed by Kaplunov and Pichugin (2009) for evaluating the nature of the boundary layer in the vicinity of the clamped section for the so-called perturbed case (with truncated higher-order boundary conditions). Kaplunov and Pichugin (2009) calculated a boundary layer solution for the dynamic response of a lattice model with direct neighboring interactions and displacement-based boundary conditions (as considered in the present study restricted to a static analysis). It is worth mentioning that this boundary layer is specific to the discreteness of the lattice, characterized by a strong oscillatory regime controlled by the size of the lattice spacing.

The exact displacement field is decomposed into a long wave and a boundary layer contribution:

$$u_i = \tilde{u}_i + \hat{u}_i \text{ with } \tilde{u}_i = A_1 + A_2 i \quad [4.41]$$

The boundary layer solution is reexpressed for convenience in the following way:

$$\hat{u}_i = (-1)^i \left[C_1 e^{\theta i - \theta n - \theta} + C_2 e^{-\theta i - \theta} \right] \quad [4.42]$$

Now, the two boundary conditions for the unsymmetrical higher-order boundary condition at the clamped section are written with respect to these two displacement fields:

$$\hat{u}(0) + \tilde{u}(0) = 0 \text{ and } \hat{u}(-a) + \tilde{u}(-a) = 0 \quad [4.43]$$

which can also be reexpressed as:

$$\tilde{u}(0) = -\hat{u}(0) \text{ and } \tilde{u}(0) - a\tilde{u}'(0) = -\hat{u}(-a) \quad [4.44]$$

The use of the boundary layer solution [4.42] in equation [4.44] leads to the two asymptotic boundary conditions:

$$\tilde{u}(0) = -\left[C_1 e^{-\theta n - \theta} + C_2 e^{-\theta} \right] \sim -C_2 e^{-\theta}$$

and

$$\tilde{u}(0) - a\tilde{u}'(0) = \left[C_1 e^{-\theta n - 2\theta} + C_2 \right] \sim C_2 \quad [4.45]$$

which means that the long-wave boundary condition is written in a single equation as:

$$\tilde{u}(0) + \left[\tilde{u}(0) - a\tilde{u}'(0) \right] e^{-\theta} = 0 \quad [4.46]$$

The boundary condition at the end associated with the application of the prescribed displacement (and implicitly the boundary layer is assumed to be negligible in this zone):

$$\tilde{u}(L) = \overline{u_n} \quad [4.47]$$

Using both equations [4.46] and [4.47] gives the constant of the so-called long-wave solution (approximated solution):

$$A_1 = \frac{\overline{u_n}}{L + a\dfrac{e^{-\theta}}{1+e^{-\theta}}}\frac{ae^{-\theta}}{1+e^{-\theta}} \text{ and } A_2 = \frac{\overline{u_n}}{L + a\dfrac{e^{-\theta}}{1+e^{-\theta}}} \quad [4.48]$$

Using the mathematical property equation [4.39], the long-wave solution (slow varying solution) is then written as:

$$\tilde{u}_i = \overline{u_n}\frac{\dfrac{1-\sqrt{\beta_1}}{2n}+\dfrac{i}{n}}{\dfrac{1-\sqrt{\beta_1}}{2n}+1} \quad [4.49]$$

For similar reasons, the boundary layer solution can be restricted asymptotically to the second decaying exponential term in equation [4.42]:

$$\hat{u}_i \sim (-1)^{i+1}\tilde{u}(0)e^{-\theta i} \quad [4.50]$$

so that an approximated solution may be obtained for the case with unsymmetrical higher-order boundary conditions:

$$\frac{u_i}{\overline{u_n}} = \frac{\dfrac{1-\sqrt{\beta_1}}{2n}\left[1-(-1)^i e^{-\theta i}\right]+\dfrac{i}{n}}{\dfrac{1-\sqrt{\beta_1}}{2n}+1} \quad [4.51]$$

Figure 4.7 shows that this approximate solution efficiently fits the exact lattice solution, especially in the vicinity of the fixed point.

It is confirmed from equation [4.51] that the boundary layer term is negligible in the case of pure direct interactions ($\beta_1=1$), whereas the contribution of the boundary layer is predominant for small values of β_1 (and large values of β_2 which controls the next-to-direct interaction contribution).

It is worth mentioning that an alternative boundary condition at the prescribed end would be based on the expansion of equation [4.11]:

$$F = EA\left[\left(\beta_1+\frac{\beta_2}{2}\right)\frac{\overline{u_n}-u_{n-1}}{a}+\frac{\beta_2}{2}\frac{\overline{u_n}-u_{n-2}}{2a}\right] \quad [4.52]$$

which can be expanded leading to the first-order approximation:

$$\tilde{u}'(L) = \frac{F}{EA} \quad [4.53]$$

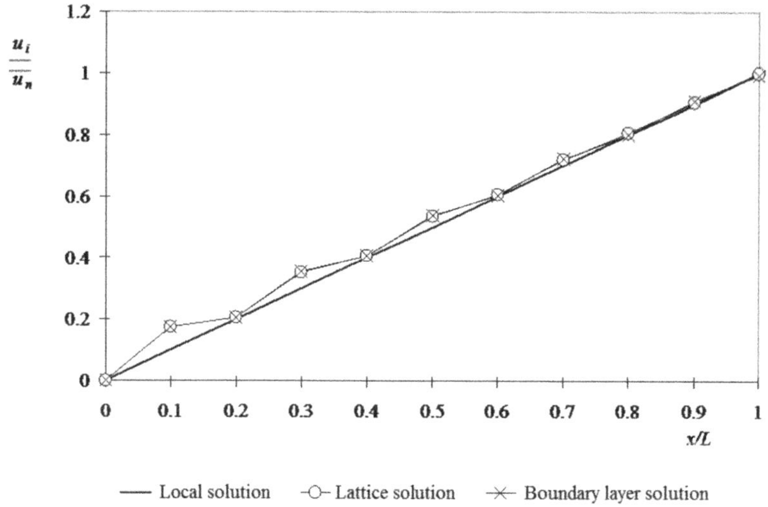

Figure 4.7. *Displacement field in the generalized lattice under uniform loading – n=10 – uniform loading – truncated higher-order boundary conditions; β_1=0.01; β_2=0.99; approximated boundary layer solution*

In this case, we obtain:

$$\tilde{u}_i = \frac{FL}{EA}\left(\frac{1-\sqrt{\beta_1}}{2n} + \frac{i}{n}\right) \quad [4.54]$$

The boundary layer contribution can be added in the total displacement formulation, leading to the same expression [4.51]. The axial stiffness is reduced, according to equation [4.54] from a factor $1\Big/\left(1+\frac{1-\sqrt{\beta_1}}{2n}\right)$, which is also consistent with the asymptotic expansion used to derive equation [4.40].

This result is also not in contradiction with the weakening found by Challamel *et al.* (2019) for the eigenfrequency of a fixed–fixed lattice with direct and next-to-direct interaction. Based on a truncated higher-order boundary condition at both sides, Challamel *et al.* (2019) found from an asymptotic expansion a weakening of the fundamental frequency from a factor $1-\frac{1-\sqrt{\beta_1}}{n}$. It is seen that the reduced stiffness factor asymptotically obtained by Challamel *et al.* (2019) in a dynamic analysis (for a two-side weakening problem) is two times the one obtained in the present static study (for a one-side weakening problem).

4.4. Conclusion

In this chapter, the static behavior of a generalized lattice composed of direct and indirect elastic interactions is studied under uniform tensile load. It is shown that the choice of higher-order boundary conditions strongly affects the macroscopic response of this microstructured system. The homogeneous response is recovered in the case of symmetrical higher-order boundary conditions. Unsymmetrical higher-order boundary conditions at the clamped section are responsible for some localization phenomena in the vicinity of the clamped section. A non-monotonic boundary layer is exhibited for the discrete displacement field of this generalized lattice. This discrete boundary layer is analyzed from an asymptotic approach, and is approximated through a simplified formulae based on the exponential law, with a change of sign from one particle to the next. Boundary layers have already been observed in continuous non-local media where some localization phenomena may be predominant (see, for instance, Chebakov *et al.* 2016). Seemingly, the discrete treatment of the boundary layers presented in this chapter is more consistent than a continuous treatment since the homogenization of the boundary layers does not appear to be fully justified, due to their short-wavelength nature.

The present results could be generalized to a nonlinear higher-order lattice, as investigated numerically by Triantafyllidis and Bardenhagen (1993). This is the generalization of FPU nonlinear lattices (Fermi *et al.* 1955) with both direct and next-to-direct interactions. Very few analytical solutions are available for the static response of nonlinear lattices (see Hérisson *et al.* (2016) for the derivation of exact analytical solutions of FPU systems under non-uniform distributed axial load). It is expected that the boundary layer may also be present for nonlinear higher-order lattices with strong localization phenomena controlled by the spacing of the lattice and the structure of the interaction model.

4.5. References

Aşkar, A. (1986). *Lattice Dynamical Foundations of Continuum Theories: Elasticity, Piezoelectricity, Viscoelasticity, Plasticity*. WorldScientific, Singapore.

Banakh, L.Y. and Kempner, M.L. (2010). *Vibrations of Mechanical Systems with Regular Structure*. Springer, New York.

Born, M. and Huang, K. (1954). *Dynamical Theory of Crystal Lattices*. Oxford University Press, Oxford.

Born, M. and von Kármán, T. (1912). On fluctuations in spatial grids. *Physikalishe Zeitschrift*, 13, 297–309.

Brillouin, L. (1946). *Wave Propagation in Periodic Structures: Electric Filters and Crystal Lattices*. McGraw-Hill, New York.

Carcaterra, A., dell'Isola, F., Esposito, R., Pulvirenti, M. (2015). Macroscopic description of microscopically strongly inhomogenous systems: A mathematical basis for the synthesis of higher gradient metamaterials. *Arch. Rational Mech. Anal.*, 218, 1239–1262.

Challamel, N., Wang, C.M., Elishakoff I. (2014). Discrete systems behave as nonlocal structural elements: Bending, buckling and vibration analysis. *Eur. J. Mech. A/Solids*, 44, 125–135.

Challamel, N., Wang, C.M., Zhang, H., Kitipornchai, S. (2018). Exact and nonlocal solutions for vibration of axial lattices with direct and indirect neighbouring interactions. *J. Eng. Mech.*, 144(5), 04018025, 1–9.

Challamel, N., Zhang, H., Wang, C.M., Kaplunov J. (2019). Scale effect and higher-order boundary conditions for generalized lattices, with direct and indirect interactions. *Mech. Res. Commun.*, 97, 1–7.

Charlotte, M. and Truskinovsky L. (2002). Linear elastic chain with a hyper-pre-stress. *J. Mech. Phys. Solids*, 50, 217–251.

Charlotte, M. and Truskinovsky, L. (2008). Towards multiscale continuum elasticity theory. *Cont. Mech. Thermodynamics*, 20, 131–161.

Chebakov, R., Kaplunov, J., Rogerson, G.A. (2016). Refined boundary conditions on the free surface of an elastic half-space taking into account non-local effects. *Proc. R. Soc. A*, 472, 20150800.

Chen, F.Y. (1970). Similarity transformation and the eigenvalue problem of certain far-coupled systems. *Am. J. Phys.*, 38(8), 1036–1039.

Chen, F.Y. (1971). On modeling and direct solution of certain free vibration systems. *J. Sound Vibration*, 14(1), 57–79.

Craster, R.V. and Kaplunov, J. (2013). *Dynamic Localization Phenomena in Elasticity, Acoustics and Electromagnetism*. Springer, Vienna.

Eaton, H.C. and Peddieson Jr., J. (1973). On continuum description of one-dimensional lattice mechanics. *J. Tenn. Acad. Sci.*, 18(3), 96–100.

Elaydi, S. (2005). *An Introduction to Difference Equations*, 3rd edition. Springer, New York.

Ericksen, J.L. (1975). Equilibrium of bars. *J. Elast.*, 5, 191–201.

Eringen, A.C. (1983). On differential equations of nonlocal elasticity and solutions of screw dislocation and surface waves. *J. Appl. Phys.*, 54, 4703–4710.

Eringen, A.C. and Kim, B.S. (1977). Relation between non-local elasticity and lattice dynamics. *Cryst. Latt. Def.*, 7, 51–57.

Fermi, E., Pasta, J., Ulam, S. (1955). Studies of nonlinear problems. Report, LA-1940. Los Alamos Scientific Lab, New Mexico

Gazis, D.C. and Wallis, R.F. (1964). Surface tension and surface modes in semi-infinite lattices. *Surface Science*, 3, 19–32.

Goldberg, S. (1958). *Introduction to Difference Equations with Illustrative Examples from Economics, Psychology and Sociology*. Dover Publications, New York.

Hencky, H. (1920). Über die angenäherte lösung von stabilitätsproblemen im raummittels der elastischen gelenkkette. *Der Eisenbau*, 11, 437–452.

Hérisson, B., Challamel, N., Picandet, V., Perrot, A. (2016). Nonlocal continuum analysis of a nonlinear uniaxial elastic lattice system under non-uniform axial load. *Physica E*, 83, 378–388.

Kaplunov, J.D. and Pichugin, A.V. (2009). On rational boundary conditions for higher-order long-wave models. In *IUTAM Symposium on Scaling in Solid Mechanics*, Borodich, F.M. (ed.). IUTAM Book, 10, 81–90.

Lagrange, J.L. (1759). Recherches sur la nature et la propagation du son. *Miscellanea Philosophico-Mathematica Societatis Privatae Taurinensis I*, 2rd Pagination, i-112, (see also *Œuvres*, Volume 1, 39–148).

Lagrange, J.L. (1853). *Mécanique analytique*, 3rd edition. Paris, 1788, Mallet-Bachelier, Gendre et successeur de bachelier, Imprimeur-libraire du bureau des longitudes, de l'école polytechnique, de l'école centrale des arts et manufactures, Paris, 367.

Maugin, G.A. (1999). *Nonlinear Waves in Elastic Crystals*. Oxford University Press, Oxford.

Michelitsch, T., Riascos, A.P., Collet, B., Nowakowski, A., Nicolleau F. (2019). *Fractional Dynamics on Networks and Lattices*. ISTE Ltd, London, and John Wiley & Sons, New York.

Mindlin, R.D. (1965). Second gradient of strain and surface-tension in linear elasticity. *Int. J. Solids Structures*, 1, 417–438.

Pipes, L.A. (1966). Circulant matrices and the theory of symmetrical components. *Matrix Tensor Quart.*, 17, 35–50.

Rosenau, P. (1987). Dynamics of dense lattices. *Phys. Rev. B*, 36(11), 5868–5876.

Tarasov, V.E. (2010). *Fractional Dynamics – Applications of Fractional Calculus to Dynamics of Particles, Fields and Media*. Springer, New York.

Toupin, R.A. and Gazis, D.C. (1964). Surface effects and initial stress in continuum and lattice models of elastic crystals. In *Proc. Int. Conf. Lattice Dynamics*, Wallis, R.F. (ed.). Pergamon Press, Copenhagen.

Triantafyllidis, N. and Bardenhagen S. (1993). On higher order gradient continuum theories in 1-D nonlinear elasticity. Derivation from and comparison to the corresponding discrete models. *J. Elast.*, 33, 259–293.

Wang, C.M., Zhang, H., Challamel, N., Duan, W.H. (2017). On boundary conditions for buckling and vibration of nonlocal beams. *Eur. J. Mech. A/Solids*, 61, 73–81.

Wang, C.M., Zhang, H., Challamel, N., Pan, W. (2020). *Hencky-Bar-Chain/Net for Structural Analysis*. World Scientific, London.

5

New Analytic Solutions for Elastic Buckling of Isotropic Plates

5.1. Introduction

In this chapter, we derive a new analytical solution for determining the elastic buckling loads of thin isotropic rectangular plates with different combinations of boundary conditions. Until very recently, an analytical solution was only available for a limited number of plates with specific combinations of boundary conditions. The solution is obtained in series form, and the coefficients are solved to match the edge conditions.

Thin plates are structural elements used in various fields, including civil, aeronautical and mechanical engineering. Such plates are often in a state in which distributed loading is applied normal to their edges and parallel to their planes, for example, in civil engineering, a floor slab with an applied lateral force due to wind, or, in aeronautical engineering, the panels of the lower skin of aircraft wings and fuselage under compression forces. These in-plane loads can lead to buckling, i.e. the loss of the plate's stability.

The first solution was obtained by Navier (1819), and his solution was for a simply supported plate along all four edges. For the cases where two opposite edges are simply supported, Levy (1899) developed an exact solution. For many years, these were the only cases that had an exact analytic solution for the buckling load and mode.

Many methods have been used for the approximate solution of buckling in plates. These methods include the finite difference method, the finite elements method, the

Chapter written by Joseph TENENBAUM, Aharon DEUTSCH and Moshe EISENBERGER.

Rayleigh–Ritz method (Dickinson 1978; Narita and Leissa 1990), the multi-term extended Kantorovich method (Shufrin and Eisenberger 2005; Shufrin *et al.* 2008a, b), the differential quadrature method and a few more. In this review, we present only exact solutions.

Exact analytical solutions for the elastic buckling of isotropic plates for Navier and Levy combinations of boundary conditions have been presented in several books (Timoshenko and Gere 1961; Bulson 1970; Reddy 1997; Szilard 2004; Wang *et al.* 2005).

Recently, analytic solutions for the buckling of rectangular plates have been presented for a very limited number of new cases. Wang *et al.* (2016) derived a symplectic superposition method for the solution of four cases of boundary conditions (CCCC, CCCS, CCSC and CCSS). The solution is a relatively complex procedure and limited to a small number of cases. Ullah *et al.* (2019a) used a double finite sine integral transform method to arrive at a solution for the same four cases. In another work, Ullah *et al.* (2019b) derived a straightforward generalized integral transform method for finding the buckling loads of plates. In this method, they used the beam functions and arrived at a solution for the same four cases. Li *et al.* (2018) solved the case of a completely free plate (FFFF) with biaxial loading by the symplectic superposition method. Liu *et al.* (2020) used an analytical spectral flexibility method to solve three cases (CCCC, SSSS and GGGG).

The complete list of the 55 possible cases for a square plate is presented in Table 5.1. This was first given by Bert and Malik (1994) for vibrations of plates. The cases include the four classical plate edge boundary conditions, which are: C – clamped, S – simply supported, F – free and G – guided. In the buckling analysis, the same 55 cases are found for unidirectional loading in the x or y direction, making a total of 110 cases. Of the 110 cases, only 100 are unique, since there are 10 cases where changing the load direction is equal to one of the cases in the other direction, and these are marked with # in Table 5.1. For cases 1–6 and 22–25, the Levy solution is known. The new solution in this chapter is only applied to cases in which all the four corners of the plate are supported, in particular, cases 7–15, 26–27, 34–35, 43–44, and 47–48. In total, we add the analytic solution for 17 cases. The cases that are not solved using the current analytic solution are marked with an * in Table 5.1.

In this chapter, we present a new method of finding the exact buckling load for isotropic thin plates by using a superposition method. The solution for the buckling load of the plates is found using a static analysis. The plate is subjected to loads parallel to its plane, both uniaxial and biaxial. The solution for the buckling load of the plate is equivalent to finding the force that will yield infinite deflection at a random point of the plate. This results from the fact that buckling of a plate causes loss of stiffness and the deflection becomes infinite. A transform method from hyperbolic sine and cosine into a combination of basic trigonometric functions was used. Such transformation allows us to obtain the appropriate equations needed to solve the partial differential equation of a plate. These equations satisfy the equation of equilibrium of a plate, for most of its boundary conditions.

1–6	7–21	22–25	26–33	34–42	43–55
1# SSSS	7# CCCC	22 SSSG	26 CSCG	34 CGSG	43 CCCG
2 SSSC	8 CCCS	23 SCSG	27 CSSG	35 CGCG	44 CCSG
3 SSSF	9 CCCF	24 SGSF	28*# SSGG	36* SGGG	45* CCGF
4 SCSC	10 CSCF	25 SGSG	29* CSGG	37* CGGG	46*# CCGG
5 SCSF	11 CFCF		30* SSGF	38* SGFG	47 CGCF
6 SFSF	12 CFSF		31* SCGF	39* CGFG	48 CGSF
	13# CCSS		32* SGGF	40* GGFG	49* CSGF
	14 CCSF		33* SFGF	41* FGFG	50* CGGF
	15 CSSF			42*# GGGG	51* CFGF
	16*# SSFF				52* SGFF
	17* CSFF				53* CGFF
	18*# CCFF				54*# GGFF
	19* SFFF				55* GFFF
	20* CFFF				
	21*# FFFF				

Table 5.1. *Possible combinations for boundary conditions*

The analytic solution can predict the "exact" values of the critical buckling loads. But even the "exact" solutions become approximate because of the truncation of the infinite series solutions for the calculation of the numerical results. The results in this paper were obtained using 35 terms in the series, and they are converged for the number of digits that are shown. These solutions help one to understand the physical behavior at the point of buckling, and also serve as benchmark values for numerical and other approximate methods.

5.2. Equilibrium equation

The equilibrium equation of in-plane stressed thin rectangular isotropic plates is (Reddy 1997)

$$D\left(\frac{\partial^4 w_0}{\partial x^4}+2\frac{\partial^4 w_0}{\partial x^2\partial y^2}+\frac{\partial^4 w_0}{\partial y^4}\right)-N_{xx}\frac{\partial^2 w_0}{\partial x^2}-N_{yy}\frac{\partial^2 w_0}{\partial y^2}=q(x,y) \quad [5.1]$$

where $w_0(x,y)$ are the out-of-plane deflections of the plate, $D = Eh^3/12(1-\nu^2)$ is the flexural stiffness of the plate, E is the modulus of elasticity, h is the plate thickness, and N_{xx} and N_{yy} are the pressure loads on the edges in direction x and y, respectively.

For the solution, we will also need the boundary condition along the edges of the plate. In thin plate theory, these are the deflection, slope, bending moment and shear force, along the edges parallel to the y and x axes, respectively, as (Reddy 1997) are

$$\Phi_x = \frac{\partial w(x,y)}{\partial x} \tag{5.2}$$

$$\Phi_y = \frac{\partial w(x,y)}{\partial y} \tag{5.3}$$

$$M_x = -D\left(\frac{\partial^2 w(x,y)}{\partial x^2} + \nu\frac{\partial^2 w(x,y)}{\partial y^2}\right) \tag{5.4}$$

$$M_y = -D\left(\frac{\partial^2 w(x,y)}{\partial y^2} + \nu\frac{\partial^2 w(x,y)}{\partial x^2}\right) \tag{5.5}$$

$$Q_x = -\left(\frac{\partial^3 w(x,y)}{\partial x^3}D + D(2-\nu)\frac{\partial^3 w(x,y)}{\partial x \partial y^2} - N_{xx}\frac{\partial w(x,y)}{\partial x}\right) \tag{5.6}$$

$$Q_y = -\left(\frac{\partial^3 w(x,y)}{\partial y^3}D + D(2-\nu)\frac{\partial^3 w(x,y)}{\partial x^2 \partial y} - N_{yy}\frac{\partial w(x,y)}{\partial y}\right) \tag{5.7}$$

These values are calculated along the edges where either the x or y coordinate is constant.

5.3. Solution

The solution for this partial differential equation is assumed as given below, and it has to satisfy the equilibrium equation all over the plate:

$$w(x,y) = \sum_{m=1}^{\infty} \bar{X}_m \cdot Y_m + \sum_{n=1}^{\infty} X_n \cdot \bar{Y}_n + w_p \tag{5.8}$$

The last term, w_p, is the solution for a simply supported plate (SSSS) with a concentrated point load P acting at coordinates (ξ, η). The solution is given in Timoshenko (1959, p. 141). The coordinates ξ, η of the application of the load P must be chosen carefully: we try to avoid choosing a point that will be on one of the nodal lines of any of the buckling modes, as, for such a location, the vertical deflection will be always zero. One such location would be, for example, at the center of the plate, where, for some combinations of boundary condition, we will have antisymmetric modes. Thus, we choose an arbitrary location for the point of application of P. The expression for w_p is

$$w_p = \sum_{m=1}^{\infty} \sum_{n=1}^{\infty} \frac{4\,P\,sin(\frac{m\pi\xi}{a})\,sin(\frac{n\pi\eta}{b})}{a\,b\,d_{mn}}\,\bar{X}_m \cdot \bar{Y}_n \tag{5.9}$$

with

$$d_{mn} = \frac{\pi^2}{a^4b^4}(D\,\pi^2b^4m^4 + 2\,D\,\pi^2a^2b^2m^2n^2 + D\,\pi^2a^4n^4 + N_{xx}\,a^4b^2\beta\,n^2 + N_{xx}\,a^2b^4m^2) \tag{5.10}$$

The two functions $\bar{X}_m$ and $\bar{Y}_n$ are trigonometric functions

$$\bar{X}_m = sin(\lambda_a x) \cdot cos\left(\frac{m\pi}{2}\right) + cos(\lambda_a x) \cdot sin\left(\frac{m\pi}{2}\right) \tag{5.11}$$

$$\bar{Y}_n = sin(\lambda_b y) \cdot cos\left(\frac{n\pi}{2}\right) + cos(\lambda_b y) \cdot sin\left(\frac{n\pi}{2}\right) \tag{5.12}$$

with

$$\lambda_a = \frac{m\pi}{a} \tag{5.13}$$

$$\lambda_b = \frac{n\pi}{b} \tag{5.14}$$

The two functions Y_m and X_n are hyperbolic functions

$$Y_m = A_m\,cosh\,(\alpha_1 y) + B_m\,cosh\,(\alpha_2 y) + C_m\,sinh\,(\alpha_1 y) + D_m\,sinh\,(\alpha_2 y) \tag{5.15}$$

$$X_n = E_n\,cosh\,(\alpha_3 x) + F_n\,cosh\,(\alpha_4 x) + G_n\,sinh\,(\alpha_3 x) + H_n\,sinh\,(\alpha_4 x) \tag{5.16}$$

with

$$\alpha_1 = \frac{\sqrt{2}\sqrt{D(2D\pi^2m^2 + N_{xx}a^2\beta - \delta_1)}}{2Da} \tag{5.17}$$

$$\alpha_2 = \frac{\sqrt{2}\sqrt{D(2D\pi^2m^2 + N_{xx}a^2\beta + \delta_1)}}{2Da} \tag{5.18}$$

$$\alpha_3 = \frac{\sqrt{2}\sqrt{D(2D\pi^2 n^2 + N_{xx}b^2 - \delta_2)}}{2Db} \quad [5.19]$$

$$\alpha_4 = \frac{\sqrt{2}\sqrt{D(2D\pi^2 n^2 + N_{xx}b^2 + \delta_2)}}{2Db} \quad [5.20]$$

where

$$\beta = \frac{N_{yy}}{N_{xx}} \quad [5.21]$$

$$\delta_1 = \sqrt{(4\,\pi^2 m^2\,(\beta - 1)\,D + \beta^2 N_{xx}\,a^2)\,a^2 N_{xx}} \quad [5.22]$$

$$\delta_2 = \sqrt{(-4\,\pi^2 n^2\,(\beta - 1)\,D + N_{xx}\,b^2)\,b^2 N_{xx}} \quad [5.23]$$

5.4. Boundary condition

The solution given in equation [5.8] is dependent on eight unknown constants for every value of m and n in equations [5.15] and [5.16]. These are $A_m, B_m, C_m, D_m, E_n, F_n, G_n$ and H_n. The plate has four edges, and on each of them we can prescribe two quantities; therefore, we obtain eight unknowns per boundary condition case.

Edge	Deflection	Rotation	Moment	Reaction
1	$w(-\frac{a}{2}, y)$	$\Phi_x(-\frac{a}{2}, y)$	$M_x(-\frac{a}{2}, y)$	$Q_x(-\frac{a}{2}, y)$
2	$w(x, -\frac{b}{2})$	$\Phi_y(x, -\frac{b}{2})$	$M_y(x, -\frac{b}{2})$	$Q_y(x, -\frac{b}{2})$
3	$w(\frac{a}{2}, y)$	$\Phi_x(\frac{a}{2}, y)$	$M_x(\frac{a}{2}, y)$	$Q_x(\frac{a}{2}, y)$
4	$w(x, \frac{b}{2})$	$\Phi_y(x, \frac{b}{2})$	$M_y(x, \frac{b}{2})$	$Q_y(x, \frac{b}{2})$

Table 5.2. *Possible edge conditions*

It is convenient to number the edges in a counter-clockwise direction, as given in Table 5.2, for all the four edges, and from every row, i.e. the edge, only two relations can be used, depending on the particular edge restraints. The expanded expressions for the edge deflection, slope, bending moment and shear can be obtained by substituting the solution [5.8] into equations [5.2]–[5.7], which are given in section 5.7, Appendix A.

As an example, we derive the equation that will be obtained when we apply zero deflection along edge 2. After substitution of $y = -b/2$ into equation [5.8] for the deflection along the edge, we have

$$w\left(x,-\frac{b}{2}\right) = \sum_{m=1}^{\infty} \bar{X}_m \times \Big[A_m\, cosh(\frac{\alpha_1 b}{2}) + B_m\, cosh(\frac{\alpha_2 b}{2}) - C_m\, sinh(\frac{\alpha_1 b}{2}) - D_m\, sinh(\frac{\alpha_2 b}{2})\Big] + \sum_{n=1}^{\infty} \bar{Y}_n \,|_{y=-\frac{b}{2}} \times \Big[E_n\, cosh(\alpha_3 x) + F_n\, cosh(\alpha_4 x) + G_n\, sinh(\alpha_3 x) + H_n\, sinh(\alpha_4 x)\Big] \quad [5.24]$$

In the second summation, we have the hyperbolic functions that are dependent on x; however, this term is multiplied by $\bar{Y}_n \,|_{y=-\frac{b}{2}}$, which is zero along the edge. Note that w_p along the four edges is zero, as it is the solution for the simply supported plate (SSSS). Then, equation [5.24] becomes, as given in equation [A.4] in section 5.7, Appendix A:

$$w\left(x,-\frac{b}{2}\right) = \sum_{m=1}^{\infty} \bar{X}_m \times \Big[A_m\, cosh(\frac{\alpha_1 b}{2}) + B_m\, cosh(\frac{\alpha_2 b}{2}) - C_m\, sinh(\frac{\alpha_1 b}{2}) - D_m\, sinh(\frac{\alpha_2 b}{2})\Big] \quad [5.25]$$

Then, as $\bar{X}_m$ is a function of x, the edge deflection can be zero if we have

$$0 = \Big[A_m\, cosh(\frac{\alpha_1 b}{2}) + B_m\, cosh(\frac{\alpha_2 b}{2}) - C_m\, sinh(\frac{\alpha_1 b}{2}) - D_m\, sinh(\frac{\alpha_2 b}{2})\Big] \quad [5.26]$$

In this equation, we have four unknowns for a given value of m. Similarly, we can derive the equations for the remaining seven edge conditions. We will obtain a total of $4(m+n)$ equations. The buckling load of the plate is found when a zero force P will generate a unit defection under the applied point load (Eisenberger and Deutsch 2015). We take a unit force $P = 1$, and calculate the unknown constants in equations [5.15] and [5.16]. The next step is to calculate the deflection at that particular point of loading using equation [5.8], or, for better convergence, we instead find the value of $(1/w(\xi,\eta))$ that becomes zero. A bisection procedure is applied to obtain the final convergence to the desired accuracy. The buckling mode is then calculated as $w(x,y)$ for the particular in-plane loading.

As an example, the equations that are needed to solve case 48, CGSF are given in Table 5.3. This case provides a good perspective of how each equation is used, and covers all types of possible boundary conditions. In this table, we make use of the equations in section 5.7, Appendix A.

Edge #	1 - C	2 - G	3 - S	4 - F
Equations 1–4	$W(-\frac{a}{2}, y) = 0$ [A.2]	$\phi_y(x, -\frac{b}{2}) = 0$ [A.8]	$W(\frac{a}{2}, y) = 0$ [A.1]	$M_y(x, \frac{b}{2}) = 0$ [A.11]
Equations 5–8	$\phi_x(-\frac{a}{2}, y) = 0$ [A.6]	$Q_y(x, -\frac{b}{2}) = 0$ [A.16]	$M_x(\frac{a}{2}, y) = 0$ [A.9]	$Q_y(x, \frac{b}{2}) = 0$ [A.15]

Table 5.3. *Equations required for solving case 48, CGSF*

5.5. Numerical results

The new exact solutions for 51 cases are presented in Table 5.4. This table contains the buckling load N_{xx} for a thin isotropic plate with various aspect ratios a/b. Some cases have the same numbering, and this number represents the case number in Table 5.1; however, the first appearance is for the loading in the x direction and the second for the loading in the y direction. All the answers are compared to the finite element results from the ANSYS commercial code. The models used the SHELL181-type elements, with a mesh containing 10,000 elements (100 divisions in both directions).

The results summarized in Table 5.4 reveal several important points regarding the exact solution. The Levy-type cases that are solved (cases 1–6, 22–25 and 34–35) all result in the known analytic solution, whereas the very detailed and fine mesh used in the finite element solution (approximately 30,000 degrees of freedom, depending on the particular case) did not fully converge to the exact result. The finite element results are, in all cases, above the correct values as the theory predicts, i.e. convergence from above.

For several cases, an important and unexpected result is obtained: the uniaxial and biaxial buckling loads are very close. The factor of reduction given here for these cases is defined as the ratio of the uniaxial buckling load divided by the biaxial load for a square plate: in cases 6, SFSF and 24, SGSF, the ratio is 1.022, in cases 25, SGSG and 34, CGSG, it is 1.0, and in case 35, it is 1.045. In these cases, the addition of loading in the perpendicular direction has a very small effect on the buckling load (i.e. below 5%).

In Table 5.5, the buckling load N_{xx} is presented for plates with biaxial loading. Five different cases with different aspect ratios a/b and various compression ratios β are provided. These cases are 7, CCCC, 11, CFCF, 13, CCSS, 23, SCSG and 48, CGSF. As in Table 5.4, we compare all our results with ANSYS, which are indicated by the numbers in parentheses.

Case number	$\frac{a}{b}$	Uniaxial		Biaxial, $\beta = 1$	
		Present study	ANSYS	Present study	ANSYS
	0.5	61.6850	61.6860	49.3480	49.3487
	1	39.4784	39.4816	19.7392	19.7410
1 SSSS	1.5	42.8368	42.8480	14.2561	14.2578
	2	39.4784	39.4840	12.3370	12.3387
	3	39.4784	39.4860	10.9662	10.9680
	0.5	67.6377	67.6410	52.9493	52.9515
	1	56.6536	56.6646	26.2798	26.2847
2 SCSS	1.5	53.6038	53.6170	22.3775	22.3835
	2	55.3250	55.3530	21.2702	21.2767
	3	53.6038	53.6297	20.6118	20.6187
	0.5	102.5104	102.5300	85.0810	85.1004
	1	47.8394	47.8480	26.2798	26.2847
2 CSSS	1.5	44.3814	44.3910	16.3656	16.3684
	2	41.8144	41.8260	13.2373	13.2395
	3	40.5468	40.5610	11.2283	11.2302
	0.5	42.9897	42.9930	39.1818	39.1832
	1	13.8332	13.8341	10.4138	10.4146
3 SFSS	1.5	8.4657	8.4660	4.9466	4.9470
	2	6.5943	6.5944	2.9359	2.9361
	3	5.2618	5.2618	1.3958	1.3959
	0.5	20.1630	20.1656	11.7435	11.7448
	1	23.3497	23.3500	10.4138	10.4146
3 SSFS	1.5	22.7479	22.7491	9.9687	9.9698
	2	22.7999	22.8020	9.7954	9.7968
	3	22.7986	22.8024	9.6790	9.6806
	0.5	75.9099	75.9190	58.4697	58.4764
	1	75.9099	75.9394	37.7996	37.8147
4 SCSC	1.5	70.2314	70.2490	37.1845	37.2041
	2	68.8070	68.8300	37.7480	37.7699
	3	69.6322	69.6980	37.1845	37.2057
	0.5	179.5024	179.6030	150.9919	151.0730
	1	66.5526	66.5830	37.7996	37.8147
4 CSCS	1.5	53.0460	53.0680	19.8349	19.8404
	2	47.8394	47.8700	14.6174	14.6206
	3	43.4900	43.5230	11.5877	11.5898
	0.5	44.0507	44.0540	39.5601	39.5608
	1	16.3096	16.3103	11.2865	11.2870
5 SFSC	1.5	12.4127	12.7440	6.2308	6.2310
	2	12.8463	13.1850	4.5266	4.5268
	3	12.7438	12.7450	3.3558	3.3559

Table 5.4. *Buckling load N_{xx} for a plate with different plate aspect ratios and boundary conditions ($\nu = 0.3$)*

Case number	$\frac{a}{b}$	Uniaxial		Biaxial, $\beta = 1$	
		Present study	ANSYS	Present study	ANSYS
5 CSFS	0.5	25.9190	25.9211	18.1064	18.1072
	1	23.6093	23.6100	11.2865	11.2870
	1.5	22.9254	22.9268	10.2134	10.2145
	2	22.8029	22.8050	9.8900	9.8914
	3	22.7990	22.8030	9.7008	9.7023
6 SFSF	0.5	38.4187	38.4240	37.6026	37.6066
	1	9.3989	9.4004	9.2005	9.2019
	1.5	4.1140	4.1147	4.0467	4.0473
	2	2.2921	2.2924	2.2646	2.2650
	3	1.0088	1.0089	1.0020	1.0022
6 FSFS	0.5	15.6142	15.6152	9.0583	9.0597
	1	20.1630	20.1637	9.2005	9.2019
	1.5	22.2690	22.2692	9.3184	9.3198
	2	22.3121	22.3135	9.4007	9.4021
	3	22.7479	22.7512	9.4977	9.4993
7 CCCC	0.5	190.8646	190.9700	154.8911	154.9770
	1	99.4259	99.4872	52.3447	52.3721
	1.5	82.4162	82.4790	40.6749	40.6975
	2	77.6449	77.7230	38.7228	38.7456
	3	72.6338	72.7003	38.1018	38.1262
8 CCSC	0.5	114.5872	114.6200	92.1244	92.1489
	1	79.8152	79.8475	42.5471	42.5645
	1.5	72.6338	72.6640	38.1018	38.1213
	2	71.1388	71.1770	37.7621	37.7833
	3	69.9107	69.9510	37.3599	37.3817
8 CCCS	0.5	184.3691	184.4730	151.0482	151.1250
	1	79.6206	79.6600	42.5471	42.5645
	1.5	69.1642	69.2190	27.0099	27.0193
	2	61.4150	61.4520	23.0311	23.0388
	3	57.4940	57.5332	21.0408	21.0480
9 CCCF	0.5	161.4800	161.6500	111.8760	112.4730
	1	45.1631	45.1877	28.5408	28.5628
	1.5	24.8789	24.8850	12.7745	12.7749
	2	19.0019	19.0040	7.4374	7.4380
	3	16.6207	16.6240	4.1766	4.1768
9 CCFC	0.5	41.7247	41.7296	29.7479	29.7482
	1	38.5512	38.5920	28.5408	28.5628
	1.5	38.1469	38.2918	28.0731	28.1481
	2	37.9796	38.2713	27.9690	28.1226
	3	37.5601	38.2785	27.7300	28.1255

Table 5.4. *(cont.) Buckling load Nxx for a plate with different plate aspect ratios and boundary conditions ($\nu = 0.3$)*

Case number	$\frac{a}{b}$	Uniaxial		Biaxial, $\beta = 1$	
		Present study	ANSYS	Present study	ANSYS
	0.5	160.6445	160.8100	111.8755	112.4730
	1	43.1486	43.1727	28.2924	28.3140
10 CSCF	1.5	21.4604	21.4670	12.6880	12.6883
	2	13.8897	13.8910	6.7676	6.7696
	3	8.4920	8.4938	2.6323	2.6326
	0.5	41.6190	41.6246	27.0710	27.0753
	1	37.9941	38.0334	28.2924	28.3140
10 SCFC	1.5	38.1353	38.2798	28.0233	28.0980
	2	37.9789	38.2706	27.9689	28.1225
	3	37.5600	38.2786	27.7300	28.1225
	0.5	156.0736	156.2300	111.1686	111.7310
	1	38.6773	38.7011	27.0569	27.0782
11 CFCF	1.5	17.0604	17.0690	10.5295	10.5302
	2	9.5363	9.5405	5.3367	5.3371
	3	4.1942	4.1963	2.0860	2.0861
	0.5	26.2953	26.2975	21.3469	21.3472
	1	36.0469	36.0802	27.0654	27.0798
11 FCFC	1.5	37.6781	37.8180	27.5693	27.6042
	2	37.7440	38.0385	27.8668	27.9370
	3	37.5502	38.2662	27.9844	28.1041
	0.5	79.3608	79.3990	72.9186	73.0130
	1	19.5658	19.5723	17.6330	17.6379
12 CFSF	1.5	8.5993	8.6020	7.5490	7.5493
	2	4.7962	4.7976	4.1161	4.1164
	3	2.1068	2.1074	1.7521	1.7521
	0.5	20.0705	20.0725	16.4639	16.4648
	1	30.1965	30.2113	17.6330	17.6380
12 FCFS	1.5	29.0975	29.1390	18.0726	18.0869
	2	30.1447	30.2285	18.2297	18.2564
	3	29.9862	30.2385	18.2623	18.3323
	0.5	107.6543	107.6800	87.6908	87.7112
	1	61.4150	61.4310	32.0524	32.0603
13 CCSS	1.5	57.9389	57.9610	24.0495	24.0563
	2	55.9436	55.9630	21.9227	21.9295
	3	54.5604	54.5870	20.7802	20.7872
	0.5	84.8584	84.8990	73.2299	73.3222
	1	26.1719	26.1780	18.4733	18.4782
14 CCSF	1.5	16.6136	16.6150	8.7530	8.7532
	2	14.5176	14.5180	5.6383	5.6388
	3	13.7841	13.7862	3.6813	3.6813

Table 5.4. *(cont.) Buckling load Nxx for a plate with different plate aspect ratios and boundary conditions ($\nu = 0.3$)*

Case number	$\frac{a}{b}$	Uniaxial		Biaxial, $\beta = 1$	
		Present study	ANSYS	Present study	ANSYS
	0.5	32.6935	32.6948	22.5533	22.5552
	1	31.0005	31.0136	18.4733	18.4782
14 CCFS	1.5	30.2031	30.2502	18.3219	18.3353
	2	30.1446	30.2414	18.3075	18.3337
	3	29.9957	30.2397	18.2651	18.3336
	0.5	83.9445	83.9830	73.2285	73.3210
	1	24.0286	24.0347	18.2502	18.2553
15 CSSF	1.5	12.9830	12.9850	7.9912	7.9920
	2	9.1295	9.1298	4.4092	4.4097
	3	6.3842	6.3846	1.8871	1.8872
	0.5	30.2075	30.2137	17.6358	17.6359
	1	30.2273	30.2390	18.2502	18.2553
15 SCFS	1.5	30.1856	30.2327	18.3172	18.3305
	2	30.1393	30.2359	18.3071	18.3334
	3	29.9956	30.2397	18.2650	18.3334
	0.5	44.5674	44.5720	41.9458	41.9502
	1	15.4213	15.4224	12.3370	12.3382
22 SSSG	1.5	10.7092	10.7100	6.8539	6.8545
	2	9.8696	9.8698	4.9348	4.9352
	3	10.7092	10.7115	3.5640	3.5643
	0.5	39.4784	39.4795	19.7392	19.7402
	1	46.3323	46.3410	12.3370	12.3376
22 SSGS	1.5	39.4784	39.4830	10.9662	10.9673
	2	41.4770	41.4910	10.4865	10.4878
	3	40.4240	40.4440	10.1438	10.1463
	0.5	45.9395	45.9440	42.9320	42.9364
	1	18.9775	18.9789	14.6174	14.6190
23 SCSG	1.5	17.5579	17.5590	10.3843	10.3854
	2	18.9775	18.9830	9.4499	9.4510
	3	17.5579	17.5581	9.2961	9.2975
	0.5	66.5526	66.5631	37.7996	37.8044
	1	52.0625	52.0760	14.6174	14.6190
23 CSGS	1.5	43.4900	43.5010	11.5877	11.5892
	2	42.1055	42.1160	10.7330	10.7346
	3	40.6060	40.6180	10.2118	10.2134
	0.5	38.9000	38.9060	38.1984	38.2025
	1	9.6047	9.6062	9.4007	9.4020
24 SGSF	1.5	4.2199	4.2206	4.1260	4.1266
	2	2.3497	2.3501	2.3001	2.3005
	3	1.0285	1.0287	1.0117	1.0118

Table 5.4. *(cont.) Buckling load Nxx for a plate with different plate aspect ratios and boundary conditions ($\nu = 0.3$)*

Case number	$\frac{a}{b}$	Uniaxial		Biaxial, $\beta = 1$	
		Present study	ANSYS	Present study	ANSYS
	0.5	25.5927	25.5946	9.2005	9.2018
	1	22.3121	22.3124	9.4007	9.4020
24 GSFS	1.5	22.8514	22.8528	9.4977	9.4992
	2	22.7977	22.7998	9.5496	9.5511
	3	22.7989	22.8027	9.5989	9.6005
	0.5	39.4784	39.4840	39.4784	39.4844
	1	9.8696	9.8712	9.8696	9.8712
25 SGSG	1.5	4.3865	4.3872	4.3865	4.3872
	2	2.4674	2.4678	2.4674	2.4678
	3	1.0966	1.0968	1.0966	1.0968
	0.5	61.6850	61.6935	9.8696	9.8724
	1	39.4784	39.4840	9.8696	9.8725
25 GSGS	1.5	42.8368	42.8490	9.8696	9.8715
	2	39.4784	39.4850	9.8696	9.8700
	3	39.4784	39.4860	9.8696	9.8712
	0.5	162.8865	163.0600	148.2573	148.4030
	1	44.8694	44.9015	37.7418	37.7700
26 CSCG	1.5	23.5197	23.5310	16.4823	16.4894
	2	16.6403	16.6450	9.4522	9.4537
	3	13.2622	13.2656	4.9599	4.9600
	0.5	84.9562	84.9613	37.8089	37.8154
	1	68.7958	68.8260	37.7418	37.7702
26 SCGC	1.5	69.6109	69.6600	37.7874	37.8175
	2	70.6640	70.7410	37.0643	37.1027
	3	68.7517	68.8360	37.1252	37.1605
	0.5	85.8036	85.8490	81.6383	81.6827
	1	25.6261	25.6340	21.2687	21.2762
27 CSSG	1.5	15.0340	15.0370	10.2804	10.2826
	2	11.9603	11.9620	6.5706	6.5712
	3	11.0954	11.0970	4.0918	4.0919
	0.5	56.6637	56.6640	26.2824	26.2836
	1	55.3216	55.3370	21.2687	21.2762
27 SCGS	1.5	56.6413	56.6640	20.6084	20.6184
	2	53.3804	53.4070	20.4096	20.4213
	3	53.7118	53.7430	20.2820	20.2928
	0.5	80.7438	80.7890	80.7437	80.7895
	1	20.1893	20.1980	20.1893	20.1977
34 CGSG	1.5	8.9740	8.9768	8.9740	8.9768
	2	5.0484	5.0494	5.0484	5.0495
	3	2.2438	2.2442	2.2438	2.2442

Table 5.4. *(cont.) Buckling load Nxx for a plate with different plate aspect ratios and boundary conditions ($\nu = 0.3$)*

Case number	$\frac{a}{b}$	Uniaxial		Biaxial, $\beta = 1$	
		Present study	ANSYS	Present study	ANSYS
	0.5	67.6472	67.6493	20.1937	20.1979
	1	56.6419	56.6670	20.1893	20.1977
34 GCGS	1.5	53.5871	53.6180	20.1874	20.1977
	2	55.3025	55.3530	20.1859	20.1978
	3	53.5664	53.6300	20.1869	20.1977
	0.5	157.8384	158.0100	150.8336	151.0750
	1	39.4725	39.5040	37.7879	37.8160
35 CGCG	1.5	17.5471	17.5570	17.5471	17.5570
	2	9.8723	9.8761	9.8723	9.8761
	3	4.3881	4.3894	4.3881	4.3894
	0.5	75.9174	75.9288	39.4894	39.5040
	1	75.8948	75.9420	37.7879	37.8158
35 GCGC	1.5	70.1767	70.2500	37.1574	37.2044
	2	68.7359	68.8310	37.7084	37.7700
	3	69.5338	69.6980	37.1552	37.2057
	0.5	164.0203	164.2000	151.0096	151.2320
	1	47.7090	47.7430	38.7141	38.7440
43 CCCG	1.5	28.7828	28.7950	18.9092	18.9163
	2	24.8573	24.8670	13.0888	13.0910
	3	20.6080	20.6160	10.1697	10.1710
	0.5	99.4369	99.4625	52.3551	52.3620
	1	77.6258	77.6770	38.7141	38.7439
43 CCGC	1.5	73.6483	73.7350	38.3798	38.4270
	2	71.2663	71.3720	37.7524	37.8094
	3	69.7470	69.8970	37.3843	37.4695
	0.5	87.0036	87.0510	82.2652	82.3140
	1	28.6451	28.6540	23.0291	23.0373
44 CCSG	1.5	20.6042	20.6080	13.2907	13.2932
	2	19.9539	19.9580	10.6374	10.6387
	3	18.1588	18.1630	9.5256	9.5270
	0.5	79.6332	79.6390	42.5495	42.5535
	1	65.0655	65.0980	23.0291	23.0373
44 CCGS	1.5	57.4752	57.5120	21.0364	21.0474
	2	55.6955	55.7430	20.5663	20.5791
	3	54.3515	54.4150	20.3214	20.3328
	0.5	156.8678	157.0100	111.8619	112.4580
	1	39.0308	39.0580	27.9114	27.9335
47 CGCF	1.5	17.2669	17.2770	12.3124	12.3142
	2	9.6709	9.6753	7.3331	7.3331
	3	4.2664	4.2673	3.9171	3.9177

Table 5.4. *(cont.) Buckling load Nxx for a plate with different plate aspect ratios and boundary conditions ($\nu = 0.3$)*

Case number	$\frac{a}{b}$	Uniaxial Present study	Uniaxial ANSYS	Biaxial, $\beta = 1$ Present study	Biaxial, $\beta = 1$ ANSYS
47 GCFC	0.5	36.0835	36.0838	29.3306	29.3328
	1	38.4653	38.5065	27.9114	27.9335
	1.5	38.1114	38.2555	28.0653	28.1403
	2	37.9793	38.2710	27.9655	28.1189
	3	37.5599	38.2785	27.7299	28.1254
48 CGSF	0.5	79.9855	80.0240	73.2185	73.3111
	1	19.8428	19.8500	18.4050	18.4101
	1.5	8.7549	8.7576	8.2860	8.2875
	2	4.8919	4.8931	4.7017	4.7025
	3	2.1502	2.1505	2.1003	2.1008
48 GCFS	0.5	30.2152	30.2175	18.8046	18.8093
	1	30.2116	30.2246	18.4050	18.4101
	1.5	30.1885	30.2353	18.3192	18.3327
	2	30.1389	30.2355	18.3046	18.3310
	3	29.9955	30.2397	18.2649	18.3334

Table 5.4. *(cont.) Buckling load Nxx for a plate with different plate aspect ratios and boundary conditions ($\nu = 0.3$)*

In Figure 5.1, the first buckling mode shapes for three cases are shown. These buckling modes are for four different aspect ratios a/b for cases 27, SCGS, 34, GCGS and 44, CCGS, all with uniaxial loading in the x direction. Here we observe that a change in the plate aspect ratio results in a very significant change in the buckling mode.

Figure 5.2 shows the modes for the same cases as Figure 5.1, but for biaxial loading with $\beta = 1$. Unlike for the unidirectional loading case in Figure 5.1, we observe here that the buckling modes remain unchanged when the aspect ratio is increased.

5.6. Conclusion

The analytical solution for the problem of isotropic plate buckling with many of the possible combinations of boundary conditions is given. This new solution does not require the separation between symmetric and antisymmetric cases; instead, they can be obtained using only one expression. Splitting the variables in the solution allows us to represent the edge conditions for deflection, slope, shear forces and bending moment in terms of common functions. Finally, when we apply the eight particular boundary conditions to the edges, we can find the set of equations that are required to determine all the terms in the complete solution. The presented results can be used as benchmark results for further studies on this topic.

Case	$\frac{a}{b}$	β				
		-1	-0.5	0	0.5	1
7 CCCC	0.5	240.3182 (240.4700)	213.3413 (213.4700)	190.8646 (190.9700)	171.7551 (171.8500)	154.8911 (154.9770)
	1	147.7043 (147.8400)	129.4421 (129.5700)	99.4259 (99.4872)	69.3919 (69.4290)	52.3447 (52.3721)
	1.5	126.3699 (126.5200)	107.6367 (107.7300)	82.4162 (82.4790)	61.2038 (61.2400)	40.6749 (40.6975)
	2	118.0669 (118.2200)	97.9599 (98.0460)	77.6449 (77.7230)	56.8042 (56.8420)	38.7228 (38.7456)
11 CFCF	0.5	156.7479 (156.8600)	156.5588 (156.6800)	156.0736 (156.2300)	150.5267 (150.9100)	111.1686 (111.7310)
	1	38.9199 (38.9440)	38.8404 (38.8650)	38.6773 (38.7011)	37.2665 (37.2900)	27.0569 (27.0782)
	1.5	17.1819 (17.1920)	17.1373 (17.1470)	17.0604 (17.0690)	16.2680 (16.2700)	10.5295 (10.5302)
	2	9.6036 (9.6089)	9.5764 (9.5817)	9.5363 (9.5405)	8.9530 (8.9542)	5.3367 (5.3371)
13 CCSS	0.5	136.9266 (136.9600)	120.7793 (120.8100)	107.6543 (107.6800)	96.8168 (96.8400)	87.6908 (87.7112)
	1	105.8388 (105.8900)	87.6193 (87.6560)	61.4150 (61.4310)	42.5475 (42.5580)	32.0524 (32.0603)
	1.5	98.6346 (98.6880)	78.0992 (78.1260)	57.9389 (57.9610)	36.6602 (36.6700)	24.0495 (24.0563)
	2	95.8631 (95.9180)	75.4577 (75.4890)	55.9436 (55.9630)	36.2018 (36.2130)	21.9227 (21.9295)
23 SCSG	0.5	49.3099 (49.3140)	47.5731 (47.5770)	45.9395 (45.9440)	44.3968 (44.4010)	42.9320 (42.9364)
	1	26.6181 (26.6190)	22.1972 (22.1980)	18.9775 (18.9789)	16.5332 (16.5350)	14.6174 (14.6190)
	1.5	29.9064 (29.9160)	26.3233 (26.3230)	17.5579 (17.5590)	13.0772 (13.0780)	10.3843 (10.3854)
	2	26.6181 (26.6190)	22.1972 (22.2010)	18.9775 (18.9830)	13.1155 (13.1170)	9.4499 (9.4510)
48 CGSF	0.5	80.2532 (80.2910)	80.1639 (80.2020)	79.9855 (80.0240)	79.3718 (79.4130)	73.2185 (73.3111)
	1	19.9495 (19.9570)	19.9116 (19.9190)	19.8428 (19.8500)	19.6597 (19.6670)	18.4050 (18.4101)
	1.5	8.8166 (8.8195)	8.7939 (8.7967)	8.7549 (8.7576)	8.6666 (8.6692)	8.2860 (8.2875)
	2	4.9322 (4.9335)	4.9167 (4.9180)	4.8919 (4.8931)	4.8431 (4.8443)	4.7017 (4.7025)

Table 5.5. *First buckling load N_{xx} for rectangular plates with different aspect ratios a/b and different compression ratios β (ν=0.3)*

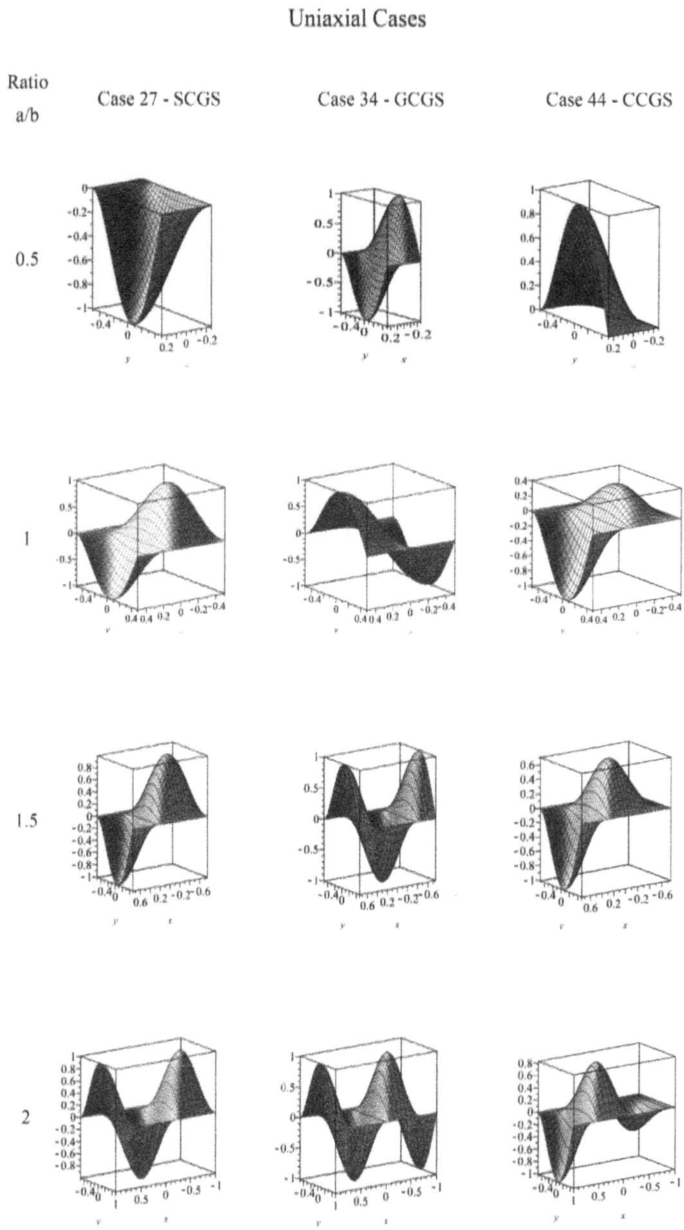

Figure 5.1. *Buckling modes for plates with different aspect ratios a/b under uniaxial loading. For a color version of this figure, see www.iste.co.uk/challamel/mechanics1.zip*

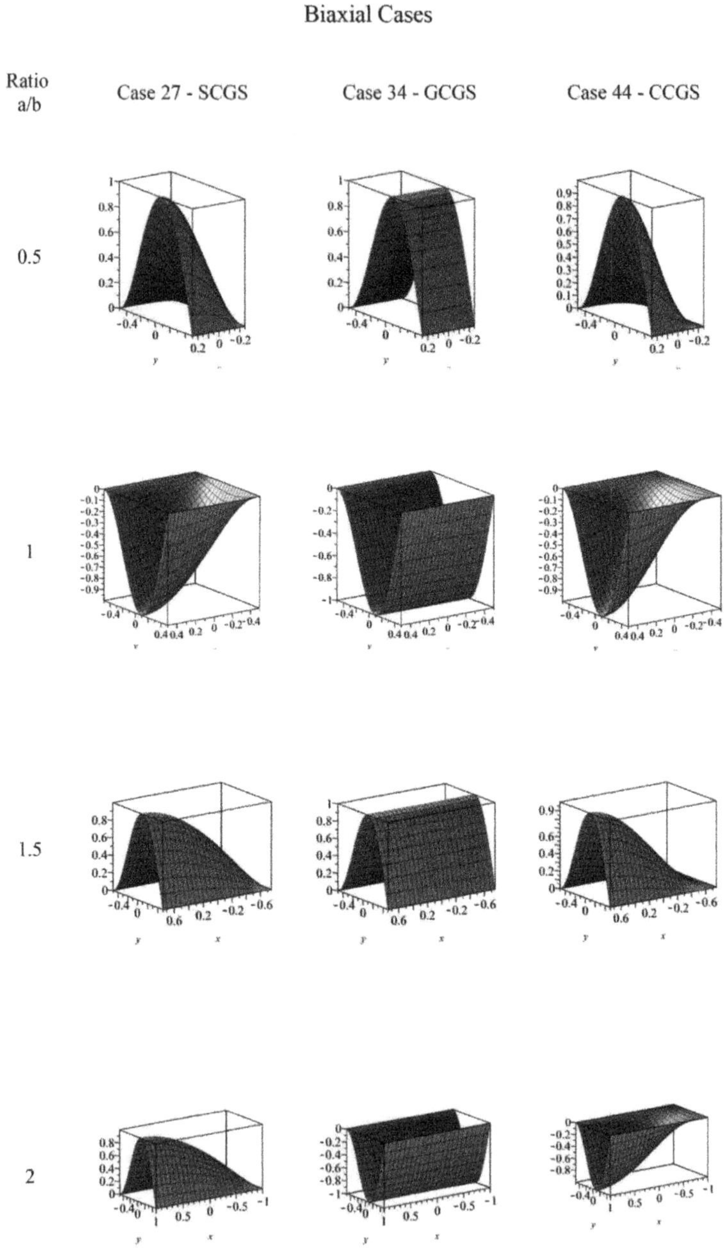

Figure 5.2. *Buckling modes for plates with different aspect ratios a/b under biaxial loading. For a color version of this figure, see www.iste.co.uk/challamel/mechanics1.zip*

5.7. Appendix A: Deflection, slopes, bending moments and shears

In this appendix, we present 16 possible edge boundary conditions for the plate.

a. The four displacement equations along the edges of the plate

$$W\left(x=\frac{a}{2},y\right)=\sum_{n=1}^{\infty}\bar{Y}_n\times\left[E_n cosh\left(\frac{\alpha_3\,a}{2}\right)+F_n cosh\left(\frac{\alpha_4\,a}{2}\right)\right.$$
$$\left.+G_n sinh\left(\frac{\alpha_3\,a}{2}\right)+H_n sinh\left(\frac{\alpha_4\,a}{2}\right)\right] \quad [A.1]$$

$$W\left(x=-\frac{a}{2},y\right)=\sum_{n=1}^{\infty}\bar{Y}_n\times\left[E_n\cosh\left(\frac{\alpha_3\,a}{2}\right)+F_n\cosh\left(\frac{\alpha_4\,a}{2}\right)\right.$$
$$\left.-G_n\sinh\left(\frac{\alpha_3\,a}{2}\right)-H_n\sinh\left(\frac{\alpha_4\,a}{2}\right)\right] \quad [A.2]$$

$$W\left(x,y=\frac{b}{2}\right)=\sum_{m=1}^{\infty}\bar{X}_m\times\left[A_m\,cosh\left(\frac{\alpha_1\,b}{2}\right)+B_m\,cosh\left(\frac{\alpha_2\,b}{2}\right)\right.$$
$$\left.+C_m\,sinh\left(\frac{\alpha_1\,b}{2}\right)+D_m\,sinh\left(\frac{\alpha_2\,b}{2}\right)\right] \quad [A.3]$$

$$W\left(x,y=-\frac{b}{2}\right)=\sum_{m=1}^{\infty}\bar{X}_m\times\left[A_m\,cosh\left(\frac{\alpha_1\,b}{2}\right)+B_m\,cosh\left(\frac{\alpha_2\,b}{2}\right)\right.$$
$$\left.-C_m\,sinh\left(\frac{\alpha_1\,b}{2}\right)-D_m\,sinh\left(\frac{\alpha_2\,b}{2}\right)\right] \quad [A.4]$$

b. The four slope equations normal to the edges of the plate

$$\phi_x\left(x=\frac{a}{2},y\right)=\sum_{n=1}^{\infty}\bar{Y}_n\times\left[E_n\alpha_3\,sinh\left(\frac{\alpha_3\,a}{2}\right)+F_n\alpha_4\,sinh\left(\frac{\alpha_4\,a}{2}\right)\right.$$
$$\left.+G_n\alpha_3\,cosh\left(\frac{\alpha_3\,a}{2}\right)+H_n\alpha_4\,cosh\left(\frac{\alpha_4\,a}{2}\right)\right]$$
$$+\sum_{m=1}^{\infty}\lambda_a\,cos(m\pi)\times(A_m cosh\,(\alpha_1\,y)$$
$$+B_m cosh\,(\alpha_2\,y)+C_m sinh\,(\alpha_1\,y)+D_m sinh\,(\alpha_2\,y))$$
$$+\sum_{m=1}^{\infty}\sum_{n=1}^{\infty}\bar{Y}_n\times\frac{4\,Psin\,(\xi\,\lambda_a)\,sin\,(\eta\,\lambda_b)\,\lambda_a}{d_{mn}\,ab}\,cos(m\pi)$$
$$[A.5]$$

$$\phi_x\left(x=-\frac{a}{2},y\right)=\sum_{n=1}^{\infty}\bar{Y}_n\times\left[-E_n\alpha_3\,sinh\left(\frac{\alpha_3\,a}{2}\right)-F_n\alpha_4\,sinh\left(\frac{\alpha_4\,a}{2}\right)\right.$$
$$\left.+G_n\alpha_3\,cosh\left(\frac{\alpha_3\,a}{2}\right)+H_n\alpha_4\,cosh\left(\frac{\alpha_4\,a}{2}\right)\right]$$
$$+\sum_{m=1}^{\infty}\lambda_a\times(A_m cosh\,(\alpha_1\,y)+B_m cosh\,(\alpha_2\,y)$$
$$+C_m sinh\,(\alpha_1\,y)+D_m sinh\,(\alpha_2\,y))$$
$$+\sum_{m=1}^{\infty}\sum_{n=1}^{\infty}\bar{Y}_n\times\frac{4\,P sin\,(\xi\,\lambda_a)\,sin\,(\eta\,\lambda_b)\,\lambda_a}{d_{mn}\,ab} \quad [A.6]$$

$$\phi_y\left(x,y=\frac{b}{2}\right)=\sum_{m=1}^{\infty}\bar{X}_m\times\left[A_m\alpha_1\,sinh\left(\frac{\alpha_1\,b}{2}\right)+B_m\alpha_2\,sinh\left(\frac{\alpha_2\,b}{2}\right)\right.$$
$$\left.+C_m\alpha_1\,cosh\left(\frac{\alpha_1\,b}{2}\right)+D_m\alpha_2\,cosh\left(\frac{\alpha_2\,b}{2}\right)\right]$$
$$+\sum_{n=1}^{\infty}\lambda_b\,cos(n\pi)\times(E_n\cosh(\alpha_3\,x)$$
$$+F_n\cosh(\alpha_4\,x)+G_n\sinh(\alpha_3\,x)+H_n\sinh(\alpha_4\,x))$$
$$+\sum_{m=1}^{\infty}\sum_{n=1}^{\infty}\bar{X}_m\times\frac{4\,P sin\,(\xi\,\lambda_a)\,sin\,(\eta\,\lambda_b)\,\lambda_b}{d_{mn}\,ab}cos(n\pi) \quad [A.7]$$

$$\phi_y\left(x,y=-\frac{b}{2}\right)=\sum_{m=1}^{\infty}\bar{X}_m\times\left[-A_m\alpha_1\,sinh\left(\frac{\alpha_1\,b}{2}\right)-\right.$$
$$B_m\alpha_2\,sinh\left(\frac{\alpha_2\,b}{2}\right)+C_m\alpha_1\,cosh\left(\frac{\alpha_1\,b}{2}\right)+$$
$$\left.D_m\alpha_2\,cosh\left(\frac{\alpha_2\,b}{2}\right)\right]$$
$$+\sum_{n=1}^{\infty}\lambda_b\times(E_n\cosh(\alpha_3\,x)+F_n\cosh(\alpha_4\,x)$$
$$+G_n\sinh(\alpha_3\,x)+H_n\sinh(\alpha_4\,x))$$
$$+\sum_{m=1}^{\infty}\sum_{n=1}^{\infty}\bar{X}_m\times\frac{4\,P sin\,(\xi\,\lambda_a)\,sin\,(\eta\,\lambda_b)\,\lambda_b}{d_{mn}\,ab} \quad [A.8]$$

c. The four bending moment equations along the edges of the plate

$$\begin{aligned}M_x\left(x=\frac{a}{2},y\right)=\sum_{n=1}^{\infty}\bar{Y}_nD\times\Bigg[-\Bigg[&E_n{\alpha_3}^2cosh\left(\frac{\alpha_3\,a}{2}\right)\\&+F_n{\alpha_4}^2cosh\left(\frac{\alpha_4\,a}{2}\right)+G_n{\alpha_3}^2sinh\left(\frac{\alpha_3\,a}{2}\right)\\&+H_n{\alpha_4}^2sinh\left(\frac{\alpha_4\,a}{2}\right)\Bigg]\\&+\nu\,{\lambda_b}^2\times\Bigg[E_ncosh\left(\frac{\alpha_3\,a}{2}\right)+F_ncosh\left(\frac{\alpha_4\,a}{2}\right)\\&+G_nsinh\left(\frac{\alpha_3\,a}{2}\right)+H_nsinh\left(\frac{\alpha_4\,a}{2}\right)\Bigg]\Bigg]\end{aligned}\qquad [A.9]$$

$$\begin{aligned}M_x\left(x=-\frac{a}{2},y\right)=\sum_{n=1}^{\infty}\bar{Y}_nD\times\Bigg[-\Bigg[&E_n{\alpha_3}^2cosh\left(\frac{\alpha_3\,a}{2}\right)\\&+F_n{\alpha_4}^2cosh\left(\frac{\alpha_4\,a}{2}\right)\\&-G_n{\alpha_3}^2sinh\left(\frac{\alpha_3\,a}{2}\right)-H_n{\alpha_4}^2sinh\left(\frac{\alpha_4\,a}{2}\right)\Bigg]\\&+\nu\,{\lambda_b}^2\times\Bigg[E_ncosh\left(\frac{\alpha_3\,a}{2}\right)+F_ncosh\left(\frac{\alpha_4\,a}{2}\right)\\&-G_nsinh\left(\frac{\alpha_3\,a}{2}\right)-H_nsinh\left(\frac{\alpha_4\,a}{2}\right)\Bigg]\Bigg]\end{aligned}\qquad [A.10]$$

$$\begin{aligned}M_y\left(x,y=\frac{b}{2}\right)=\sum_{m=1}^{\infty}\bar{X}_mD\times\Bigg[-\Bigg[&A_m{\alpha_1}^2cosh\left(\frac{\alpha_1\,b}{2}\right)\\&+B_m{\alpha_2}^2cosh\left(\frac{\alpha_2\,b}{2}\right)\\&+C_m{\alpha_1}^2sinh\left(\frac{\alpha_1\,b}{2}\right)+D_m{\alpha_2}^2sinh\left(\frac{\alpha_2\,b}{2}\right)\Bigg]\\&+\nu\,{\lambda_a}^2\times\Bigg[A_mcosh\left(\frac{\alpha_1\,b}{2}\right)+B_mcosh\left(\frac{\alpha_2\,b}{2}\right)\\&+C_msinh\left(\frac{\alpha_1\,b}{2}\right)+D_msinh\left(\frac{\alpha_2\,b}{2}\right)\Bigg]\Bigg]\end{aligned}\qquad [A.11]$$

$$M_y\left(x, y=-\frac{b}{2}\right)=\sum_{m=1}^{\infty} \bar{X}_m D \times\left[-\left[A_m {\alpha_1}^2 cosh\left(\frac{\alpha_1 b}{2}\right)\right.\right.$$
$$+B_m {\alpha_2}^2 cosh\left(\frac{\alpha_2 b}{2}\right)$$
$$\left.-C_m {\alpha_1}^2 sinh\left(\frac{\alpha_1 b}{2}\right)-D_m {\alpha_2}^2 sinh\left(\frac{\alpha_2 b}{2}\right)\right]$$
$$+\nu {\lambda_a}^2 \times\left[A_m cosh\left(\frac{\alpha_1 b}{2}\right)+B_m cosh\left(\frac{\alpha_2 b}{2}\right)\right.$$
$$\left.\left.-C_m sinh\left(\frac{\alpha_1 b}{2}\right)-D_m sinh\left(\frac{\alpha_2 b}{2}\right)\right]\right] \qquad [A.12]$$

d. The four reaction equations along the edges of the plate

$$Q_x\left(x=\frac{a}{2}, y\right)=\sum_{m=1}^{\infty} D\lambda_a^3 cos(m\pi) \times\left(A_m cosh\left(\alpha_1 y\right)\right.$$
$$\left.+B_m cosh\left(\alpha_2 y\right)+C_m sinh\left(\alpha_1 y\right)+D_m sinh\left(\alpha_2 y\right)\right)$$
$$-\sum_{n=1}^{\infty} D \bar{Y}_n \times\left(E_n {\alpha_3}^3 sinh\left(\frac{\alpha_3 a}{2}\right)+\right.$$
$$F_n {\alpha_4}^3 sinh\left(\frac{\alpha_4 a}{2}\right)$$
$$\left.+G_n {\alpha_3}^3 cosh\left(\frac{\alpha_3 a}{2}\right)+H_n {\alpha_4}^3 cosh\left(\frac{\alpha_4 a}{2}\right)\right)$$
$$+D(\nu-2)\left(\sum_{m=1}^{\infty} \lambda_a cos(m\pi) \times\left(A_m {\alpha_1}^2 cosh\left(\alpha_1 y\right)\right.\right.$$
$$+B_m {\alpha_2}^2 cosh\left(\alpha_2 y\right)+C_m {\alpha_1}^2 sinh\left(\alpha_1 y\right)$$
$$\left.+D_m {\alpha_2}^2 sinh\left(\alpha_2 y\right)\right)$$
$$-\sum_{n=1}^{\infty} \lambda_b^2 \bar{Y}_n \times\left[E_n \alpha_3 sinh\left(\frac{\alpha_3 a}{2}\right)+F_n \alpha_4 sinh\left(\frac{\alpha_4 a}{2}\right)\right.$$
$$\left.\left.+G_n \alpha_3 cosh\left(\frac{\alpha_3 a}{2}\right)+H_n \alpha_4 cosh\left(\frac{\alpha_4 a}{2}\right)\right]\right)$$

$$+ N_{xx}\left(\sum_{m=1}^{\infty} \lambda_a cos(m\pi) \times (A_m cosh\,(\alpha_1\, y)\right.$$
$$+B_m cosh\,(\alpha_2\, y) + C_m sinh\,(\alpha_1\, y)$$
$$+D_m sinh\,(\alpha_2\, y)) + \sum_{n=1}^{\infty} \bar{Y}_n \times \left[E_n \alpha_3\, sinh\left(\frac{\alpha_3\, a}{2}\right)\right.$$
$$+F_n \alpha_4\, sinh\left(\frac{\alpha_4\, a}{2}\right) + G_n \alpha_3\, cosh\left(\frac{\alpha_3\, a}{2}\right)$$
$$\left.\left.+H_n \alpha_4\, cosh\left(\frac{\alpha_4\, a}{2}\right)\right]\right)$$
$$-\sum_{m=1}^{\infty}\sum_{n=1}^{\infty} \bar{Y}_n \times \frac{4P sin\,(\xi\,\lambda_a)\, sin\,(\eta\,\lambda_b)\,\lambda_a}{d_{mn}\, ab} cos(m\pi)$$
$$\times \left(D\,\lambda_b{}^2 \nu - D\,\lambda_a{}^2 - 2\,D\,\lambda_b{}^2 - N_{xx}\right) \qquad [A.13]$$

$$Q_x\left(x = -\frac{a}{2}, y\right) = \sum_{m=1}^{\infty} D\lambda_a^3 \times (A_m cosh\,(\alpha_1\, y) + B_m cosh\,(\alpha_2\, y)$$
$$+C_m sinh\,(\alpha_1\, y) + D_m sinh\,(\alpha_2\, y))$$
$$-\sum_{n=1}^{\infty} D\,\bar{Y}_n \times \left(-E_n \alpha_3{}^3 sinh\left(\frac{\alpha_3\, a}{2}\right)\right.$$
$$-F_n \alpha_4{}^3 sinh\left(\frac{\alpha_4\, a}{2}\right)$$
$$\left.+G_n \alpha_3{}^3 cosh\left(\frac{\alpha_3\, a}{2}\right) + H_n \alpha_4{}^3 cosh\left(\frac{\alpha_4\, a}{2}\right)\right)$$
$$+ D(\nu - 2)\left(\sum_{m=1}^{\infty} \lambda_a \times \left(A_m \alpha_1{}^2 cosh\,(\alpha_1\, y)\right.\right.$$
$$+ B_m \alpha_2{}^2 cosh\,(\alpha_2\, y) + C_m \alpha_1{}^2 sinh\,(\alpha_1\, y)$$
$$\left.+ D_m \alpha_2{}^2 sinh\,(\alpha_2\, y)\right)$$
$$-\sum_{n=1}^{\infty} \lambda_b^2 \bar{Y}_n \times \left[-E_n \alpha_3\, sinh\left(\frac{\alpha_3\, a}{2}\right)\right.$$
$$-F_n \alpha_4\, sinh\left(\frac{\alpha_4\, a}{2}\right) + G_n \alpha_3\, cosh\left(\frac{\alpha_3\, a}{2}\right)$$
$$\left.\left.+H_n \alpha_4\, cosh\left(\frac{\alpha_4\, a}{2}\right)\right]\right)$$

$$+ N_{xx}\left(\sum_{m=1}^{\infty} \lambda_a \left(A_m cosh\left(\alpha_1 y\right) + B_m cosh\left(\alpha_2 y\right)\right.\right.$$

$$+C_m sinh\left(\alpha_1 y\right)$$

$$\left.+D_m sinh\left(\alpha_2 y\right)\right) + \sum_{n=1}^{\infty} \bar{Y}_n \times \left[-E_n \alpha_3\, sinh\left(\frac{\alpha_3\, a}{2}\right)\right.$$

$$-F_n \alpha_4\, sinh\left(\frac{\alpha_4\, a}{2}\right) + G_n \alpha_3\, cosh\left(\frac{\alpha_3\, a}{2}\right)$$

$$\left.\left.+H_n \alpha_4\, cosh\left(\frac{\alpha_4\, a}{2}\right)\right]\right)$$

$$- \sum_{m=1}^{\infty}\sum_{n=1}^{\infty} \bar{Y}_n \times \frac{4P sin\left(\xi\, \lambda_a\right) sin\left(\eta\, \lambda_b\right) \lambda_a}{d_{mn}\, ab}$$

$$\times \left(D\, {\lambda_b}^2 \nu - D\, {\lambda_a}^2 - 2\, D\, {\lambda_b}^2 - N_{xx}\right) \qquad \text{[A.14]}$$

$$Q_y\left(x, y = \frac{b}{2}\right) = -\sum_{m=1}^{\infty} D\bar{X}_m \times \left(A_m {\alpha_1}^3 sinh\left(\frac{\alpha_1\, b}{2}\right)\right.$$

$$+B_m {\alpha_2}^3 sinh\left(\frac{\alpha_2\, b}{2}\right)$$

$$\left.+C_m {\alpha_1}^3 cosh\left(\frac{\alpha_1\, b}{2}\right) + D_m {\alpha_2}^3 cosh\left(\frac{\alpha_2\, b}{2}\right)\right)$$

$$-\sum_{n=1}^{\infty} D\lambda_b^3 cos(n\pi) \times \left(E_n cosh\left(\alpha_3\, x\right)\right.$$

$$\left.+F_n cosh\left(\alpha_4\, x\right) + G_n sinh\left(\alpha_3\, x\right) + H_n sinh\left(\alpha_4\, x\right)\right)$$

$$+ D(\nu - 2)\left(-\sum_{m=1}^{\infty} \lambda_a^2 \bar{X}_m \times \left[A_m \alpha_1\, sinh\left(\frac{\alpha_1\, b}{2}\right)\right.\right.$$

$$+B_m \alpha_2\, sinh\left(\frac{\alpha_2\, b}{2}\right) + C_m \alpha_1\, cosh\left(\frac{\alpha_1\, b}{2}\right)$$

$$\left.+D_m \alpha_2\, cosh\left(\frac{\alpha_2\, b}{2}\right)\right]$$

$$+\sum_{n=1}^{\infty} \lambda_b\, cos(n\pi) \times \left(E_n {\alpha_3}^2 cosh\left(\alpha_3\, x\right)\right.$$

$$+F_n{\alpha_4}^2cosh\left(\alpha_4\,x\right)$$

$$\left.+G_n{\alpha_3}^2sinh\left(\alpha_3\,x\right)+H_n{\alpha_4}^2sinh\left(\alpha_4\,x\right)\right)\Bigg)$$

$$+\,N_{yy}\Bigg(\sum_{m=1}^{\infty}\bar{X}_m\times\left[A_m\alpha_1\,sinh\left(\frac{\alpha_1\,b}{2}\right)+\right.$$

$$B_m\alpha_2\,sinh\left(\frac{\alpha_2\,b}{2}\right)$$

$$\left.+C_m\alpha_1\,cosh\left(\frac{\alpha_1\,b}{2}\right)+D_m\alpha_2\,cosh\left(\frac{\alpha_2\,b}{2}\right)\right]$$

$$+\sum_{n=1}^{\infty}\lambda_b cos(n\pi)\times\left(E_ncosh\left(\alpha_3\,x\right)\right.$$

$$\left.+F_ncosh\left(\alpha_4\,x\right)+G_nsinh\left(\alpha_3\,x\right)+H_nsinh\left(\alpha_4\,x\right)\right)\Bigg)$$

$$-\sum_{m=1}^{\infty}\sum_{n=1}^{\infty}\bar{X}_m\times\frac{4Psin\left(\xi\,\lambda_a\right)sin\left(\eta\,\lambda_b\right)\lambda_b}{d_{mn}\,ab}cos(n\pi)$$

$$\times\left(D\,{\lambda_a}^2\nu-D\,{\lambda_b}^2-2\,D\,{\lambda_a}^2-N_{yy}\right)\qquad[A.15]$$

$$Q_y\left(x,y=-\frac{b}{2}\right)=-\sum_{m=1}^{\infty}D\bar{X}_m\times\left(-A_m{\alpha_1}^3sinh\left(\frac{\alpha_1\,b}{2}\right)\right.$$

$$-B_m{\alpha_2}^3sinh\left(\frac{\alpha_2\,b}{2}\right)$$

$$\left.+C_m{\alpha_1}^3cosh\left(\frac{\alpha_1\,b}{2}\right)+D_m{\alpha_2}^3cosh\left(\frac{\alpha_2\,b}{2}\right)\right)$$

$$+\sum_{n=1}^{\infty}D\lambda_b^3\times\left(E_ncosh\left(\alpha_3\,x\right)\right.$$

$$\left.+F_ncosh\left(\alpha_4\,x\right)+G_nsinh\left(\alpha_3\,x\right)+H_nsinh\left(\alpha_4\,x\right)\right)$$

$$+\,D(\nu-2)\Bigg(-\sum_{m=1}^{\infty}\lambda_a^2\bar{X}_m\times\left[-A_m\alpha_1\,sinh\left(\frac{\alpha_1\,b}{2}\right)\right.$$

$$-B_m\alpha_2\,sinh\left(\frac{\alpha_2\,b}{2}\right)+C_m\alpha_1\,cosh\left(\frac{\alpha_1\,b}{2}\right)$$

$$\left.+D_m\alpha_2\,cosh\left(\frac{\alpha_2\,b}{2}\right)\right]$$

$$+\sum_{n=1}^{\infty}\lambda_b\times\left(E_n{\alpha_3}^2cosh\left(\alpha_3\,x\right)+F_n{\alpha_4}^2cosh\left(\alpha_4\,x\right)\right.$$

$$\left.+G_n{\alpha_3}^2sinh\left(\alpha_3\,x\right)+H_n{\alpha_4}^2sinh\left(\alpha_4\,x\right)\right)\Bigg)$$

$$+\,N_{yy}\left(\sum_{m=1}^{\infty}\bar{X}_m\times\left[-A_m\alpha_1\,sinh\left(\frac{\alpha_1\,b}{2}\right)\right.\right.$$

$$-B_m\alpha_2\,sinh\left(\frac{\alpha_2\,b}{2}\right)$$

$$\left.+C_m\alpha_1\,cosh\left(\frac{\alpha_1\,b}{2}\right)+D_m\alpha_2\,cosh\left(\frac{\alpha_2\,b}{2}\right)\right]$$

$$+\sum_{n=1}^{\infty}\lambda_b\times\left(E_ncosh\left(\alpha_3\,x\right)+F_ncosh\left(\alpha_4\,x\right)\right.$$

$$\left.\left.+G_nsinh\left(\alpha_3\,x\right)+H_nsinh\left(\alpha_4\,x\right)\right)\right)$$

$$-\sum_{m=1}^{\infty}\sum_{n=1}^{\infty}\bar{X}_m\times\frac{4Psin\left(\xi\,\lambda_a\right)sin\left(\eta\,\lambda_b\right)\lambda_b}{d_{mn}\,ab}$$

$$\times\left(D\,{\lambda_a}^2\nu-D\,{\lambda_b}^2-2\,D\,{\lambda_a}^2-N_{yy}\right)\qquad\text{[A.16]}$$

5.8. Appendix B: Function transformation

In this appendix, the expansions of equations [5.15] and [5.16] are derived. By rewriting X_n, we obtain

$$X_n=E_n\;cosh\left(\alpha_3x\right)+F_n\;cosh\left(\alpha_4x\right)+G_n\;sinh\left(\alpha_3x\right)+H_n\;sinh\left(\alpha_4x\right)\qquad\text{[B.1]}$$

We are interested in expanding the four $cosh$ and $sinh$ hyperbolic functions in terms of the trigonometric functions in $\bar{X}_m$, which are as follows:

$$cosh(\alpha_ix)=\sum_{m=2,4...}^{\infty}R_msin\left(\frac{m\pi x}{a}\right)cos\left(\frac{m\pi}{2}\right)$$

$$+\sum_{m=1,3...}^{\infty}S_mcos\left(\frac{m\pi x}{a}\right)sin\left(\frac{m\pi}{2}\right)\qquad\text{[B.2]}$$

with

$$R_m = \frac{2}{a}\int_{-\frac{a}{2}}^{\frac{a}{2}} cosh(\alpha_i x) sin\left(\frac{m\pi x}{a}\right) cos\left(\frac{m\pi}{2}\right) dx = 0 \tag{B.3}$$

$$S_m = \frac{2}{a}\int_{\frac{-a}{2}}^{\frac{a}{2}} cosh(\alpha_i x) cos\left(\frac{m\pi x}{a}\right) sin\left(\frac{m\pi}{2}\right) dx = \frac{4\,\pi\, m\, cosh(\frac{\alpha_i a}{2})}{m^2\,\pi^2 + \alpha_i^2\, a^2} \tag{B.4}$$

and using trigonometric identities, we finally arrive at the following, simpler, expression:

$$cosh(\alpha_i x) = \sum_{m=1,2,3\ldots}^{\infty} sin\left(\frac{m\pi}{2}\right)^2 \frac{4\,\pi\, m\, cosh(\frac{\alpha_i a}{2})}{m^2\,\pi^2 + \alpha_i^2\, a^2} \bar{X}_m, \tag{B.5}$$

Similarly, we obtain, for $sinh(\alpha_i x)$

$$sinh(\alpha_i x) = \sum_{m=1,2,3\ldots}^{\infty} -cos\left(\frac{m\pi}{2}\right)^2 \frac{4\,\pi\, m\, sinh(\frac{\alpha_i a}{2})}{m^2\,\pi^2 + \alpha_i^2\, a^2} \bar{X}_m, \tag{B.6}$$

and

$$cosh(\alpha_i y) = \sum_{n=1,2,3\ldots}^{\infty} sin\left(\frac{n\pi}{2}\right)^2 \frac{4\,\pi\, n\, cosh(\frac{\alpha_i b}{2})}{n^2\,\pi^2 + \alpha_i^2\, b^2} \bar{Y}_n, \tag{B.7}$$

$$sinh(\alpha_i y) = \sum_{n=1,2,3\ldots}^{\infty} -cos\left(\frac{n\pi}{2}\right)^2 \frac{4\,\pi\, n\, sinh(\frac{\alpha_i b}{2})}{n^2\,\pi^2 + \alpha_i^2\, b^2} \bar{Y}_n, \tag{B.8}$$

Here we can rewrite the expressions for the hyperbolic functions $sinh$ and $cosh$ in a compact form as

$$sinh(\alpha_1 y) = \sum_{n=1}^{\infty} s_1 \bar{Y}_n, \tag{B.9}$$

$$sinh(\alpha_2 y) = \sum_{n=1}^{\infty} s_2 \bar{Y}_n, \tag{B.10}$$

$$sinh(\alpha_3 x) = \sum_{m=1}^{\infty} s_3 \bar{X}_m, \tag{B.11}$$

$$sinh(\alpha_4 x) = \sum_{m=1}^{\infty} s_4 \bar{X}_m, \qquad [B.12]$$

$$cosh(\alpha_1 y) = \sum_{n=1}^{\infty} c_1 \bar{Y}_n, \qquad [B.13]$$

$$cosh(\alpha_2 y) = \sum_{n=1}^{\infty} c_2 \bar{Y}_n, \qquad [B.14]$$

$$cosh(\alpha_3 x) = \sum_{m=1}^{\infty} c_3 \bar{X}_m, \qquad [B.15]$$

$$cosh(\alpha_4 x) = \sum_{m=1}^{\infty} c_4 \bar{X}_m, \qquad [B.16]$$

$$s_1 = -4\,\pi\,n\,\frac{cos^2(\frac{n\pi}{2})\,sinh(\frac{\alpha_1 b}{2})}{n^2\pi^2 + {\alpha_1}^2\,b^2}, \qquad [B.17]$$

$$s_2 = -4\,\pi\,n\,\frac{cos^2(\frac{n\pi}{2})\,sinh(\frac{\alpha_2 b}{2})}{n^2\pi^2 + {\alpha_2}^2\,b^2}, \qquad [B.18]$$

$$s_3 = -4\,\pi\,m\,\frac{cos^2(\frac{m\pi}{2})\,sinh(\frac{\alpha_3 a}{2})}{m^2\pi^2 + {\alpha_3}^2\,a^2}, \qquad [B.19]$$

$$s_4 = -4\,\pi\,m\,\frac{cos^2(\frac{m\pi}{2})\,sinh(\frac{\alpha_4 a}{2})}{m^2\pi^2 + {\alpha_4}^2\,a^2}, \qquad [B.20]$$

$$c_1 = 4\,\pi\,n\,\frac{sin^2(\frac{n\pi}{2})\,cosh(\frac{\alpha_1 b}{2})}{n^2\pi^2 + {\alpha_1}^2\,b^2}, \qquad [B.21]$$

$$c_2 = 4\,\pi\,n\,\frac{sin^2(\frac{n\pi}{2})\,cosh(\frac{\alpha_2 b}{2})}{n^2\pi^2 + {\alpha_2}^2\,b^2}, \qquad [B.22]$$

$$c_3 = 4\,\pi\,m\,\frac{sin^2(\frac{m\pi}{2})\,cosh(\frac{\alpha_3 a}{2})}{m^2\pi^2 + {\alpha_3}^2\,a^2}, \qquad [B.23]$$

$$c_4 = 4\,\pi\,m\,\frac{sin^2(\frac{m\pi}{2})\,cosh(\frac{\alpha_4 a}{2})}{m^2\pi^2 + {\alpha_4}^2\,a^2}. \qquad [B.24]$$

5.9. References

Bert, C.W. and Malik, M. (1994). Frequency equations and modes of free vibrations of rectangular plates with various edge conditions. *Proceedings of the Institution of Mechanical Engineers, Part C: Journal of Mechanical Engineering Science*, 208, 308–319.

Bulson, P.S. (1970). *The Stability of Flat Plates*. Chatto & Windus, London.

Dickinson, S. (1978). The buckling and frequency of flexural vibration of rectangular isotropic and orthotropic plates using Rayleigh's method. *Journal of Sound and Vibration*, 61(1), 1–8.

Eisenberger, M. and Deutsch, A. (2015). Static analysis for exact vibration analysis of clamped plates. *International Journal of Structural Stability and Dynamics*, 15, 1540030.

Levy, M. (1899). Sur l'équilibre élastique d'une plaque rectangulaire. *Comptes rendus de l'Académie des Sciences de Paris*, 129(1), 535–539.

Li, R., Zheng, X., Wang, H., Xiong, S., Yan, K., Li, P. (2018). New analytic buckling solutions of rectangular thin plates with all edges free. *International Journal of Mechanical Sciences*, 144, 67–73.

Liu, X., Liu, X., Xie, S. (2020). A highly accurate analytical spectral flexibility formulation for buckling and wrinkling of orthotropic rectangular plates. *International Journal of Mechanical Sciences*, 168, 105311.

Narita, Y. and Leissa, A.W. (1990). Buckling studies for simply supported symmetrically laminated rectangular plates. *International Journal of Mechanical Sciences*, 32(11), 909–924.

Navier, L. (1819). *Résumé des leçons de mécanique*. École Polytechnique, Paris.

Reddy, J.N. (1997). *Mechanics of Laminated Composite Plates: Theory and Analysis*. CRC Press, Boca Raton.

Shufrin, I. and Eisenberger, M. (2005). Stability and vibration of shear deformable plates – First order and higher order analyses. *International Journal of Solids and Structures*, 42(3), 1225–1251.

Shufrin, I., Rabinovitch, O., Eisenberger, M. (2008a). Buckling of laminated plates with general boundary conditions under combined compression, tension, and shear – A semi-analytical solution. *Thin-Walled Structures*, 46(7), 925–938.

Shufrin, I., Rabinovitch, O., Eisenberger, M. (2008b). Buckling of symmetrically laminated rectangular plates with general boundary conditions – A semi-analytical approach. *Composite Structures*, 82(4), 521–531.

Szilard, R. (2004). *Theories and Application of Plate Analysis*. John Wiley & Sons, New York.

Timoshenko, S.P. (1959). *Theory of Plates and Shells*. McGraw-Hill, Reading.

Timoshenko, S.P. and Gere, J.M. (1961). *Theory of Elastic Stability*. McGraw-Hill, New York.

Ullah, S., Zhang, J., Zhong, Y. (2019a). Accurate buckling analysis of rectangular thin plates by double finite sine integral transform method. *Structural Engineering and Mechanics*, 72, 491–502.

Ullah, S., Zhong, Y., Zhang, J. (2019b). Analytical buckling solutions of rectangular thin plates by straightforward generalized integral transform method. *International Journal of Mechanical Sciences*, 152, 535–544.

Wang, C.M., Wang, C.Y., Reddy, J.N. (2005). *Exact Solutions for Buckling of Structural Members*. CRC Press, Boca Raton.

Wang, B., Li, P., Li, R. (2016). Symplectic superposition method for new analytic buckling solutions of rectangular thin plates. *International Journal of Mechanical Sciences*, 119, 432–441.

6

Buckling and Post-Buckling of Parabolic Arches with Local Damage

6.1. Introduction

The birth of fracture mechanics is attributed to Griffith (1920), who based it on the solutions provided by Kirsch (1898) and Inglis (1913). Many efforts to enhance this new branch of solid mechanics followed (Irwin 1957a,b, 1960a,b; Dugdale 1960; Wells 1963; Rice 1968; Rice and Levy 1972). Applications of fracture mechanics to one-dimensional continua were possible after the introduction of the stress intensity factor (SIF) in Irwin (1957a,b). This, together with Castigliano's theorem, allows crack-like damages to be modeled by springs, and the response of beam frames can be described with no need to find the stress field around the crack tip. The structural response includes the displacement profiles due to benchmark load distributions, as well as the frequencies and mode shapes of natural free vibration, which encompass information about geometry, material properties and boundary conditions. A decade before the introduction of SIFs, Kirshmer (1944) and Thomson (1949) focused on simple structures with crack-type damages using a reduced area. Even after the introduction of SIFs, these crack models were used by many researchers.

Dimarogonas (1981, 1982, 1996); Dimarogonas and Paipetis (1983); Anifantis and Dimarogonas (1983a,b,c) successfully used the spring analogy by means of SIFs, especially on rotating machines. These papers paved the way to many more, which focus on building enhanced models for either the crack or the structure and on examining the effects of cracks on the static and dynamic behavior of beam-type structures (Gounaris and Dimarogonas 1988; Ostachowicz and Krawwczuk 1992;

Chapter written by Uğurcan EROĞLU, Giuseppe RUTA, Achille PAOLONE and Ekrem TÜFEKCI.

Chati *et al.* 1997; Khiem and Lien 2001; Sinha and Friswell 2002; Viola *et al.* 2002; Krawczuk *et al.* 2003; Kisa and Arif Gurel 2007; Orhan 2007; Mazanoglu *et al.* 2009; Caddemi and Morassi 2013; Caddemi *et al.* 2017; Caliò *et al.* 2017; Cannizzaro *et al.* 2017; Eroglu and Tufekci 2017b; Eroglu *et al.* 2019, 2020), without pretending to be exhaustive. A thorough investigation of the structural response suggested that monitoring the mechanical behavior of cracked elements may lead to damage identification by suitable methods. Thus, many papers focused on experimental settings, optimization techniques and their effects on identification procedures (Cawley and Adams 1979; Ju and Minovich 1986; Cerri and Ruta 2004; Cerri *et al.* 2008; Pau *et al.* 2010; Ciambella and Vestroni 2015; Capecchi *et al.* 2016; Eroglu and Tufekci 2016, 2017a; Caddemi and Morassi 2007, 2011; Greco and Pau 2011). A survey of the entire literature is out of our scope, yet, we may refer to the review of Doebling *et al.* (1996) for details.

Most studies on cracked one-dimensional structural elements deal with their statics and free dynamics, while their stability is only given marginal consideration, especially arches. A possible reason for this is that stability analysis is usually related to structural design and functioning, and is not meant for structural monitoring purposes. We may quote Karaagac *et al.* (2011), who examined the static and dynamic stability of circular arches by the finite element technique. They used the SIFs provided in Müller *et al.* (1993) and assumed the crack on the inner surface of the arch.

We investigate buckling and post-buckling of parabolic arches with crack-like damages. The latter may be simplified models of either the failure of connections of large structures and actual cracks in relatively small-size applications. We model the arch as a one-dimensional curved beam and study its fundamental and bifurcated path under reasonable kinematic and constitutive assumptions. The crack is modeled by the spring analogy and the SIFs, following literature and previous works by the authors. The novel points here are: a) the fundamental path is non-trivial, in that we linearize the finite field equations and find a non-zero linear elastic response to a uniform load with respect to the arch span; b) the critical values of the load multiplier are found in closed form; c) the germ of the post-buckling path is found by an additional perturbation of the finite field equations about the bifurcation point, providing information on the quality of the bifurcated path. Thorough comments and mechanical interpretations, in light of classic results on stability, are provided, together with hints on possible applications and future developments.

6.2. A one-dimensional model for arches

We model arches as initially curved fully deformable beams, i.e. one-dimensional structured continua that are assumed flexible, extensible and shear-deformable. Then, we suppose that their reference shape is well defined starting from a segment of a regular planar curve, called the arch axis. The curve segment lies on a plane. On the latter we fix a Cartesian frame x, y, to which a consistent basis of unit vectors $\{\mathbf{e}_x, \mathbf{e}_y\}$ is associated. A third unit vector $\mathbf{e}_z$, which coincides with one of the unit normals to the plane of the arch axis, completes a vector basis for the vector space associated with the three-dimensional Euclidean ambient space. The position of any point P of the axis is described by the vector field $\mathbf{r}_0$

$$\mathbf{r}_0(x) = x\mathbf{e}_x + y(x)\mathbf{e}_y, \quad 0 \le x \le \bar{x},\ y(0) \le y \le y(\bar{x}) \tag{6.1}$$

The tangent vector to the arch axis at any point P and its unit counterpart $\mathbf{l}(s)$ are

$$\begin{gathered}\frac{d\mathbf{r}_0(x)}{dx} = \mathbf{e}_x - \frac{dy(x)}{dx}\mathbf{e}_y, \qquad ds = \sqrt{\frac{d\mathbf{r}_0(x)}{dx} \cdot \frac{d\mathbf{r}_0(x)}{dx}}dx,\\ \mathbf{l}(x(s)) = \frac{d\mathbf{r}_0(x(s))}{dx(s)}\frac{dx(s)}{ds},\end{gathered} \tag{6.2}$$

where ds is the arc length, which is a measure of an element of the axis compatible with the given Cartesian metrics. Equation $[6.2]_2$ yields the intrinsic abscissa s in terms of x and vice versa; hence, $\mathbf{r}_0$ and all the fields depending on P may be expressed in terms of either x or s; thus, equation $[6.2]_3$ gives the field of unit vectors tangent to the axis.

The s-derivative of the field $\mathbf{l}(x(s))$, its magnitude and its unit counterpart are

$$\begin{gathered}\frac{d\mathbf{l}(x(s))}{ds} = \frac{d\mathbf{l}(x(s))}{dx(s)}\frac{dx(s)}{ds} =: k(x(s))\mathbf{m}(x(s)),\\ \mathbf{m}(x(s)) = \frac{1}{k(x(s))}\frac{d\mathbf{l}(x(s))}{ds}\end{gathered} \tag{6.3}$$

The quantity k is the local curvature of the axis at P; the Frénet–Serret local basis consists of the triad $\{\mathbf{l}, \mathbf{m}, \mathbf{n}\}$, $\mathbf{n} = \mathbf{l} \times \mathbf{m}$. Since the axis is a plane curve, the unit tangents and normals depend on P, while the bi-normal does not and coincides, with the exception of the a sign depending on the location of the osculating circle at P, with $\mathbf{e}_z$.

The reference shape is a three-dimensional region that is completed by attaching the copies of a prototype plane figure to any point of the axis. These are called transverse cross-sections and, being all equal, model an arch with uniform geometry with respect to the axis. With no loss in generality, the cross-sections are attached to the axis orthogonally to the unit tangent to the axis. That is, the initial setting of the cross-sections is determined by the Frénet–Serret local triad of unit base vectors for any point of the axis: the arch element is along the unit tangent $\mathbf{l}$, the corresponding cross-section lies in the plane spanned by the unit normal $\mathbf{m}$ and the unit bi-normal $\mathbf{n}$.

For simplicity, the dependence on $x(s)$ will be dropped if no confusion arises.

6.2.1. *Finite kinematics and balance, linear elastic law*

The reference configuration is supposed to be stress-free (i.e. the arch is in an initial natural state). A different shape is provided by the fields $\mathbf{d}$, the vector of the axis displacement (equivalently, by the new position $\mathbf{r} = \mathbf{r}_0 + \mathbf{d} \; \forall P$), and $\mathbf{R}$, the orthogonal tensor of the cross-section's rotation; both are functions of P, hence of x (or s) and of an evolution parameter (time, for instance).

Neutral, or rigid, changes of shape are characterized by a uniform value of $\mathbf{R}$ and by a tangent vector in the new shape that is simply the $\mathbf{R}$-transformed of $\mathbf{l}$. If primes denote s-derivatives to shorten the notation, finite deformation measures in the actual configuration are thus (see Antman (1973, 1995) and Pignataro *et al.* (2008))

$$\mathbf{v} = (\mathbf{r}_0 + \mathbf{d})' - \mathbf{R}\mathbf{l}, \quad \mathbf{V} = \mathbf{R}'\mathbf{R}^{\top} \qquad [6.4]$$

The components of the vector field $\mathbf{v}$ are the elongation ε of the axis and the shearing strains γ between the axis and the cross-sections; the non-zero components of the skew-symmetric tensor field $\mathbf{V}$ are the variation of curvature χ of the axis.

If we suppose the loads and the change of shape take place in the same plane of the referential configuration, $\mathbf{R}$ is expressed in terms of a single angle ϑ, providing the rotation of the cross-section at P about the bi-normal $\mathbf{n}$, and the displacement field has just two components:

$$(\mathbf{R}) = \begin{pmatrix} \cos\vartheta & -\sin\vartheta \\ \sin\vartheta & \cos\vartheta \end{pmatrix}, \qquad \mathbf{d} = u\mathbf{l} + v\mathbf{m} \qquad [6.5]$$

Then, we have only two and one components of the strain measures $\mathbf{v}, \mathbf{V}$, respectively

$$\varepsilon = 1 - \cos\vartheta + kv + u', \qquad \gamma = -\sin\vartheta + ku + v', \qquad \chi = \vartheta' \qquad [6.6]$$

The environment acts on the arch by: a vector force field, power dual of the incremental displacement of the axis; and a skew-symmetric tensor couple field, power dual of the incremental rotation of the cross-sections. Force and couple may be either distributed (actions at a distance) and localized at the end cross-sections (actions by contact); they are denoted $\mathbf{b}, \mathbf{B}, \mathbf{f}, \mathbf{F}$, respectively. The parts of the arch interact by contact via a vector and a skew symmetric tensor field, power duals of the incremental strains, denoted $\mathbf{t}, \mathbf{T}$, respectively. All the action fields are functions of the place P along the axis and of the evolution parameter, in general.

The balance equations in the configuration at a given value of the evolution parameter, also called actual shape, may be derived either by applying the virtual work principle or by assuming balance of force and torque as a basic principle. In both cases, they read (see Antman (1973, 1995) and Pignataro *et al.* (2008))

$$\begin{gathered}\mathbf{t}' + \mathbf{b} = \mathbf{0}, \qquad \mathbf{T}' + (\mathbf{r}_0 + \mathbf{d})' \wedge \mathbf{t} + \mathbf{B} = \mathbf{0} \quad \forall x \in (-l, l), \\ \mathbf{t} = \pm\mathbf{t}_{\pm l}, \qquad \mathbf{T} = \pm\mathbf{T}_{\pm l} \quad x = \pm l\end{gathered} \qquad [6.7]$$

with $\wedge$ the external product of vectors, providing skew-symmetric tensors. The balance of torque has a parallel expression in terms of axial vectors of skew-symmetric tensors and the cross-product between them. The two components of the inner force and the sole component of the inner couple, with respect to the local basis in the actual configuration, will be denoted $\tilde{N}, \tilde{T}, \tilde{M}$, respectively; they are usually named normal and transverse (or shearing) force and bending couple, respectively.

The components of the inner actions are resultants and resultant moments of Cauchy stresses (i.e., surface densities of contact force in the actual shape). It is customary in structural mechanics, however, to express them in terms of the inner actions in the reference configuration, with components denoted N, T, M, respectively (this is what is usually done by Cauchy and Piola stresses in three-dimensional continuum mechanics). The two sets of components are related via the rotation tensor $\mathbf{R}$ in equation $[6.5]_1$, ruling the change of unit vectors solidal to the cross-sections.

Then, the scalar consequences of equation [6.7] in the presence of planar external actions with components n, t, m, with respect to the local basis in the actual shape, read

$$\begin{gathered}N' - kT + n = 0, \qquad T' + kN + t = 0, \\ M' - N(\sin\vartheta + \gamma) + T(\cos\vartheta + \varepsilon) + m = 0\end{gathered} \qquad [6.8]$$

Here, the local curvature k appears due to the Frénet–Serret formulas for the s-derivatives of the intrinsic local basis.

Even though kinematics is in a finite setting and balance is written in the actual configuration in terms of the inner actions in the reference shape, we admit that the inner actions are linearly related to the strain measure components by elastic uncoupled constitutive laws, according to

$$\tilde{N} = EA\varepsilon, \qquad \tilde{T} = GA_s\gamma, \qquad \tilde{M} = EI\chi \qquad [6.9]$$

where E, G are Young's and transverse elastic moduli of the material of the arch, respectively, and A, A_s, I are the cross-sectional area, shearing area and second moment of area relative to a principal axis of inertia parallel to $\mathbf{n}$, respectively. The positions in equation [6.9] are reasonable in that we will operate a perturbation of the finite field equations, and linear (incremental) constitutive equations suffice to describe the arch response. The uncoupled laws in equation [6.9], similar to those provided by Saint-Venant's theory, are reasonable enough for compact cross-sections with main dimensions that are small compared with the local radius of the osculating circle (see Timoshenko and Goodier (1951)).

Then, taking into account that the local frame attached to the cross-section changes according to the orthogonal tensor $\mathbf{R}$ in equation [6.5], the constitutive relations in equation [6.9] can be given in terms of the components of the inner action in the reference shape, providing

$$\varepsilon = N\left(\frac{\cos^2\vartheta}{EA} + \frac{\sin^2\vartheta}{GA_s}\right) + T\sin\vartheta\cos\vartheta\left(\frac{1}{EA} - \frac{1}{GA_s}\right),$$
$$\gamma = T\left(\frac{\cos^2\vartheta}{GA_s} + \frac{\sin^2\vartheta}{EA}\right) + N\sin\vartheta\cos\vartheta\left(\frac{1}{EA} - \frac{1}{GA_s}\right), \quad \chi = \frac{M}{EI} \qquad [6.10]$$

In the following sections, we describe how we use the set of field equations for the arch in order to follow its response to an external load uniformly distributed with respect to the metrics of the span and monotonically increasing. A perturbation approach will follow a linearized solution in the neighborhood of the referential configuration (fundamental path), until a critical value of the load multiplier is reached. This corresponds to a bifurcation of the static equilibrium; in order to follow the jet of the bifurcated path, a second perturbation of the field equations in the neighborhood of the buckled shape will be performed.

6.2.2. *Non-trivial fundamental equilibrium path*

The field equations [6.6], [6.8] and [6.10] form a system of ordinary differential equations with respect to the spatial coordinate of any point P of the arch axis. To find the static response to any set of distributed external actions n, t, m in the actual shape, we are supposed to solve this nonlinear differential system, supplemented by the relevant boundary conditions. Let all external actions be rescaled by a load multiplier: then, the solutions of the field equations provide a one-parameter family of new shapes, called the fundamental path, which is described in exact, nonlinear, finite form. It goes without saying that, bar very special cases, the resolution of the field equations in finite form is computationally very demanding.

However, in structural engineering, we usually do not face finite displacements, and we may assume the structural response to be limited to "small" displacements and rotations. Hence, linearization of the field equations is possible, and the search for the fundamental path is greatly simplified. To this aim, let us introduce an evolution parameter $\eta \in [0, 1]$ such that $\eta = 0$, $\eta = 1$ characterize the reference and the actual shape, respectively. A perturbation approach, i.e., a formal power series expansion of all the fields of interest in terms of η, provides an approximate response of the structure about a given value of the evolution parameter (see Pignataro *et al.* (1991)). If we thus let $\eta = 0$ and limit the formal expansion at the first order, we study a non-trivial neighborhood of the reference configuration, supposed to represent the natural state of the arch. Then, if the suffix 0 denotes quantities evaluated in the reference shape,

$$\mathbf{R}_0 = \mathbf{I} \Leftrightarrow \vartheta_0 = 0, \qquad \mathbf{d}_0 = \mathbf{0}, \qquad N_0 = T_0 = M_0 = 0 \qquad [6.11]$$

Hence, the linear expansion of equations [6.6], [6.8] and [6.10] about the unstressed reference state provides

$$\begin{aligned} &EA(\dot{u}'_f - k\dot{v}_f) = \dot{N}_f, \quad GA_s(\dot{v}'_f - \dot{\vartheta}_f + k\dot{u}_f) = \dot{T}_f, \quad EI\dot{\vartheta}'_f = \dot{M}_f, \\ &\dot{N}'_f - k\dot{T}_f + \dot{n}_f = 0, \quad \dot{T}'_f + k\dot{N}_f + \dot{t}_f = 0, \quad \dot{M}'_f + \dot{T}_f + \dot{m}_f = 0, \end{aligned} \qquad [6.12]$$

where the overdot denotes η-derivatives of functions evaluated at $\eta = 0$; the suffix f indicates fields along the fundamental path. As is well known, the fundamental path is a single-valued curve expressing a characteristic displacement (or strain) versus the load multiplier when the latter monotonically grows from zero. When the load multiplier attains a so-called critical value, the equilibrium path becomes a multi-valued curve and bifurcation of the static solution occurs (see Pignataro *et al.* (1991)).

6.2.3. *Bifurcated path*

To investigate the static bifurcation phenomenon in a neighborhood of the critical point, we assume that any function g of interest in this problem may be written in the additive form $g = g_f + g_b$, where the subscripts f and b denote the values of g in the fundamental and bifurcated paths, respectively. If we let $g_b = 0$, we investigate the fundamental path only; if $g_b \neq 0$ and we add it to g_f, we obtain the value attained by g in the bifurcated path when departing from the fundamental one.

Let us investigate a neighborhood of the bifurcation point, found by evaluating the critical value of the load multiplier along the fundamental path. Then, we suppose that any function of interest along the bifurcated path regularly depends on another evolution parameter β, equal to zero at the bifurcation point and growing along the bifurcated path. We may thus perform a power series expansion of g_b in terms of β about the critical point $\beta = 0$ (see Pignataro *et al.* (1991)); if overdots now stand for the β-derivatives evaluated at $\beta = 0$ and $g_b(\beta = 0) = 0$, i.e. $g(\beta = 0) = g_f$, we have

$$g_b = 0 + \beta\dot{g} + (\beta^2/2)\ddot{g} + \dots \qquad [6.13]$$

Performing such an expansion for the system of equations [6.6], [6.8] and [6.10], where all functions of interest are now evaluated along the bifurcated path, we obtain a hierarchy of systems of ordinary differential equations, one at each order of the β-formal power series expansion (see Pignataro *et al.* (1991)).

The first set of the hierarchy provides the first-order germ of the bifurcated path and accounts for the quantities of the fundamental path at the critical point

$$\begin{aligned} -k\dot{v}_b + \sin\vartheta_f\dot{\vartheta}_b + \dot{u}_b' &= \frac{\dot{N}_b}{EA} + T_f\vartheta_b\left(\frac{1}{EA} - \frac{1}{GA_s}\right) = \dot{\varepsilon}_b, \\ k\dot{u}_b - \cos\vartheta_f\dot{\vartheta}_b + \dot{v}_b' &= \frac{\dot{T}_b}{GA_s} + N_f\vartheta_b\left(\frac{1}{EA} - \frac{1}{GA_s}\right) = \dot{\gamma}_b, \\ \dot{\vartheta}_b' = \frac{\dot{M}_b}{EI}, \qquad & \dot{N}_b' - k\dot{T}_b + \dot{n}_b = 0, \qquad \dot{T}_b' + k\dot{N}_b + \dot{t}_b = 0, \\ & \dot{M}_b' - N_f(\gamma_b + \cos\vartheta_f\dot{\vartheta}_b) - \dot{N}_b(\gamma_f + \sin\vartheta_f) + \\ & + T_f(\dot{\varepsilon}_b - \sin\vartheta_f\dot{\vartheta}_b) + \dot{T}_b(\cos\vartheta_f + \varepsilon_f) + \dot{m}_b = 0 \end{aligned} \qquad [6.14]$$

The terms along the fundamental path may be replaced with those evaluated by the η-linearization discussed in the previous section, since we assumed that the fundamental path, however non-trivial, consists of shapes adjacent to the reference configuration. Thus, equation [6.14] becomes

$$\begin{aligned} u_b' - kv_b &= \frac{N_b}{EA} + T_f\vartheta_b\left(\frac{1}{EA} - \frac{1}{GA_s}\right) - \vartheta_f\vartheta_b \\ v_b' + ku_b - \vartheta_b &= \frac{T_b}{GA_s} + N_f\vartheta_b\left(\frac{1}{EA} - \frac{1}{GA_s}\right), \\ EI\vartheta_b' = M_b \qquad & N_b' - kT_b + n_b = 0, \qquad T_b' + kN_b + t_b = 0, \\ M_b' - N_f(\gamma_b + \vartheta_b) - & N_b(\gamma_f + \vartheta_f) + T_f(\varepsilon_b - \vartheta_f\vartheta_b) + T_b(1 + \varepsilon_f) + m_b = 0 \end{aligned} \qquad [6.15]$$

where all overdots are omitted for a simpler notation: all quantities are thus actually first-order increments (with the exception of the curvature k in the reference configuration): those with subscript f are with respect to the reference configuration, and those with subscript b with respect to the critical state. The solution to the set in equation [6.15] provides the first-order bifurcated path with respect to β, accounting for the strains and inner actions attained at the bifurcation point.

Next, we show a couple of meaningful benchmark examples to show how the proposed two-parameter perturbation expansions work and provide known results.

6.2.4. *Special benchmark examples*

Here, we provide qualitative evidence on the generality of the differential system [6.15]. We present two benchmark examples from the literature; since shearing strain is neglected in these, we admit the same inner constraint, without loss of generality.

Straight compressed beams. In a straight ($k = 0$) and slender ($\gamma = 0$) beam subjected only to a compressive force P, it is immediate to see that the non-trivial linear response, describing the fundamental path, is

$$\gamma_f = 0, \quad \varepsilon_f = -P/EA, \quad \vartheta_f = 0, \quad N_f = -P, \quad T_f = M_f = 0 \qquad [6.16]$$

Substituting equation [6.16] into equation [6.15] leads to the set of linear differential equations

$$\begin{aligned} EAu_b' &= N_b, \qquad v_b' = \vartheta_b(1-\varepsilon_f), \qquad EI\vartheta_b' = M_b, \\ N_b' &= T_b' = 0, \qquad M_b' + P\vartheta_b + T_b(1-\varepsilon_f) = 0 \end{aligned} \qquad [6.17]$$

which can be rearranged as

$$EAu_b'' = 0, \qquad EIv_b^{IV} + P\,(1-\varepsilon_f)\,v_b'' = 0 \qquad [6.18]$$

Equations [6.18] are identical to the first two of (31) in Pignataro and Ruta (2003), and we also find them in Pflüger (1964). If we neglect the initial contraction, i.e. $\varepsilon_f = 0$, equations [6.18] coincide with the well-known differential equations for Euler buckling.

Circular arch under radial load. A portion of a slender ($\gamma = 0$) circular arch of radius R has uniform initial curvature $k = 1/R$ and is loaded by a uniform radial distribution of magnitude q with the same direction throughout the motion (i.e. "dead" or conservative). If, as is usual in the literature, the arch is supported by rollers, the only non-zero fields of the fundamental path follow "Mariotte's formula"

$$\varepsilon_f = -(qR)/(EA), \quad N_f = -qR \qquad [6.19]$$

Substituting equation [6.19] into equation [6.15] leads to the set of linear differential equations

$$\begin{aligned} u_b' - \frac{v_b}{R} &= \frac{N_b}{EA}, \qquad v_b' - \vartheta_b\left(1-\frac{qR}{EA}\right) + \frac{u_b}{R} = 0, \\ \vartheta_b' &= \frac{M_b}{EI}, \qquad N_b' - \frac{T_b}{R} - q\vartheta_b = 0, \\ T_b' + \frac{N_b}{R} &= 0, \qquad M_b' + qR\vartheta_b + T_b\left(1-\frac{qR}{EA}\right) = 0 \end{aligned} \qquad [6.20]$$

If the axial strain is negligible, i.e. $1-(qR/EA) \approx 1$, the system [6.20] becomes

$$\begin{aligned} EA\left(u_b' - \frac{v_b}{R}\right) - qR^2\left(v_b'' + \frac{u_b'}{R}\right) - REI\left(v_b^{IV} + \frac{u_b'''}{R}\right) = 0 \\ EI\left(v_b''' + \frac{u_b''}{R}\right) + REA\left(u_b'' - \frac{v_b'}{R}\right) - qR\left(v_b' + \frac{u_b}{R}\right) = 0 \end{aligned} \qquad [6.21]$$

Equation [6.21] is identical to those of Simitses and Hodges (2006), after suitably changing the independent variable and the orientation of the radial displacement.

From these two benchmark studies, we deduce that the two-parameter perturbation expansions of the finite field equations proposed here are a reliable tool for investigating the static stability of non-trivial equilibria. In the next section, we investigate parabolic arches, which are the main objectives of our study.

6.3. Parabolic arches

Parabolic arches are curved beams, the axes of which are described by a segment of parabola, symmetric with respect to its vertex. We fix the coordinate frame as in Figure 6.1; thus, the position vector of the points of the axis is

$$\mathbf{r}_0(x) = x\mathbf{e}_x + f[1 - (x^2/l^2)]\mathbf{e}_y, \quad -l \leq x \leq l,\ 0 \leq y \leq f \qquad [6.22]$$

in terms of the Cartesian abscissa x along the span, $2l$ long; f is the height of the arch crown with respect to the line of the span, connecting the two ends.

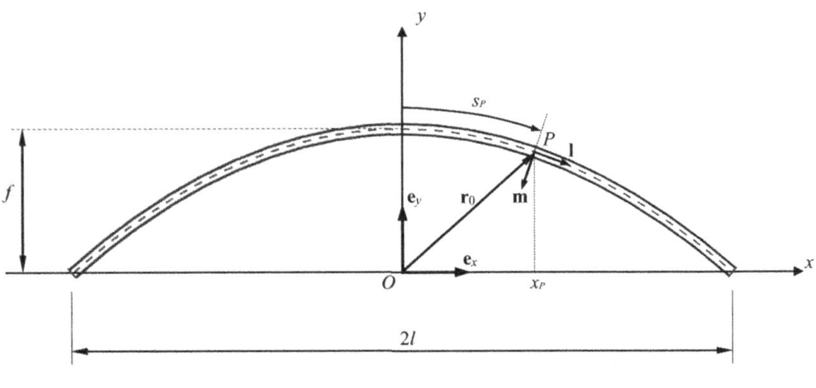

Figure 6.1. *Reference shape of a parabolic arch with cross-sections orthogonal to the axis*

By equations [6.2] and [6.3], the tangent vectors, their magnitude (the arc length), their unit counterpart $\mathbf{l}$, the local curvature and unit normals are

$$\frac{d\mathbf{r}_0}{dx} = \mathbf{e}_x - f\frac{2x}{l^2}\mathbf{e}_y, \quad ds = \sqrt{1 + \frac{4f^2x^2}{l^4}}dx, \quad \mathbf{l} = \frac{l^2\mathbf{e}_x - 2fx\mathbf{e}_y}{\sqrt{l^4 + 4f^2x^2}},$$
$$k = \frac{2fl^4}{(l^4 + 4f^2x^2)^{3/2}}, \qquad \mathbf{m} = -\frac{2fx\mathbf{e}_x + l^2\mathbf{e}_y}{\sqrt{l^4 + 4f^2x^2}} \qquad [6.23]$$

The Frénet–Serret triad in equation [6.23] is completed by the bi-normal $\mathbf{n} = \mathbf{l} \times \mathbf{m} = -\mathbf{e}_z$ (the sign is due to the assigned curvature, with osculating circle below the axis $\forall P$).

To abstract from the particular values of the geometrical and physical quantities in a problem, it is customary to operate non-dimensional analyses; let us then define

$$(\bar{x}, \bar{s}, \bar{u}, \bar{v}) = \frac{(x, s, u, v)}{l}, \quad \alpha = \frac{f}{l}, \quad \lambda = l\sqrt{\frac{A}{I}}, \quad \bar{A} = \frac{GA_s}{EA},$$
$$(\bar{N}, \bar{T}) = \frac{(N, T)l^2}{EI}, \quad \bar{M} = \frac{Ml}{EI}, \quad (\bar{n}, \bar{t}) = \frac{(n, t)l^3}{EI}, \quad \bar{m} = \frac{ml^2}{EI} \qquad [6.24]$$

Equation $[6.23]_2$ provides (at least locally) the functions $x(s)$ and $s(x)$; thus, we may also deduce that $\bar{x} = \bar{x}(\bar{s})$ and equation [6.24] provides further non-dimensional quantities

$$\frac{ds}{dx} = \frac{d\bar{s}}{d\bar{x}} = L(\bar{x}) = \sqrt{1 + 4\alpha^2\bar{x}^2}, \qquad \bar{k}(\bar{x}) = kl = \frac{2\alpha}{L^3(\bar{x})}. \qquad [6.25]$$

Substituting equations [6.24] and [6.25] into equation [6.12], the equations for the non-dimensional η-first-order fundamental path are written in matrix form

$$\frac{d\mathbf{y}_f}{d\bar{x}} = \mathbf{A}_f\mathbf{y}_f + \mathbf{q}_f, \qquad \mathbf{y}_f = \left\{\bar{u}_f\ \bar{v}_f\ \vartheta_f\ \bar{N}_f\ \bar{T}_f\ \bar{M}_f\right\}^T,$$
$$\mathbf{q}_f = -L\{0\,0\,0\,\bar{n}_f\ \bar{t}_f\ \bar{m}_f\}^T, \qquad \mathbf{A}_f = L\begin{pmatrix} 0 & k & 0 & \frac{1}{\lambda^2} & 0 & 0 \\ -k & 0 & 1 & 0 & \frac{1}{\bar{A}\lambda^2} & 0 \\ 0 & 0 & 0 & 0 & 0 & 1 \\ 0 & 0 & 0 & 0 & k & 0 \\ 0 & 0 & 0 & -k & 0 & 0 \\ 0 & 0 & 0 & 0 & -1 & 0 \end{pmatrix}. \qquad [6.26]$$

A force load $-q\mathbf{e}_y$ that is uniform with respect to the span length implies that $m_f = 0 \Rightarrow \bar{m}_f = 0$; its components with respect to the arc length in the local basis are given by the force equivalence

$$-q\mathbf{e}_y\,dx = (n\mathbf{l} + t\mathbf{m})ds \qquad [6.27]$$

whence their non-dimensional counterparts according to equations [6.24].

The solution of equation [6.26] is (Hubbard and West 1995; Tufekci *et al.* 2017)

$$\mathbf{y}_f(\bar{x}) = \mathbf{Y}_f(\bar{x})\left[\mathbf{y}_f(\bar{x}_0) + \int_0^{\bar{x}} \mathbf{Y}_f^{-1}(\xi)\,\mathbf{q}(\xi)\,d\xi\right] \qquad [6.28]$$

where $\mathbf{Y}(\bar{x})$ is called the principal matrix, or *matricant* (Pease 1965), or *transfer matrix* (Pestel and Leckie 1963), of the homogeneous equation [6.26] about $\bar{x} = 0$. The entries of the principal matrix for a planar curved beam in integral form can be found in Tufekci and Arpaci (2006). They are provided in Eroglu *et al.* (2020) for parabolic arches of uniform cross-sections; variable cross-sections are considered in Eroglu and Ruta (2020). The state vector $\mathbf{y}$ in equation [6.28] accounts for distributed external loads; concentrated ones at any point $\bar{x}_i \in [-1, 1]$ were recently discussed in Eroglu *et al.* (2020) through continuity of displacement and local balance of actions.

We admit that the given load is "dead"; then, it is easy to prove that $n_b = -t_f\vartheta_b, t_b = n_f\vartheta_b$. Substituting equations [6.24] and [6.25] into equations [6.15] provides the equations for the non-dimensional β-first-order bifurcated path in matrix form

$$\frac{d\mathbf{y}_b}{d\bar{x}} = \mathbf{A}_b\mathbf{y}_b\,, \qquad \mathbf{y}_b = \left\{\bar{u}_b\ \bar{v}_b\ \vartheta_b\ \bar{N}_b\ \bar{T}_b\ \bar{M}_b\right\}^T$$
$$\mathbf{A}_b = L\begin{pmatrix} 0 & k & -\vartheta_f + T_f(\frac{1}{\lambda^2} - \frac{1}{\bar{A}\lambda^2}) & \frac{1}{\lambda^2} & 0 & 0 \\ -k & 0 & 1 + N_f(\frac{1}{\lambda^2} - \frac{1}{\bar{A}\lambda^2}) & 0 & \frac{1}{\bar{A}\lambda^2} & 0 \\ 0 & 0 & 0 & 0 & 0 & 1 \\ 0 & 0 & t_f & 0 & k & 0 \\ 0 & 0 & -n_f & -k & 0 & 0 \\ 0 & 0 & N_f & \frac{T_f}{\bar{A}\lambda^2} + \vartheta_f & -1 - \frac{N_f}{\bar{A}\lambda^2} & 0 \end{pmatrix} \qquad [6.29]$$

In a homogeneous arch, the mass per unit length and the bending stiffness are uniform, but the terms in equation [6.29] depend on $\bar{x}$, and $\mathbf{A}_b$ cannot be reduced to an upper triangular form. Hence, equation [6.29], in general, does not have closed-form solutions. Thus, Eroglu *et al.* (2019) propose to find approximate solutions by the Peano series (Peano 1888) and Volterra's multiplicative integral (Pease 1965; Slavik 2007)

$$\begin{aligned} &\mathbf{y}_b\,(\bar{x}) = \mathbf{Y}_b(\bar{x}, \bar{x}_0)\mathbf{y}_b(\bar{x}_0) \\ &\mathbf{Y}_b(\bar{x}, \bar{x}_0) = \prod_{\imath=1}^{n} \mathbf{Y}_2(\bar{x}_0 + \imath\Delta\bar{x}, \bar{x}_0 + (\imath - 1)\Delta\bar{x}), \\ &\mathbf{Y}_2(\bar{x}_2, \bar{x}_1) \approx \mathbf{I} + \mathbf{A}_b(\bar{x}_1)(\bar{x}_2 - \bar{x}_1) + \\ &+ \left(\frac{1}{2}\left.\frac{d\mathbf{A}_b}{dx}\right|_{\bar{x}_1} + \frac{1}{4}\left.\frac{d^2\mathbf{A}_b}{dx^2}\right|_{\bar{x}_1} + \frac{1}{2}\mathbf{A}_b^2(\bar{x}_1)\right)(\bar{x}_2 - \bar{x}_1)^2 \end{aligned} \qquad [6.30]$$

having split $\{\bar{x}, \bar{x}_0\}$ in n intervals with $\Delta\bar{x} = (\bar{x} - \bar{x}_0)/n$. As n increases, $\Delta\bar{x} \to 0$ and equation [6.30] turns into Volterra's integral (Slavik 2007). We keep n finite, yet large enough to ensure convergence.

Equation [6.30]$_1$ requires a point $\bar{x}_0$ where the values of the displacements and contact actions are known. This occurs at the boundary, where non-dimensional natural conditions are (see equation [6.24]):

$$\begin{aligned} \text{clamped end}: \bar{u} &= 0, \quad \bar{v} = 0, \quad \vartheta = 0, \\ \text{pinned end}: \bar{u} &= 0, \quad \bar{v} = 0, \quad \bar{M} = 0, \\ \text{free end}: \bar{N} &= 0, \quad \bar{T} = 0, \quad \bar{M} = 0. \end{aligned} \qquad [6.31]$$

They are three such natural conditions at each end, yielding six linear equations in the six components $\mathbf{y}_b$

$$\mathbf{T}(q)\, \mathbf{y}_b(\bar{x}_0) = \mathbf{0} \qquad [6.32]$$

with $\mathbf{T}(q)$ a square 6×6 matrix of coefficients. Non-trivial solutions $\mathbf{y}_b(\bar{x}_0)$ require the singularity of $\mathbf{T}(q)$ in terms of the load multiplier q, yielding a highly nonlinear equation, the solutions of which are the critical loads for the arch,

$$\det[\mathbf{T}(q)] = 0 \Rightarrow q = q_{cr} \qquad [6.33]$$

6.4. Crack models for one-dimensional elements

If the arch has a small plane crack on the cross-section at $\bar{x}_c$, its damaging effect is described by considering two regular arch chunks connected at x_c by a set of springs; as in Eroglu and Tufekci (2017a), their compliances are related to the depth of the crack through its strain energy U_c, given by Tada *et al.* (2000)

$$U_c = \int_{A_c} \frac{1}{E'} \left[\left(\sum K_{I\jmath} \right)^2 + \left(\sum K_{II\jmath} \right)^2 \right] dA, \quad c_{\imath\jmath} = \frac{\partial^2 U_c}{\partial \imath \partial \jmath}, \imath, \jmath = N, T, M \qquad [6.34]$$

where A_c is the cross-section as reduced by the crack; $E' = E/(1 - \nu^2)$, with ν Poisson's ratio of the material; $K_{I\jmath}, K_{II\jmath}$ are the stress intensity factors in opening and shearing modes, respectively, due to the $\jmath$-th contact action; and $c_{\imath\jmath}$ is the compliance of the spring representing the effect of the crack on the $\jmath$-th contact action due to a unit value of the $\imath$-th kinematic descriptor.

In a rectangular cross-section of height h with a crack of depth a at opposite sides, with respect to the center of curvature, consider the additional non-dimensional quantities related to the crack

$$\begin{aligned} &\bar{a} = \frac{a}{h}, \quad \bar{K}_{\imath\jmath} = \frac{K_{\imath\jmath}}{E\sqrt{h}}, \quad \bar{U}_c = \frac{U_c}{hEA}, \quad \bar{c}_{\bar{N}\bar{N}} = \frac{EI}{l^3} c_{NN}, \\ &\bar{c}_{\bar{T}\bar{T}} = \frac{EI}{l^3} c_{TT}, \quad \bar{c}_{\bar{M}\bar{M}} = \frac{EI}{l} c_{MM}, \quad \bar{c}_{\bar{M}\bar{N}} = \bar{c}_{\bar{N}\bar{M}} = \frac{EI}{l^2} c_{MN} \end{aligned} \qquad [6.35]$$

Following Tada *et al.* (2000), the shape functions f_i of the crack magnitude ratio

$$f_1(\bar{a}) = \sqrt{\frac{2\tan(\pi\bar{a}/2)}{\pi\bar{a}}}\,\frac{0.752 + 2.02\bar{a} + 0.37[1 - \sin(\pi\bar{a}/2)]^3}{\cos}(\pi\bar{a}/2),$$

$$f_2\,(\bar{a}) = \frac{1.22 - 0.561\bar{a} + 0.085\bar{a}^2 + 0.18\bar{a}^3}{\sqrt{1-\bar{a}}},$$

$$f_3(\bar{a}) = \sqrt{\frac{2\tan(\pi\bar{a}/2)}{\pi\bar{a}}}\,\frac{0.923 + 0.199\,[1 - \sin\,(\pi\bar{a}/2)]^4}{\cos\,(\pi\bar{a}/2)} \qquad [6.36]$$

and equation [6.34] yield the non-dimensional compliances

$$\begin{aligned}
\bar{c}_{\bar{N}\bar{N}} &= 4\pi\sqrt{3}(1-\nu^2)\int_0^{\bar{a}} \frac{\xi}{\lambda^3} f_1^2\,(\xi)\,d\xi,\\
\bar{c}_{\bar{T}\bar{T}} &= 16\pi\sqrt{3}\frac{1-\nu}{1+\nu}\int_0^{\bar{a}} \frac{\xi}{\bar{A}^2\lambda^3} f_2^2(\xi)d\xi,\\
\bar{c}_{\bar{M}\bar{M}} &= 12\pi\sqrt{3}(1-\nu^2)\int_0^{\bar{a}} \frac{\xi}{\lambda} f_3^2\,(\xi)\,d\xi,\\
\bar{c}_{\bar{N}\bar{M}} &= 12\pi(1-\nu^2)\int_0^{\bar{a}} \frac{\xi}{\lambda^2} f_1\,(\xi)\,f_3\,(\xi)\,d\xi.
\end{aligned} \qquad [6.37]$$

Henceforth, only non-dimensional quantities will be used; with another abuse of notation, overbars will be omitted for simplicity, except when confusion may arise.

The crack at x_c implies a jump of the axis displacement in the actual shape; the jump depends on the contact actions at the cross-sections of the left and right chunks of the arch (whence the superscripts l, r) facing each other at x_c. The inner contact actions at the ends of the fictitious springs will, of course, be balanced. These vector conditions are written in the local basis of the left chunk, pulled by the operator $\mathbf{R}$:

$$\mathbf{R}\,(\vartheta_r - \vartheta_l)\,\tilde{\mathbf{d}}^r - \tilde{\mathbf{d}}^l = \mathbf{C}^*\tilde{\mathbf{f}}^l, \qquad \mathbf{R}\,(\vartheta_r - \vartheta_l)\,\tilde{\mathbf{f}}^r = \tilde{\mathbf{f}}^l \qquad [6.38]$$

$$\mathbf{C}^* = \begin{pmatrix} c_{NN} & 0 & pc_{NM} \\ 0 & c_{TT} & 0 \\ pc_{NM} & 0 & c_{MM} \end{pmatrix}$$

The tilde here denotes vectors in actual configuration. For easier calculations, inner actions $\mathbf{f} = \{N, T, M\}^T$ and displacement are written with respect to their components in the reference frame; hence, equation [6.38] becomes

$$\begin{aligned}
\mathbf{R}(\vartheta_r - \vartheta_l)\mathbf{R}^T(\vartheta_r)\mathbf{d}^r - \mathbf{R}^T(\vartheta_l)\mathbf{d}^l &= \mathbf{C}^*\mathbf{R}^T(\vartheta_l)\mathbf{f}^l,\\
\mathbf{R}(\vartheta_r - \vartheta_l)\mathbf{R}^T(\vartheta_r)\mathbf{f}^r &= \mathbf{R}^T(\vartheta_l)\mathbf{f}^l
\end{aligned} \qquad [6.39]$$

Since $\mathbf{R}(\vartheta_r - \vartheta_l)\mathbf{R}^T(\vartheta_r) = \mathbf{R}^T(\vartheta_l)$, equation [6.39] may be written in matrix form

$$\mathbf{y}^r(x_c) = \mathbf{C}(a,p)\mathbf{y}^l(x_c),$$
$$\mathbf{C}(a,p) = \begin{pmatrix} \mathbf{I}\ \mathbf{R}(\vartheta_l(x_c))\,\mathbf{C}^*\mathbf{R}^T(\vartheta_l(x_c)) \\ \mathbf{0} \qquad\qquad \mathbf{I} \end{pmatrix}, \tag{6.40}$$

where $\mathbf{0}, \mathbf{I}$ are the 3×3 null and identity matrices, respectively, and $p = \pm 1$ is a non-material parameter indicating that the crack is at the top or bottom of the cross-section with respect to the center of curvature, respectively. Equation [6.40] is nonlinear and can be subjected to the same perturbation performed for field equations. When it is linearized about the initial unstressed configuration, the condition $\mathbf{y}^r(x_c) = \mathbf{C}_f\mathbf{y}^l(x_c)$, identical to that in Eroglu *et al.* (2020), is recovered. A linearization about the non-trivial fundamental path leads to the matrix $\mathbf{C}$ depending on ϑ_b, rearranged as

$$\mathbf{y}^r(x_c) = \mathbf{C}_b\mathbf{y}^l(x_c) \tag{6.41}$$

where $\mathbf{C}_b$ is too long to be reported and will be omitted for the sake of space.

For both right and left chunks, equation [6.28], solution for the static problem, holds; the unknown initial values are now 12, and adding the jump to the boundary conditions yields 12 equations for calculating them. It is then better to present the jump conditions in terms of these initial values; if $\hat{\mathbf{I}}$ is the 6×6 identity, equation [6.40] becomes

$$\mathbf{y}_f^r(x_0) = \mathbf{Y}_f^{-1}(x_c)\left(\mathbf{C}_f\mathbf{Y}_f(x_c)\mathbf{y}_f^l(x_0) + \left(\mathbf{C}_f - \hat{\mathbf{I}}\right)\int_0^{x_c}\mathbf{Y}_f^{-1}(\xi)\,\mathbf{q}(\xi)\,d\xi\right) \tag{6.42}$$

For the eigenvalue problem equation [6.33] we will then update the principal matrix for a given crack location x_c inside the $\jmath$-th interval between x_0 and x as follows:

$$\mathbf{Y}_b(x,x_0) = \mathbf{Y}_b(x, x_0 + \jmath\Delta x)\,\mathbf{Y}_{bc}\mathbf{Y}_b(x_0 + (\jmath-1)\Delta x, x_0)$$
$$\mathbf{Y}_{bc} = \mathbf{Y}_2(x_0 + \jmath\Delta x, x_c)\,\mathbf{C}_b\mathbf{Y}_2(x_c, x_0 + (\jmath-1)\Delta x) \tag{6.43}$$

6.5. An application

Consider a fixed parabolic arch with geometric aspect $\alpha = 0.4$ and slenderness ratio $\varsigma = 100$, under the described load; if the cross-section is rectangular, $\bar{A} = 0.3$. The arch is assumed to have a transverse crack at $x = x_c$ with depth ratio $a = 0.6$

(recall that, to simplify the notation, we dropped the overbars for the non-dimensional quantities related to the crack). We refer to Eroglu *et al.* (2020) for a detailed examination of the static problem providing the first-order strain increments, which are assumed to be a sufficiently accurate representation of the actual deformed shape.

If the crack is not at the crown, one of the arch chunks is more "slender" than the other. Then, it is possible that when the former buckles, the latter goes on along its fundamental path; thus, two different principal matrices should be used. In general, both chunks buckle in correspondence to the same load multiplier; in any case, we consider two different scenarios: both chunks bifurcate for the same load multiplier ("global buckling"), or the more slender reaches bifurcation, while the other remains on its fundamental path ("local buckling"). However, recall that the two sub-arches are linked by relatively stiff springs, not by a smooth hinge; i.e. the crack affects a very small portion of the cross-section and does not actually cut the arch in two.

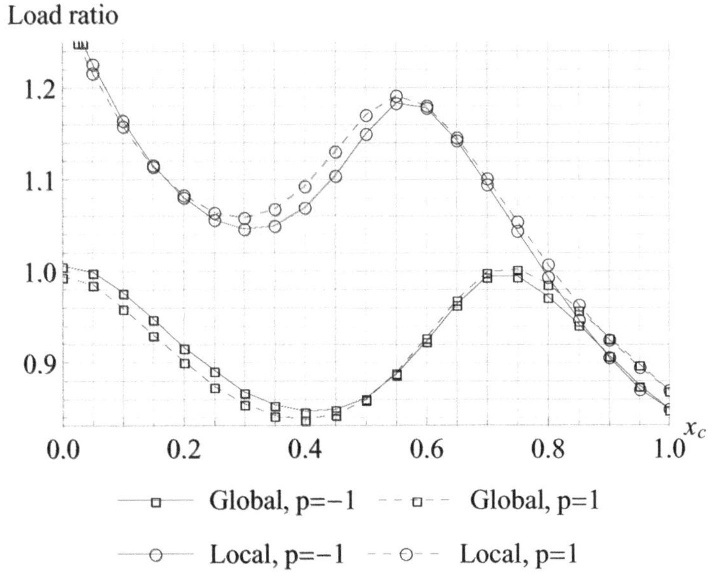

Figure 6.2. *Critical load ratio for a fixed arch with a crack characterized by* $a = 0.6$

Figure 6.2 presents the plot of the critical load ratio (critical load of the cracked arch divided by that of the intact arch) for an arch damaged by a crack with aspect ratio $a = 0.6$ versus the crack location x_c. The plot considers one half of the arch, because of symmetry, i.e. we plot what happens when the crack is located at the right half of the considered arch. We present the results for both "global" (square markers) and "local" buckling (round markers) and for both crack locations at the opposite sides of the cross-section with respect to the center of curvature (i.e. for the non-material parameter $p = 1$ (full lines) and $p = -1$ (dashed lines)).

We see that "local" buckling requires a greater critical load than the "global" one, except for $x_c \gtrapprox 0.8$, i.e. in the vicinity of the right constraint. This has a clear physical interpretation: when the crack is near the fixed end, the difference in slenderness of the two sub-arches reaches its top; thus, the more slender chunk attains buckling before the more stubby sub-arch. The critical load ratio of this "local" buckling is very high at the beginning, then decreases as x_c approaches the constrained end, resembling that of "global" buckling. This also has a physical interpretation, since for $x_c \approx 0.8$, the stubby chunk actually acts as an extended clamp, then makes the system globally stiffer; on the other hand, when $x_c \to 1$, the system is practically equivalent to the entire arch.

Actually, for other values of x_c, the global buckling prevails, which is also physically reasonable: if both chunks have comparable slenderness, they are similarly prone to buckling and bifurcate as one. We also remark that the effect of the crack position on opposite sides of the cross-section has limited effect on critical load, due to considerations similar to those provided in Eroglu *et al.* (2019). For more shallow arches, we may expect a stronger dependence of the critical load on p.

We remark that at $x_c \approx 0$ and $x_c \approx 0.75$, the critical load ratio attains values greater than unity, i.e. the critical load multiplier of the damaged arch is greater than that of the intact one. This is a seemingly paradoxical behavior, since it was highlighted by Timoshenko and Gere (1961) that a parabolic arch with one hinge at the crown, in general, buckles under a lower critical load with respect to an entire fixed arch. However, we may check that this phenomenon occurs for cracks located at places around which the incremental bending moment vanishes. Therefore, the coupling between axial force and bending moment (accounted for by the parameter p) creates a bending moment, which acts against the buckling strain, providing a stiffening effect. Note that for the problems considered herein, the mode shape is still skew-symmetric, albeit the crack induces some slight deviations from it (see Figure 6.3).

Figure 6.3 shows the buckling modes about the non-trivial fundamental path. The deformed configuration is plotted with an amplification factor of 10, to better reflect the effect of crack location on the section. Effects of crack position on the bifurcated paths are clearly shown; moreover, the question is raised about the stability of bifurcated paths and its dependence on the crack location on the section, which needs to be investigated in future contributions.

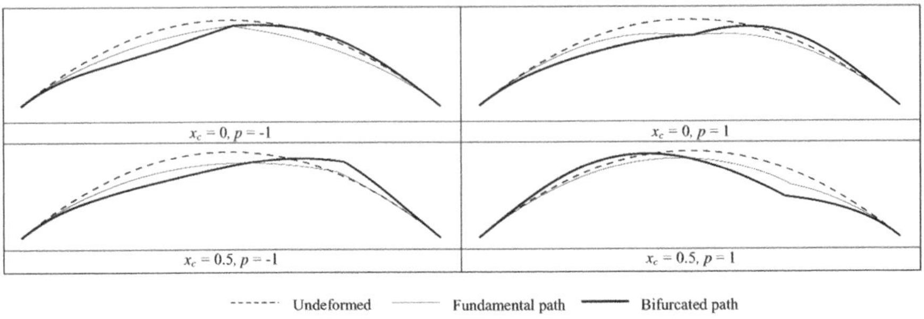

Figure 6.3. *Buckling modes about the deformed shape for* $a = 0.6$

6.5.1. *A comparison*

For comparison, we consider a symmetric circular arch, consisting of two regular identical chunks connected with a torsional spring with stiffness k_T, as in Wang *et al.* (2005). This shape approximates that of a parabolic arch when its opening angle is moderate. The arch has radius R and opening angle φ. This is a primitive model of possible damage at the apex, neglecting axial, and shear compliances as well as axial-bending coupling. For this simplified model, in our treatment, we have a single non-zero component in the compliance matrix, c_{MM}, which equals $1/k_T$.

φ	$\pi/6$	$\pi/6$	$\pi/3$	$\pi/3$	$\pi/2$	$\pi/2$	$2\pi/3$	$2\pi/3$
$\frac{k_T R}{EI}$	W	Present	W	Present	W	Present	W	Present
0	108.36*	108.21*	27.077*	27.052*	12.025*	12.008*	6.7578*	6.755*
2	129.48*	129.30*	36.794	36.737	17.956	17.926	10.806	10.788
4	147.61	147.40	43.979	43.909	21.751	21.712	13.060	13.035
6	163.21	162.97	49.366	49.282	24.277	24.231	14.418	14.388

Table 6.1. *Critical loads for the symmetric mode shapes of a hinged circular arch with a weakened section at the apex. Stars denote that the symmetric mode has a lower critical load than the corresponding skew-symmetric mode; W=Wang* et al. *(2005)*

In the exact treatment in Wang *et al.* (2005), the arch is assumed to be inextensible and shear rigid. Table 6.1 lists the critical loads for symmetric mode shapes: indeed, note that for skew-symmetric mode shape, the bending couple at the apex vanishes and the torsion spring does not affect the behavior. The weakened section only affects the critical load for symmetric mode shapes.

6.6. Final remarks

We presented two perturbation expansions of the finite governing equations for arches modeled as curved beams, in order to investigate their non-trivial fundamental path and their post-buckling path. Kinematics is finite, balance is in the actual configuration and only constitutive equations are supposed linear elastic. The perturbation approach lets us investigate the effect of a deformed and pre-stressed shape on the buckling load and the relevant equilibria following possible static bifurcation. We showed that this approach recovers well-known benchmark results and applied it to the analysis of buckling scenarios of parabolic arches under a "vertical" load uniformly distributed with respect to the span, affected by small cracks. The effect of the latter is suitably modeled as a set of linear elastic springs, following previous investigations; the solution of the governing equations was found numerically by using the so-called principal matrix of the differential system. We obtained results for both "local" and "global" buckling and commented on them, highlighting how seemingly paradoxical results actually have a clear physical interpretation. Future investigations will be about the quality of the post-buckling path and on linear vibration about non-trivial pre-stressed states, in order to detect the effect of local damages for monitoring purposes.

6.7. Acknowledgments

This work began when Uğurcan Eroğlu was a visiting research student at the Dipartimento di Ingegneria Strutturale e Geotecnica of the Sapienza University of Rome, the support of which is gratefully acknowledged. Giuseppe Ruta acknowledges the support of institutional grants from Sapienza University of Rome for the year 2019.

6.8. References

Anifantis, N. and Dimarogonas, A.D. (1983a). Buckling of rings and tubes with longitudinal cracks. *Mechanics Research Communications*, 18, 693–702.

Anifantis, N. and Dimarogonas, A.D. (1983b). Post buckling behavior of transverse cracked columns. *Composite Structures*, 12(2), 351–356.

Anifantis, N. and Dimarogonas, A.D. (1983c). Stability of columns with a single crack subjected to follower and vertical loads. *International Journal of Solids and Structures*, 19, 281–291.

Antman, S. (1973). *The Theory of Rods*. Springer, Berlin.

Antman, S. (1995). *Nonlinear Problems of Elasticity*. Springer-Verlag, New York.

Caddemi, S. and Morassi, A. (2007). Crack detection in elastic beams by static measurements. *International Journal of Solids and Structures*, 44, 5301–5315.

Caddemi, S. and Morassi, A. (2011). Detecting multiple open cracks in elastic beams by static tests. *Journal of Engineering Mechanics*, 137, 113–124.

Caddemi, S. and Morassi, A. (2013). Multi-cracked Euler–Bernoulli beams: Mathematical modeling and exact solutions. *International Journal of Solids and Structures*, 50, 944–956.

Caddemi, S., Caliò, I., Cannizzaro, F. (2017). The dynamic stiffness matrix (DSM) of axially loaded multi-cracked frames. *Mechanics Research Communications*, 84, 90–97.

Caliò, I., D'Urso, D., Greco, A. (2017). The influence of damage on the eigen-properties of Timoshenko spatial arches. *Computers and Structures*, 190, 13–24.

Cannizzaro, F., Greco, A., Caddemi, S., Caliò, I. (2017). Closed form solutions of a multi-carcked circular arch under static loads. *International Journal of Solids and Structures*, 121, 191–200.

Capecchi, D., Ciambella, J., Pau, A., Vestroni, F. (2016). Damage identification in a parabolic arch by means of natural frequencies, modal shapes and curvatures. *Meccanica*, 51, 2847–2859.

Cawley, P. and Adams, R. (1979). The location of defects in structures from measurements of natural frequencies. *Journal of Strain Analysis*, 14(2), 49–57.

Cerri, M. and Ruta, G. (2004). Detection of localised damage in plane circular arches by frequency data. *Journal of Sound and Vibration*, 270, 39–59.

Cerri, M., Dilena, M., Ruta, G. (2008). Vibration and damage detection in undamaged and cracked circular arches: Experimental and analytical results. *Journal of Sound and Vibration*, 314, 83–94.

Chati, M., Rand, R., Mukherjee, S. (1997). Modal analysis of a cracked beam. *Journal of Sound and Vibration*, 207(2), 249–270.

Ciambella, J. and Vestroni, F. (2015). The use of modal curvatures for damage localization in beam-type structures. *Journal of Sound and Vibration*, 340, 126–137.

Dimarogonas, A.D. (1981). Buckling of rings and tubes with longitudinal cracks. *Mechanics Research Communications*, 8, 179–186.

Dimarogonas, A.D. (1982). Crack identification in aircraft structures. *Proceedings of the First National Aircraft Conference*, pp. 11–22.

Dimarogonas, A.D. (1996). Vibration of cracked structures: A state of the art review. *Engineering Fracture Mechanics*, 55(5), 831–857.

Dimarogonas, A.D. and Paipetis, S. (1983). *Analytical Methods in Rotor Dynamics*. Applied Science, Paris.

Doebling, S., Farrar, C., Prime, M., Shevitz, D. (1996). Damage identification and health monitoring of structural and mechanical systems from changes in their vibration characteristics: A literature review. *Los Alamos Nat. Lab.* Report, LA-13070-MS, pp. 1–129.

Dugdale, D. (1960). Yielding of steel sheets containing slits. *Journal of Mechanical Physics and Solids*, 8, 100–104.

Eroglu, U. and Ruta, G. (2020). Closed-form solutions for elastic tapered parabolic arches under uniform thermal gradients. *Meccanica*, 55, 1135–1152.

Eroglu, U. and Tufekci, E. (2016). Exact solution based finite element formulation of cracked beams for crack detection. *International Journal of Solids and Structures*, 96, 240–253.

Eroglu, U. and Tufekci, E. (2017a). Crack modeling and identification in curved beams using differential evolution. *International Journal of Mechanical Sciences*, 131–132, 435–450.

Eroglu, U. and Tufekci, E. (2017b). Free vibration of damaged frame structures considering the effects of axial extension, shear deformation and rotatory inertia: Exact solution. *International Journal of Structural Stability and Dynamics*, 17(10), 1750111.

Eroglu, U., Ruta, G., Tufekci, E. (2019). Natural frequencies of parabolic arches with a single crack on opposite cross-section sides. *Journal of Vibration and Control*, 25(7), 1313–1325.

Eroglu, U., Paolone, A., Ruta, G., Tufekci, E. (2020). Exact closed-form static solutions for parabolic arches with concentrated damage. *Archive of Applied Mechanics*, 90, 673–689.

Gounaris, G. and Dimarogonas, A. (1988). A finite element of a cracked prismatic beam for structural analysis. *Computers and Structures*, 28(3), 309–313.

Greco, A. and Pau, A. (2011). Detection of a concentrated damage in a parabolic arch by measured static displacements. *Structural Engineering and Mechanics*, 39(6), 751–765.

Griffith, A. (1920). The phenomena of rapture and flow in solids. *Philosophical Transactions of the Royal Society of London, Series A*, 221(582–593), 163–198.

Hubbard, J. and West, B. (1995). *Differential Equations: A Dynamical Systems Approach*. Springer-Verlag, New York.

Inglis, C. (1913). Stresses in plates due to the presence of cracks and sharp corners. *Transactions of the Institute of Naval Architects*, 55, 219–241.

Irwin, G. (1957a). Analysis of stresses and strains near the end of a crack traversing a plate. *Journal of Applied Mechanics*, 24, 361–364.

Irwin, G. (1957b). Relation of stresses near a crack to the crack extension force. *Proceedings of the 9th Congress of Applied Mechanics*, Brussels, p. 245.

Irwin, G. (1960a). *Fracture Mechanics in Structural Mechanics*. Pergamon Press, Oxford.

Irwin, G. (1960b). Plastic zone near a crack and fracture toughness. *Proceedings of the 7th Sagamore Ordnance Materials Conference*, Syracuse University, New York.

Ju, F. and Minovich, M. (1986). Modal frequency method in diagnosis of fracture damage in structures. *Proceedings of the 4th International Modal Analysis Conference*, Los Angeles, CA.

Karaagac, C., Ozturk, H., Sabuncu, M. (2011). Crack effects on the in-plane static and dynamic stabilities of a curved beam with an edge crack. *Journal of Sound and Vibration*, 330(8), 1718–1736.

Khiem, N. and Lien, T. (2001). A simplified method for natural frequency analysis of a multiple cracked beam. *Journal of Sound and Vibration*, 245(4), 737–751.

Kirsch, E. (1898). Die theorie der elastizität und die bedürfnisse der festigkeitslehre. *Zeitschrift des Vereines deutscher Ingenieure*, 42, 797–807.

Kirshmer, P. (1944). The effect of discontinuities on the natural frequency of beams. *Proceedings of ASTM International*, 44, 897–904.

Kisa, M. and Arif Gurel, M. (2007). Free vibration analysis of uniform and stepped cracked beams with circular cross sections. *International Journal of Engineering Science*, 45(2–8), 364–380.

Krawczuk, M., Palacz, M., Ostachowicz, W. (2003). The dynamic analysis of a cracked Timoshenko beam by the spectral element method. *Journal of Sound and Vibration*, 264(5), 1139–1153.

Mazanoglu, K., Yesilyurt, I., Sabuncu, M. (2009). Vibration analysis of multiple-cracked non-uniform beams. *Journal of Sound and Vibration*, 320(4–5), 977–989.

Müller, W., Herrmann, G., Gao, H. (1993). A note on curved cracked beams. *International Journal of Solids and Structures*, 30, 1527–1532.

Orhan, S. (2007). Analysis of free and forced vibration of a cracked cantilever beam. *NDT & E International*, 40(6), 443–450.

Ostachowicz, W. and Krawwczuk, M. (1992). Coupled torsional and bending vibrations of a rotor with an open crack. *Archive of Applied Mechanics*, 62, 191–201.

Pau, A., Greco, A., Vestroni, F. (2010). Numerical and experimental detection of concentrated damage in a parabolic arch by measured frequency variations. *Journal of Vibration and Control*, 17(4), 605–614.

Peano, G. (1888). Intégration par séries des équations différentielles linéaires. *Mathematische Annalen*, 32(3), 450–456.

Pease, M. (1965). *Methods of Matrix Algebra*. Academic Press, New York.

Pestel, E. and Leckie, F. (1963). *Matrix Methods in Elastomechanics*. McGraw-Hill, New York.

Pflüger, A. (1964). *Stabilitätsprobleme der Elastostatik*. Springer-Verlag, Berlin.

Pignataro, M. and Ruta, G. (2003). Coupled instabilities in thin-walled beams: A qualitative approach. *European Journal of Mechanics – A/Solids*, 22(1), 139–149.

Pignataro, M., Rizzi, N., Luongo, A. (1991). *Stability, Bifurcation and Post-critical Behaviour of Elastic Structures*. Elsevier, Amsterdam.

Pignataro, M., Rizzi, N., Ruta, G. (2008). A beam model for the flexural-torsional buckling of thin-walled members. *Thin Walled Structures*, 46, 816–822.

Rice, J. (1968). A path independent integral and the approximate analysis of strain concentration by notches and cracks. *ASME Journal of Applied Mechanics*, 35, 379–386.

Rice, J. and Levy, N. (1972). The part-through surface crack in an elastic plate. *Journal of Applied Mechanics*, 39, 185–194.

Simitses, G. and Hodges, D. (2006). *Fundamentals of Structural Stability*. Butterworth-Heinemann, Burlington.

Sinha, J.K. and Friswell, M.I. (2002). Simulation of the dynamic response of a cracked beam. *Computers and Structures*, 80(18–19), 1473–1476.

Slavik, A. (2007). *Product Integration, its History and Applications*. Matfyz Press, Prague.

Tada, H., Paris, P., Irwin, G. (2000). *The Stress Analysis of Cracks Handbook*, 3rd edition. ASME Press, New York.

Thomson, W. (1949). Vibration of slender bars with discontinuities in stiffness. *Journal of Applied Mechanics*, 17, 203–207.

Timoshenko, S. and Gere, J. (1961). *Theory of Elastic Stability*. McGraw-Hill, New York.

Timoshenko, S. and Goodier, J. (1951). *Theory of Elasticity*, 2nd edition. McGraw Hill, New York.

Tufekci, E. and Arpaci, A. (2006). Analytical solutions of in-plane static problems for non-uniform curved beams including axial and shear deformations. *Structural Engineering and Mechanics*, 22(2), 131–150.

Tufekci, E., Eroglu, U., Aya, S. (2017). A new two-noded curved beam finite element formulation based on exact solution. *Engineering with Computers*, 33(2), 261–273.

Viola, E., Nobile, L., Federici, L. (2002). Formulation of cracked beam element for structural analysis. *Journal of Engineering Mechanics*, 128(2), 220–230.

Wang, C., Wang, C., Reddy, J. (2005). *Exact Solutions for Buckling of Structural Members*. CRC Press, Boca Raton.

Wells, A. (1963). Application of fracture mechanics at and beyond general yielding. *British Welding Journal*, 11, 563–570.

7

Inelastic Microbuckling of Composites by Wave-Buckling Analogy

7.1. Introduction

Microbuckling in fiber-reinforced elastic composite materials, associated with the composite compressive strength, has been extensively studied over the years. A two-dimensional problem of a periodically layered bi-material, in which the stiff and soft layers represent the fibers and matrix, respectively, has been frequently used to model the composite. In the classical investigation by Rosen (1964), which has been summarized by Jones (1975), the two lower buckling modes were identified to be: the shear buckling mode characterized by in-phase deformation of fiber and matrix layers, and the transverse buckling mode with fiber and matrix layers exhibiting anti-phase deformation. Based on energy considerations, Rosen established an expression for the stress buckling in shear mode with very long wavelength (when compared to the thickness of the stiff fiber layer). Typically, in many studies of this problem, the stiff layer was modeled as an Euler–Bernoulli beam (see Waas *et al.* (1990), for example). This approach was also used by Parnes and Chiskis (2002), where a comprehensive list of references can be found. In this study, the infinite fiber layer was considered to be embedded in an elastic matrix, with the interactions between the layers deduced on the basis of the mechanics of materials. The results show that the long-wavelength shear buckling mode is adequate for composites of a relatively large fiber volume fraction, while composites of fiber volume fractions smaller than a specific transition value buckle in a finite wavelength shear mode. Guz (1969), Waas (1992), Nestorovic and Triantafyllidis (2004), Aboudi and Gilat (2006) and Gilat (2010), among others, used a full elasticity analysis of layered and fibrous composites, including the effects of combined loading as well as material anisotropy.

Chapter written by Rivka GILAT and Jacob ABOUDI.

To better adjust theoretical predictions to experimentally measured composite compressive strength, models taking optional failure mechanisms, non-uniform pre-buckling stress distribution, fibers misalignment and waviness into account, as well as nonlinear and inelastic material behavior, were suggested by Budiansky and Fleck (1993), Kyriakides *et al.* (1995), Drapier *et al.* (1999), Guz *et al.* (2005) and Yokozeki *et al.* (2005), where further literature is referenced.

In previous work, Aboudi and Gilat (2006) used the analogy between the governing equations for the analysis of microbuckling and the elastodynamic equations for wave propagation in solids, in order to predict the elastic microbuckling within a periodic layered bi-material and fiber-reinforced composites subjected to a uniaxial compressive load, parallel to the layer/fiber direction. In the present investigation, this wave-buckling analogy is used for the analysis of inelastic bifurcation microbuckling in layered composites. To this end, the instantaneous stiffness tensor of the inelastic layer is herein used in the bifurcation buckling analysis. The evolution of the instantaneous stiffness tensor of initially isotropic material under axial loading conditions, reveals that plasticity induces the softening and orthotropy of the material. In order to take this developing orientation dependency into account, the wave-buckling analogy is generalized for the case of orthotropic materials. It is applied to the rate version of the linearized buckling equation, in an incremental manner and in conjunction with constitutive relations reflecting the degradation of material stiffnesses. Assuming that the composite consists of elastic–viscoplastic materials, its behavior could have been conveniently described by the unified (Bodner and Partom 1975) model. However, while this viscoplastic constitutive law is well performing for proportional loading conditions, it cannot follow the plasticity induced decrease of shear modulus under uniaxial loading, and thus predicts this value as if it remains intact. As the importance of shear in the onset of instability in layered composites was indicated by the extensive work of the above-mentioned research and others, the modified constitutive law proposed by Rubin and Bodner (1995) for non-proportional buckling is adopted. This enables an adequate update of all the instantaneous stiffnesses ruling the onset of instability.

7.2. Buckling-wave propagation analogy

Consider a periodic bi-material layered composite as shown in Figure 7.1(a). The repeating layered unit cell consists of two layers and is described with respect to global coordinates X_1, X_3, where X_3 is the layering direction. In addition, a local coordinate $x_3^{(\beta)}$, located at the bottom surface of each layer, $\beta = 1, 2$, is introduced. In a previous article by Aboudi and Gilat (2006), a wave-buckling analogy was used for the analysis of the microbuckling within such a periodic bi-material subjected to a uniaxial compressive load parallel to the layers direction (Figure 7.1(b)). According to this analogy, the buckling load can be directly extracted from the corresponding dispersion relation of the layered material, which provides the phase velocity of a wave in terms of the wavelength.

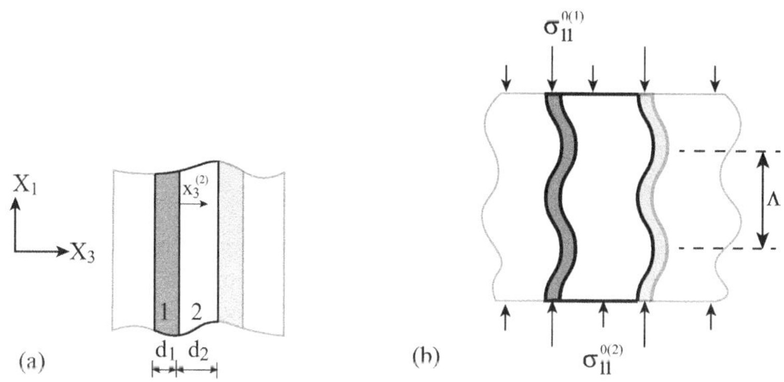

Figure 7.1. *(a) The repeating unit cell and (b) shear mode microbuckling of a periodic layered bi-material composite*

The linear equations of motion governing the propagation of harmonic waves in a continuum are

$$\sigma_{jm,j} = \rho u_{m,tt} \qquad j, m = 1, 2, 3 \tag{7.1}$$

where σ_{jm} and u_m are the components of the stress tensor and the displacements vector, respectively, ρ is the mass density, t is the time, and the comma represents the differentiation. For the investigation of sagittal harmonic waves propagating in the layered composite, it is assumed that any field variable $s^{(\beta)}$, in the β layer, has the form:

$$s^{(\beta)} = S^{(\beta)} \exp[ik(X_1 + \eta^{(\beta)} x_3^{(\beta)} - ct)] \tag{7.2}$$

where $S^{(\beta)}$ is an amplitude factor, k is the wave number associated with the wavelength $\Lambda = 2\pi/k$, c is the phase velocity, $x_3^{(\beta)}$ is the local layer coordinate in the layering direction, $\eta^{(\beta)}$ is an unknown parameter, and i is the imaginary unit.

By using the form [7.2] in [7.1], it follows that the derivative with respect to X_1 should be replaced by ik, the derivative with respect to x_3 should be replaced by $ik\eta$ and the derivative with respect to the time t is replaced by $-ikc$. Consequently, the following equations are established:

$$\begin{aligned} ik\sigma_{11}^{(\beta)} + ik\eta^{(\beta)}\sigma_{31}^{(\beta)} + \rho^{(\beta)}(kc)^2 u_1^{(\beta)} &= 0 \\ ik\sigma_{13}^{(\beta)} + ik\eta^{(\beta)}\sigma_{33}^{(\beta)} + \rho^{(\beta)}(kc)^2 u_3^{(\beta)} &= 0 \end{aligned} \tag{7.3}$$

On the contrary, the linearized equilibrium equations governing the bifurcation buckling of a layer under pre-buckling uniaxial layer stress $\sigma_{11}^{0(\beta)}$ are (Whitney 1987)

$$\sigma_{jm,j}^{(\beta)} - \sigma_{jn}^{0(\beta)} u_{m,nj}^{(\beta)} = 0 \quad j, m, n = 1, 3 \qquad [7.4]$$

For a classical cylindrical buckling, it is assumed that any field variable $b^{(\beta)}$ in the layer has the form

$$b^{(\beta)} = B^{(\beta)} \exp[ik(X_1 + \eta^{(\beta)} x_3^{(\beta)})] \qquad [7.5]$$

where $B^{(\beta)}$ is an amplitude factor and k is the buckling wave number which is associated with the wavelength $\Lambda = 2\pi/k$ in the X_1 direction. By substituting the form [7.5] in [7.4], the derivative with respect to X_1 should be replaced by ik, and the derivative with respect to x_3 should be replaced by $ik\eta^{(\beta)}$. Accordingly, for the material occupying a layer β, [7.4] takes the form

$$\begin{aligned} &ik\sigma_{11}^{(\beta)} + ik\eta^{(\beta)}\sigma_{31}^{(\beta)} + \sigma_{11}^{0(\beta)} k^2 u_1^{(\beta)} = 0 \\ &ik\sigma_{13}^{(\beta)} + ik\eta^{(\beta)}\sigma_{33}^{(\beta)} + \sigma_{11}^{0(\beta)} k^2 u_3^{(\beta)} = 0 \end{aligned} \qquad [7.6]$$

The comparison between [7.3] and [7.6] reveals that an analogy between wave propagation and buckling can be established and

$$\rho^{(\beta)} c^2 \Longleftrightarrow \sigma_{11}^{0(\beta)} \qquad [7.7]$$

Hence, the buckling load can be directly extracted from the corresponding dispersion relation of the layered material, which provides the phase velocity of a wave in terms of the wavelength.

7.3. Microbuckling in elastic orthotropic composites

Aiming at the analysis of layered composites consisting of inelastic materials, we note that the properties of such materials develop orientation dependency upon loading. Focusing on buckling due to the overall uniaxial compression in the X_1 direction, orthotropic constitutive relations should be taken into account for the layer material, with the material symmetry axes coinciding with the load direction. Hence, the stress–strain relations for the material occupying the β layer are

$$\sigma_{jn}^{(\beta)} = C_{jnpq}^{(\beta)} \epsilon_{pq}^{(\beta)}, \qquad \mathbf{C}^{(\beta)} = \begin{bmatrix} C_{11} & C_{12} & C_{13} & 0 & 0 & 0 \\ C_{21} & C_{22} & C_{23} & 0 & 0 & 0 \\ C_{31} & C_{32} & C_{33} & 0 & 0 & 0 \\ 0 & 0 & 0 & C_{44} & 0 & 0 \\ 0 & 0 & 0 & 0 & C_{55} & 0 \\ 0 & 0 & 0 & 0 & 0 & C_{66} \end{bmatrix}^{(\beta)} \qquad [7.8]$$

and the strains $\epsilon_{pq}^{(\beta)}$ are related to the displacements through the following linear relations:

$$\epsilon_{pq}^{(\beta)} = \frac{1}{2}(u_{p,q}^{(\beta)} + u_{q,p}^{(\beta)}) \qquad [7.9]$$

To obtain the critical stress, the buckling curve is first obtained by following the approach suggested by Nayfeh (1995) for establishing the dispersion curves. The substitution of [7.8] and [7.9], in conjunction with the form [7.5] for the displacement components, into [7.6] reduces the latter to the following two algebraic equations:

$$K_{nj}^{(\beta)} U_j^{(\beta)} = 0, \qquad n, j = 1, 3 \qquad [7.10]$$

with

$$\begin{aligned}
K_{11}^{(\beta)} &= C_{11}^{(\beta)} + C_{55}^{(\beta)}(\eta^{(\beta)})^2 + \sigma_{11}^{0(\beta)} \\
K_{13}^{(\beta)} &= \eta^{(\beta)}(C_{13}^{(\beta)} + C_{55}^{(\beta)}) \\
K_{33}^{(\beta)} &= C_{55}^{(\beta)} + C_{33}^{(\beta)}(\eta^{(\beta)})^2 + \sigma_{11}^{0(\beta)} \\
K_{nj}^{(\beta)} &= K_{jn}^{(\beta)}
\end{aligned} \qquad [7.11]$$

where $U_j^{(\beta)}$ are the displacement amplitudes. For a non-trivial solution, $U_j^{(\beta)}$, the determinant of $\mathbf{K}^{(\beta)}$ must vanish, thus forming a characteristic equation of order four, defining four values of $\eta^{(\beta)}$ for each layer

$$\eta_{1,3}^{(\beta)} = \pm\sqrt{\frac{-a_2^{(\beta)} \pm \sqrt{a_2^{(\beta)2} - 4a_1^{(\beta)}a_3^{(\beta)}}}{2a_1^{(\beta)}}}, \qquad \eta_2^{(\beta)} = -\eta_1^{(\beta)}, \qquad \eta_4^{(\beta)} = -\eta_3^{(\beta)}$$

$$\begin{aligned}
a_1^{(\beta)} &= C_{33}^{(\beta)}C_{55}^{(\beta)}, \qquad a_3^{(\beta)} = (C_{55}^{(\beta)} + \sigma_{11}^{0(\beta)})(C_{11}^{(\beta)} + \sigma_{11}^{0(\beta)}) \\
a_2^{(\beta)} &= C_{55}^{(\beta)}(C_{55}^{(\beta)} + \sigma_{11}^{0(\beta)}) + C_{33}^{(\beta)}(C_{11}^{(\beta)} + \sigma_{11}^{0(\beta)}) - (C_{13}^{(\beta)} + C_{55}^{(\beta)})^2
\end{aligned} \qquad [7.12]$$

For each η_k, $k = 1, 2, ...4$, [7.10] can be used to define

$$\delta_q^{(\beta)} = \left(\frac{U_{3q}}{U_{1q}}\right)^{(\beta)} = -\left(\frac{K_{11}}{K_{13}}\right)^{(\beta)} \qquad [7.13]$$

By using this definition in conjunction with the constitutive relations [7.8] and the kinematic relations [7.9], the displacements and stresses in a layer can be expressed as follows:

$$\mathbf{P}^{(\beta)} = \begin{bmatrix} u_1 \\ u_3 \\ \sigma_{33} \\ \sigma_{31} \end{bmatrix}^{(\beta)} = \mathbf{X}^{(\beta)}\mathbf{D}^{(\beta)} \begin{bmatrix} U_{11} \\ U_{12} \\ U_{13} \\ U_{14} \end{bmatrix}^{(\beta)} \exp[ikX_1] \qquad [7.14]$$

where $\mathbf{X}^{(\beta)}$ and $\mathbf{D}^{(\beta)}$ are 4×4 matrices with the entries

$$
\begin{aligned}
&X_{1q}^{(\beta)} = 1, && X_{2q}^{(\beta)} = \delta_q^{(\beta)} \\
&X_{3q}^{(\beta)} = C_{13}^{(\beta)} + C_{33}^{(\beta)} \eta_q^{(\beta)} \delta_q^{(\beta)}, && X_{4q}^{(\beta)} = C_{55}^{(\beta)} (\eta_q^{(\beta)} + \delta_q^{(\beta)}) \\
&D_{pq}^{(\beta)} = 0, && p \neq q \\
&D_{pq}^{(\beta)} = \exp[ik\eta_q^{(\beta)} x_3^{(\beta)}], && p = q, \qquad p, q = 1, 2, ...4
\end{aligned}
\qquad [7.15]
$$

The continuity constraints at the interface between adjacent layers of the representative unit and periodicity conditions can be cast to have the following form:

$$
\mathbf{P}^{(2)}_{(x_3^{(2)}=d_2)} = \mathbf{A}^{(2)} \mathbf{A}^{(1)} \mathbf{P}^{(1)}_{(x_3^{(1)}=0)} = \mathbf{P}^{(1)}_{(x_3^{(1)}=0)} \qquad [7.16]
$$

with

$$
\mathbf{A}^{(\beta)} = \mathbf{X}^{(\beta)} \mathbf{D}^{(\beta)}_{(x_3^{(\beta)}=d_\beta)} (\mathbf{X}^{(\beta)})^{-1} \qquad [7.17]
$$

For a non-trivial solution, it is required that

$$
det[\mathbf{A}^{(2)} \mathbf{A}^{(1)} - \mathbf{I}] = 0 \qquad [7.18]
$$

Recalling that $\mathbf{A}^{(\beta)} = \mathbf{A}^{(\beta)}(k, \sigma_{11}^{0(\beta)})$ and that under the uniform strain loading condition

$$
\sigma_{11}^{0(\beta)} = \sigma_{11}^{0} \frac{\bar{C}_{11}^{(\beta)} (d_1 + d_2)}{\bar{C}_{11}^{(1)} d_1 + \bar{C}_{11}^{(2)} d_2}, \qquad \bar{C}_{11}^{(\beta)} = \frac{C_{11}^{(\beta)} C_{33}^{(\beta)} - C_{13}^{(\beta)} C_{13}^{(\beta)}}{C_{33}^{(\beta)}} \qquad [7.19]
$$

[7.18] constitutes the relations between k and σ_{11}^0. For each wave number k, the smallest root of [7.18] is $\tilde{\sigma}_{11}^0$, which is the buckling stress. The corresponding eigenvector $U_j^{(\beta)}$ represents the buckling mode of wavelength $\Lambda = 2\pi/k$. The results of numerous previous works, showing that this buckling mode is the shear mode shown in Figure 7.1(b), make our choice of the considered representative unit legitimate. Microbuckling occurs under the minimal value σ_{11cr}^0, with wavelength Λ_{cr}, minimizing $\tilde{\sigma}_{11}^0(\Lambda)$.

7.4. Inelastic microbuckling

To consider inelastic bifurcation microbuckling in layered composites, the layer material is assumed to be an elastic–viscoplastic material and a constitutive law, suitable for the description of its behavior, is adopted.

To explain the consideration guiding our choice of a material constitutive law, we first mention that simple expressions for the microbuckling stress in layered materials were proposed in the pioneering and numerous subsequent works dealing with the subject. Relating to a bi-material composite with a repeating unit cell consisting of one layer occupied by a stiffer material and termed fiber (subscript f) and a second softer layer termed matrix (subscript m), the long-wavelength limit from Rosen (1964) for elastic materials is

$$\sigma^0_{11cr} = \frac{\mu_m}{1 - v_f} \qquad [7.20]$$

where μ_m is the matrix shear modulus and v_f is the volume fraction of the fiber material. One of the modifications suggested by Budiansky and Fleck (1993) for composites including a plastic matrix layer is given by

$$\sigma^0_{11cr} = \frac{\mu_m}{1 - v_f}\left(\frac{E_{ms}E_m}{E_{ms}E_m + 3\mu_m(E_m - E_{ms})}\right) = \frac{\mu^I_m}{1 - v_f} \qquad [7.21]$$

where E_m and E_{ms} are the matrix elastic and the secant moduli, respectively, and the superscript I stands for inelastic.

Both the above expressions testify the significant role of the matrix shear modulus in the microbuckling of layered composites. Furthermore, inelastic material behavior reveals softening, which can be viewed as the reduction of the material mechanical properties. Hence, it is concluded that the effect of plasticity on the material shear modulus might be of importance when analyzing the inelastic microbuckling behavior of composites. In order for this effect to be reflected, the model proposed by Rubin and Bodner (1995) for non-proportional buckling is used. This model, which is based on the assumption that the plastic strain is explicitly related to the total strain rate, is a modification of the Bodner and Partom (1975) elastic–viscoplastic model. While in the framework of the latter, the shear modulus is not affected by plasticity evolved under axial loading, the modified model is capable of predicting the plasticity induced reduction of the shear modulus due to such a loading. The adopted constitutive law is

$$\dot{\sigma}_{ij} = (L^E_{ijkl} + L^{VP}_{ijkl})\dot{\epsilon}_{kl} - \Gamma\sigma'_{ij} \qquad [7.22]$$

where $\dot{\epsilon}_{ij}$ is the total strain rate, σ'_{ij} is the deviatoric stress, and Γ is a scalar function involved in the flow rule. L^E_{ijkl} is the elastic part of the stiffness tensor and L^{VP}_{ijkl} is its viscoplastic part, which enables the detection of changes in the shear stiffness in the absence of direct shear stress and which is unique for the present model. This constitutive relation can be written in the following common form:

$$\dot{\sigma}_{ij} = C^I_{ijkl}\dot{\epsilon}_{kl} \qquad [7.23]$$

such that C^I_{ijkl} represents the instantaneous stiffness of the elastic–viscoplastic material. In order to determine the inelastic microbuckling, we use the rate version of the buckling equations [7.4]

$$\dot{\sigma}^{(\beta)}_{jm,j} - \sigma^{0(\beta)}_{jn}\dot{u}^{(\beta)}_{m,nj} = 0 \quad j,m,n = 1,3, \quad \beta = 1,2 \qquad [7.24]$$

Following the procedure for elastic buckling, the inelastic buckling stress is obtained in an incremental manner as follows. At each loading increment, namely for each incrementally increased value of σ^0_{11}, the instantaneous stiffnesses of the inelastic phase are computed. For each value of k, the instantaneous matrix $\mathbf{K}^{(\beta)\mathbf{I}}$, [7.11] is constructed, and the instantaneous characteristic equation $det[\mathbf{K}^{(\beta)\mathbf{I}}] = \mathbf{0}$ is solved to determine the values of $\eta^{(\beta)}_q$, $\beta = 1,2$, $q = 1,...4$. The instantaneous matrices $\mathbf{A}^{(\beta)\mathbf{I}}$ are then built and the instantaneous buckling condition [7.18]

$$det[\mathbf{A}^{(2)I}\mathbf{A}^{(1)I} - \mathbf{I}] = 0 \qquad [7.25]$$

is checked. The smallest stress for which the instantaneous determinant vanishes is the buckling stress $\tilde{\sigma}^0_{11}$ with a mode of wavelength $\Lambda = 2\pi/k$. The critical inelastic bifurcation microbuckling stress is the minimum of the latter values over k.

7.5. Results and discussion

To demonstrate the validity of the present approach, results for the elastic case are first examined. In the following, the volume fraction v_f defines the portion occupied by the material of layer 1. The effect of the stiffness ratio $E^{(1)}/E^{(2)}$ (where $E^{(\beta)}$ is the β layer material Young's modulus) on the critical buckling strain is shown in Figure 7.2. The present results are close to those extracted from the works of Waas (1992) and Parnes and Chiskis (2002).

Further results assessing the validity of the wave-buckling analogy and the adoption of Nayfeh's wave propagation analysis can be found in Aboudi and Gilat (2006) and Gilat (2010).

The inelastic microbuckling in a composite consisting of elastic stiff layers and elasto-viscoplastic soft layers is presented in Figures 7.3–7.5. Specifically, the B/Al composite is considered with elastic shear moduli $\mu^{(1)} = 153.8GPa$ and $\mu^{(2)} = 26.6GPa$ and Poisson's ratios $\nu^{(1)} = 0.3$ and $\nu^{(2)} = 0.335$, for boron and aluminum, respectively. The inelastic material parameters, characterizing the inelastic aluminum layer (see Rubin and Bodner (1995)), are $\Gamma_0 = 10^8 s^{-1}$, $\kappa_0 = 0.4GPa$, $n = 1.$, $m_1 = 1000GPa^{-1}$, $m_2 = 40000GPa^{-1}$, $Z_1 = 0.53GPa$, $Z_3 = 0.3GPa$, $b = 10^{-6}s^{-1}$, $f_1 = 0.23$ and $f_2 = 0.7$.

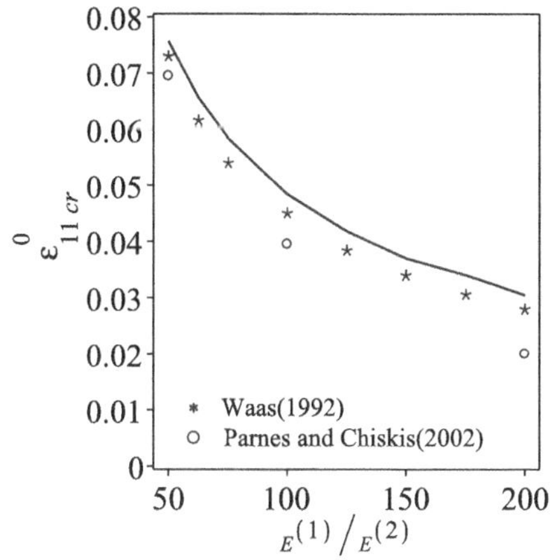

Figure 7.2. *Effect of layer stiffness ratio on the elastic critical bifurcation microbuckling strain, $v_f = 0.05$. For a color version of this figure, see www.iste.co.uk/challamel/mechanics1.zip*

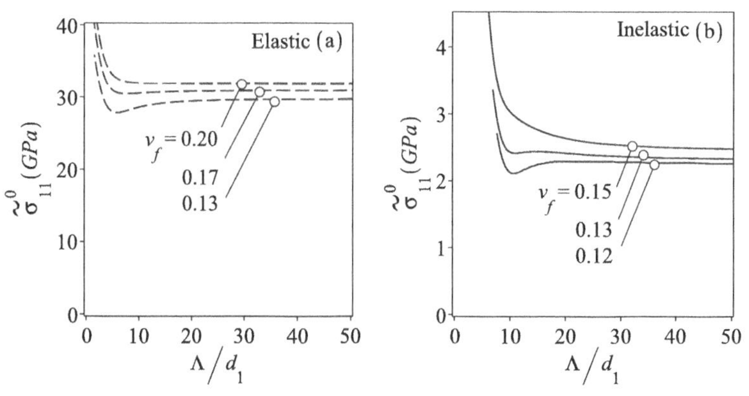

Figure 7.3. *Buckling curves exhibiting the volume fraction-dependent variation of microbuckling stress with the buckling wavelength for B/Al resulting from (a) elastic (b) inelastic analyses. For a color version of this figure, see www.iste.co.uk/challamel/mechanics1.zip*

Figure 7.3 presents the buckling curves corresponding to the dispersion curves of the analogical wave propagation problem. To reveal the effect of plasticity, curves resulting from elastic and inelastic analyses are given for various values of volume fractions. For each buckling wavelength, these curves show the corresponding smallest buckling stress (smallest root of [7.25]). For each volume fraction, the minimum point of the curve defines the critical bifurcation microbuckling stress and wavelength. As expected, for both elastic and elasto-plastic cases, composites of a small volume fraction undergo small wavelength microbuckling, while those of a large volume fraction yield long-wavelength buckling. Plasticity is observed to decrease the volume fraction at which the transition occurs from short-wave buckling to long-wave buckling. It also dramatically reduces the critical microbuckling stresses.

The dependence of the critical buckling stress and wavelength on the composite volume fraction is presented in Figure 7.4. In addition to the present elastic results (dashed line), the long-wave approximation of Rosen (1964) [7.20] is also shown. The present inelastic buckling stresses (solid line) are somewhat higher than those predicted by [7.21], Budiansky and Fleck (1993). A similar trend can be deduced from the results presented by Yokozeki *et al.* (2005) for the special case of almost aligned linear elastic fibers embedded in an elastic–plastic matrix. Both present elastic and inelastic curves exhibit a tendency change at the point of transition from finite to infinite buckling wavelength.

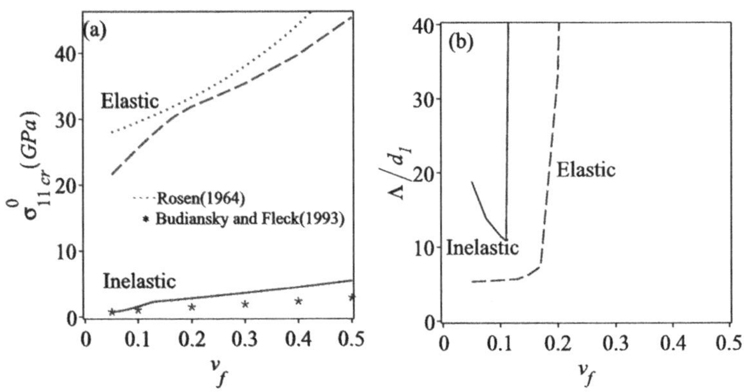

Figure 7.4. *Effect of the volume fraction on (a) the critical microbuckling stress and (b) the critical buckling wavelength of B/Al. For a color version of this figure, see www.iste.co.uk/challamel/mechanics1.zip*

The presence of this transition point can also be identified by the solid curve presented in Figure 7.5(a), showing the variation of the critical microbuckling strain with the volume fraction, as predicted by the presently suggested approach. This is in contrast with the qualitative behavior given by [7.21] which, being based on Rosen's expression, cannot reflect the essence of finite wavelength buckling in composites of a low volume fraction.

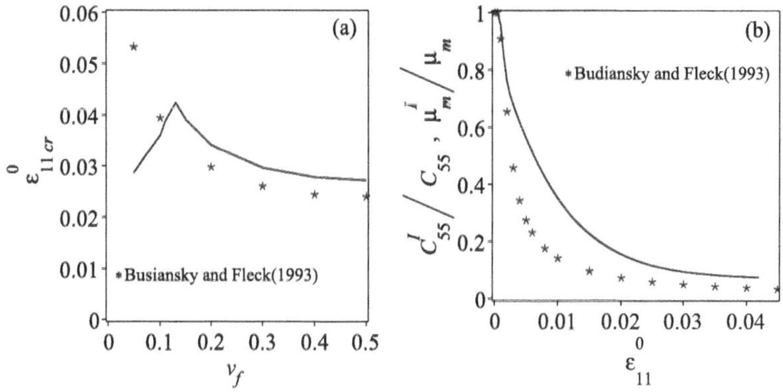

Figure 7.5. *(a) Effect of the volume fraction on the critical microbuckling strain in the B/Al composite and (b) plasticity-induced reduction of the Al shear modulus. For a color version of this figure, see www.iste.co.uk/challamel/mechanics1.zip*

The plasticity-induced change of the inelastic phase shear modulus, defined by the presently adopted constitutive law, is indicated by the solid line shown in Figure 7.5(b). This shows a significant reduction at the level of critical buckling strain. As observed with respect to the buckling stress, the values of the presently predicted modified shear modulus are less conservative than those defined by [7.21].

To conclude, we examine the behavior of a composite in which the stiffer layer is inelastic while the softer is assumed to be linearly elastic. Specifically, Al/epoxy is considered with $\mu^{(1)} = 26.6GPa$, $\mu^{(2)} = 1.96GPa$, $\nu^{(1)} = 0.335$ and $\nu^{(2)} = 0.328$ and with the inelastic material properties as given above. The behavior of such composites is presented in Figure 7.6. Figure 7.6(a) reveals that in contrast to the qualitative similarity between the elastic and inelastic behaviors exhibited in Figure 7.3, in the present case, there is a qualitative difference between the elastic and inelastic behaviors.

According to Figure 7.6, inelasticity of the stiffer layer eliminates the short-wavelength buckling to long-wavelength buckling transition. This can be attributed to the plasticity-induced decrease of the inelastic material stiffness, decreasing the layer stiffness ratio $E^{(1)}/E^{(2)}$. As shown by Parnes and Chiskis (2002), for smaller stiffness ratios, the short- to long-wavelength transition occurs at a larger volume fraction. As indicated by Figure 7.2, such a reduction of the layers stiffness ratio yields buckling under larger critical buckling strains. It should be noted that under large strains, the assumption of linearly elastic behavior of the softer layer is questionable.

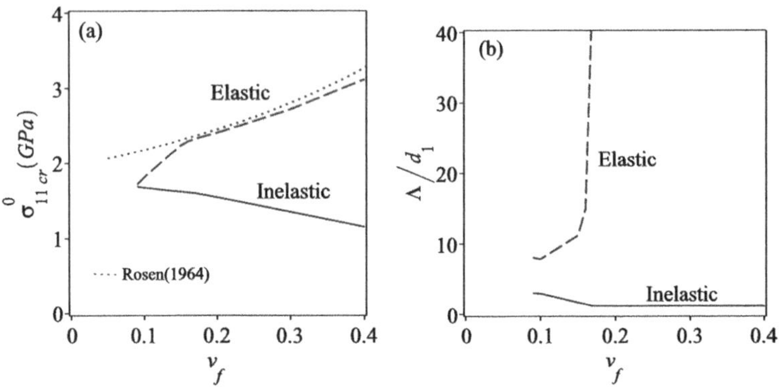

Figure 7.6. *Effect of the volume fraction on (a) the critical microbuckling stress and (b) the critical buckling wavelength of Al/epoxy. For a color version of this figure, see www.iste.co.uk/challamel/mechanics1.zip*

In summary, in the present work, the inelastic bifurcation microbuckling in a bi-material periodic layered composite was formulated as a micromechanical problem. A solution was obtained by using the analogy between the buckling and wave propagation problems, in conjunction with an incremental approach, while adopting an elastic–viscoplastic constitutive law indicating plasticity reduction of all the involved moduli.

7.6. References

Aboudi, J. and Gilat, R. (2006). Buckling analysis of fibers in composite materials by wave propagation analogy. *Int. J. Solids Struct.*, 43, 5168–5181.

Bodner, S.R. and Partom, Y. (1975). Constitutive equations for elastic-viscoplastic strain-hardening materials. *ASME J. Appl. Mech.*, 42, 385–389.

Budiansky, B. and Fleck, N.A. (1993). Compressive failure of fibre composites. *J. Mech. Phys. Solids*, 41, 183–211.

Drapier, S., Grandidier, J., Poiter-Ferry, M. (1999). Towards a numerical model of the compressive strength for long fibre composites. *Eur. J. Mech. A/Solids*, 18, 69–92.

Gilat, R. (2010). A 3D thermoelastic analysis of the buckling of a layer bonded to a compliant substrate and related problems. *Int. J. Solids Struct.*, 47, 2533–2543.

Guz, A.N. (1969). On setting up a stability theory of unidirectional fibrous materias. *Institute of Mechanics, Academy of Sciences of the Ukrainian SSR*, Prikladnaya Mekhanika, 5, 62–70.

Guz, A.N., Rodger, A., Guz, I. (2005). Developing a compressive failure theory for nanocomposites. *Int. Appl. Mech.*, 41, 233–255.

Jones, R. (1975). *Mechanics of Composite Materials*. Scripta Book Company, Washington, DC.

Kyriakides, S., Arseculeratne, R., Perry, E.J., Liechti, K.M. (1995). On the compressive failure of fiber reinforced composites. *Int. J. Solids Struct.*, 32, 689–738.

Nayfeh, A. (1995). *Wave Propagation in Layered Anisotropic Media with Applications to Composites*. Elsevier, Amsterdam.

Nestorovic, M. and Triantafyllidis, N. (2004). Onset of failure in finitely strained layered composites subjected to combined normal and shear loading. *J. Mech. Phys. Solids*, 52, 941–974.

Parnes, R. and Chiskis, A. (2002). Buckling of nano-fiber reinforced composites: A re-examination of elastic buckling. *J. Mech. Phys. Solids*, 50, 855–879.

Rosen, B. (1964). Mechanisms of Composite Strengthening. In *Fiber Composite Materials*. American Society of Metals, Cleveland, OH.

Rubin, M.B. and Bodner, S.R. (1995). An incremental elastic–viscoplastic theory indicating a reduced modulus for non-proportional buckling. *Int. J. Solids Struct.*, 32, 2967–1987.

Waas, A.M. (1992). Effect of interphase on compressive strength of unidirectional composites. *J. Appl. Mech.*, 59, s183–s188.

Waas, A.M., Babcock, C.D., Knauss, W.G. (1990). A mechanical model for elastic fiber microbuckling. *ASME J. Appl. Mech.*, 57, 138–149.

Whitney, J. (1987). *Structural Analysis of Laminated Anisotropic Plates*. Pergamon Press, Oxford.

Yokozeki, T., Ogasawara, T., Ishikawa, T. (2005). Effects of fiber nonlinear properties on the compressive strength prediction of unidirectional carbon–fiber composites. *Compos. Sci. Technol.*, 65, 2140–2147.

8

Quasi-Bifurcation of Discrete Systems with Unstable Post-Critical Behavior under Impulsive Loads

The huge economic and environmental consequences due to explosions in the oil industry have recently motivated the study of thin-walled cylindrical shells subjected to impulsive waves due to an explosion. One of the main challenges in this field is the identification of suitable dynamic buckling criteria for this class of loads. This paper focuses on a simple geometrically nonlinear two degree-of-freedom system subjected to impulsive loads of very short duration with respect to their fundamental periods. Similar to what occurs in shell structures, the model includes both membrane and bending stiffness components and displays an unstable static behavior at the critical state. The description of the motion is formulated using Lagrange equations and they are integrated numerically by means of an implicit code. By analogy with the static case, the dynamic path that follows the motion of a degree of freedom at a given load level is defined as the fundamental motion; any perturbed path that departs from the fundamental path is interpreted as a bifurcation of the original motion and termed quasi-bifurcation by Lee. Following the original criterion due to Lee, a coefficient calculated as the projection of the perturbed motion on the corresponding perturbed acceleration is investigated, and instability is said to occur for positive values of the coefficient. The present results show that the quasi-bifurcation criterion, as originally presented by Lee, is a necessary condition but may yield low values of instability loads because a positive projection may occur for a short time before the structure returns to a stable condition with a negative coefficient. A modification is proposed in which the time integration of the inner product due to Lee is carried out, and the analysis shows that this allows a

Chapter written by Mariano P. AMEIJEIRAS and Luis A. GODOY.

clearer identification of bounds in the response that are associated with dynamic buckling.

8.1. Introduction

Explosions have been identified as the most frequent accidents affecting the oil industry (Kang *et al.* 2016). Interest in the behavior of oil storage tanks under blast loads caused by explosions has intensified over the last decade following the notorious accidents at a tank farm located in Buncefield, UK, in 2005 and another one in Bayamon, Puerto Rico, in 2009.

Understanding and modeling this class of problems requires computation of the nonlinear dynamic response under short duration impulsive loads. Although such analysis is part of the state of the art in analytical and finite element approaches (see, for example, Nayfeh and Mook (1979); Nayfeh and Balachandran (2004)), such studies may sometimes be obscured by the complexity of the problem under consideration, whereas the interpretation of results and the evaluation of stability to identify dynamic buckling are topics still subject to discussion. Thus, to improve our understanding of the mechanics of behavior of shells under blast loads, it may be convenient to perform studies on much simpler analogous structural systems, for which the response can be modeled using analytical tools.

The buckling of structures under step or impulsive loading has been the subject of research since the 1960s. Reviews of dynamic buckling criteria were presented, for example, in the books by Simitses (1990), Amabili (2008) and Kubiak (2013). This area of research falls within the field of non-classical problems in the theory of stability, as defined by Elishakoff *et al.* (2001).

By analogy with the static case, we will define that a fundamental motion is the trajectory in a plot of a representative degree of freedom (DOF) with respect to time, computed at a given load level. Two forms of dynamic buckling, which are relevant to the present study, have been identified in the literature:

(a) Instability in the same mode of the fundamental motion, associated with large amplitude oscillations. Forced vibrations are investigated in this approach using nonlinear dynamics, and a criterion is employed to assess the occurrence of unstable oscillations. The most common criterion employed in the literature was originally proposed by Budiansky and Roth (1962), in which the dynamic response is first computed for various load levels and the load versus amplitude of a representative DOF is plotted. By analogy with the static case, this plot is here called a pseudo-equilibrium path. Dynamic buckling is identified whenever there is a jump in the pseudo-equilibrium path, i.e. when a small increment in load causes a non-proportional (large) increment in transient displacements. This criterion was originally developed for step loads, but its application to impulsive loads (such as those associated with explosions) is less evident.

(b) Instability in a mode which is orthogonal to the fundamental motion. This behavior is known as quasi-bifurcation from a fundamental trajectory and was originally identified by Lee (1977, 1981). As stated before, the motion of an undisturbed system shows oscillations in a mode here identified as the fundamental mode. But there are cases in which a second mode grows at a critical time, leading to a quasi-bifurcation; at first, this takes the form of small random oscillations, but a divergent motion is finally achieved. Lee (1981) also identified that the dynamic post-bifurcation mode may change in time.

The identification of a quasi-bifurcation state may be carried out either: (a) by establishing conditions at which dynamic bifurcation should occur, such as in the works of Kounadis *et al.* (1989) and Simitses (1990), or (b) by following stability indicators along the fundamental motion, such as frequencies of vibration, static eigenvalue problems, tangent matrix or static stability coefficients. Various authors in Europe (notably Kroplin and Dinkler 1986; Kleiber *et al.* 1987; Burmeister and Ramm 1990; Kratzig and Eller 1992) implemented strategies to evaluate stability along the fundamental motion.

The dynamic buckling of a simple, two DOF model is investigated in this work under an impulsive load in order to explore the type of analysis that should be considered when addressing shell problems under blast pressures due to explosions. The framework of the present analysis is that given by Lee, and shortcomings are highlighted together with ways to improve the identification of stability of a motion. The chapter focuses on quasi-bifurcation phenomena, which is also believed to occur in shells under impulsive loads due to an explosion and which may be instrumental in explaining the failure modes observed in oil infrastructure.

8.2. Case study of a two DOF system with unstable static behavior

In order to investigate the dynamic instability criterion developed by Lee (1977), a simple nonlinear geometric and elastic two DOF system is considered in this paper, for which a semi-analytical formulation may be implemented. The specific mechanical system was originally studied by Croll and Walker (1972) while addressing static stability. Interest in this model is twofold: first, it contains bending and membrane mechanisms to equilibrate the load, in much the same way as in shell structures; and second, its static behavior is characterized by an unstable critical state with imperfection sensitivity, as shown by Croll and Walker (1972).

The three rigid bars shown in Figure 8.1 have length L and their initial position is given by α_0 and β_0. Bending stiffness is provided by rotational springs K_f, whereas membrane stiffness is controlled by the extensional spring K_m. The mass of the system is lumped at the joints with value M. The generalized DOF of the problem are rotations $\alpha(t)$ and $\beta(t)$, also shown in Figure 8.1.

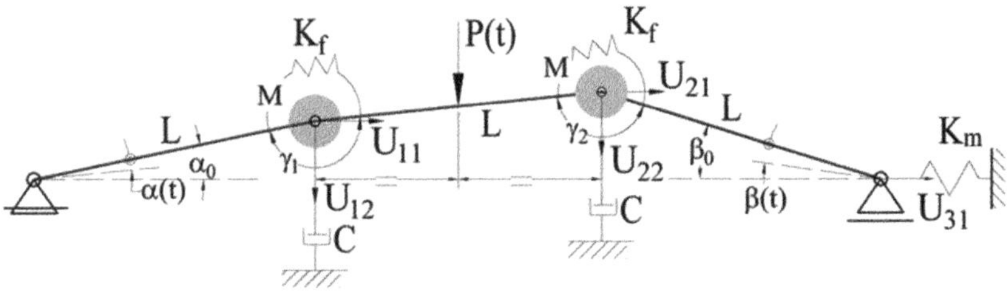

Figure 8.1. *Two DOF model investigated in this work*

The kinetic energy T, elastic potential V, dissipation D and generalized forces Q (excluding viscous damping) are given by:

$$
\begin{aligned}
T &= \frac{1}{2}M\left(\dot{U}_{11}^{\,2}+\dot{U}_{12}^{\,2}+\dot{U}_{21}^{\,2}+\dot{U}_{22}^{\,2}\right)\\
V &= \frac{1}{2}\left(K_m U_{31}^{\,2}+K_f\Delta\gamma_1^2+K_f\Delta\gamma_2^{\,2}\right)\\
D &= \frac{1}{2}C\left(\dot{U}_{12}^{\,2}+\dot{U}_{21}^{\,2}\right)\\
Q_1^{nc} &= \frac{P}{2}\left(\frac{\partial U_{12}}{\partial\alpha}+\frac{\partial U_{22}}{\partial\alpha}\right)\\
Q_2^{nc} &= \frac{P}{2}\left(\frac{\partial U_{12}}{\partial\beta}+\frac{\partial U_{22}}{\partial\beta}\right)
\end{aligned}
\qquad [8.1]
$$

with displacement components given by:

$$U_{11} = L\left[-\cos\left(\beta_0-\beta_0\xi\right)+\cos\left(\beta_0-\beta_0\xi-\alpha\right)\right] \qquad [8.2a]$$

$$U_{12} = L\left[\sin\left(\beta_0-\beta_0\xi\right)-\sin\left(\beta_0-\beta_0\xi-\alpha\right)\right] \qquad [8.2b]$$

$$U_{21} = -L\cos(\beta_0 - \beta_0\xi) + L\cos(\beta_0 - \beta_0\xi - \alpha) - L\left\{1 - \left[\sin(\beta_0) - \sin(\beta_0 - \beta_0\xi)\right]^2\right\}^{\frac{1}{2}} + L\left\{1 - \left[-\sin(\beta_0 - \beta_0\xi - \alpha) + \sin(\beta_0 - \beta)\right]^2\right\}^{\frac{1}{2}} \quad [8.2c]$$

$$U_{22} = L\left[\sin(\beta_0) - \sin(\beta_0 - \beta)\right] \quad [8.2d]$$

The scalar ξ denotes the amplitude of geometric imperfection, and is given by:

$$\xi = \frac{|\alpha_0 - \beta_0|}{\beta_0} \quad [8.3]$$

The angle changes at the hinges between bars are given by:

$$\Delta\gamma_1 = -\arctan\left[\frac{\sin(\beta_0) - \sin(\beta_0 - \beta_0\xi)}{\left\{1 - \left[\sin(\beta_0) - \sin(\beta_0 - \beta_0\xi)\right]^2\right\}^{\frac{1}{2}}}\right] + \arctan\left[\frac{-L\sin(\beta_0 - \beta_0\xi - \alpha) + L\sin(\beta_0 - \beta)}{L\left\{1 - \left[-\sin(\beta_0 - \beta_0\xi - \alpha) + \sin(\beta_0 - \beta)\right]^2\right\}^{\frac{1}{2}}}\right] + \alpha \quad [8.4a]$$

$$\Delta\gamma_2 = \arctan\left[\frac{\sin(\beta_0) - \sin(\beta_0 - \beta_0\xi)}{\left\{1 - \left[\sin(\beta_0) - \sin(\beta_0 - \beta_0\xi)\right]^2\right\}^{\frac{1}{2}}}\right] - \arctan\left[\frac{-L\sin(\beta_0 - \beta_0\xi - \alpha) + L\sin(\beta_0 - \beta)}{L\left\{1 - \left[-\sin(\beta_0 - \beta_0\xi - \alpha) + \sin(\beta_0 - \beta)\right]^2\right\}^{\frac{1}{2}}}\right] + \beta \quad [8.4b]$$

The equations of motion for this system are derived by use of Lagrange's formulation (see, for example, Baruh 1999):

$$\frac{d}{dt}\frac{\partial T}{\partial \dot{q}_i}-\frac{\partial T}{\partial q_i}+\frac{\partial D}{\partial \dot{q}_i}+\frac{\partial V}{\partial q_i}=Q_i^{nc}, \quad i=\overline{1,n} \qquad [8.5]$$

in which q_i are the generalized coordinates. For quasi-static systems, equation [8.5] takes the form of the principle of minimum potential energy and might be used to define the static equilibrium path including geometric nonlinearities.

The resulting dynamic equations are cumbersome and for reasons of brevity have not been included in this work. The full version of the analytical formulation may be found in Ameijeiras (2020). The equations of motion are integrated numerically through the IDA code in the SUNDIALS package (Hindmarsh *et al.* 2005) within the Mathematica software (Wolfram Mathematica 2019).

8.3. Exploring the static and dynamic behavior of the two DOF system

The static and dynamic response of the two DOF model is explored in this section by assuming the data adopted by Croll and Walker (1972): α_0=0.2 rad, β_0=0.2 rad, L_0=1 m, K_f=1 MN/m, K_m=1000 MN/m, M=10 MNs^2/m, and no viscous damping has been assumed (C=0) in the results presented in this chapter. The load $P(t)$ is given by a triangular impulse:

$$P(t)=\begin{cases} P_0\left(1-\dfrac{t}{t_0}\right) & \text{for} \quad 0\le t\le t_0 \\ 0 & \text{Otherwise} \end{cases} \qquad [8.6]$$

where t_0 is the duration of the positive phase of the impulsive load (assumed in this case as 0.05 s) and P_0 is the peak force. As reference values, the natural periods of vibration of the system are computed as T_1 = 6.71 s, T_2 = 2.22 s, so the duration of the impulse is very small in comparison with the period of both modes, i.e. t_0/T_1=0.0075 and t_0/T_2=0.0225.

Considering the static response, the case with ξ=0 displays a linear bifurcation at P^B=1.18 MN, α^B=β^B=0.00764 rad, and the energy stored at this bifurcation state is V=0.0046 MNm. The geometrically nonlinear static response of the system with imperfections has been computed using Riks' algorithm and the equilibrium paths are shown in Figure 8.2(a), where for increasing values of imperfection amplitude ξ the equilibrium paths show a maximum in each curve. As stated before, interest in this system is associated with its unstable static behavior at the critical state, which is what occurs in some shell structures. The imperfection-sensitivity plot is shown in Figure 8.2(b) for the values of maximum in each curve normalized with respect to the bifurcation load, and this illustrates a behavior that is typical of some shell structures.

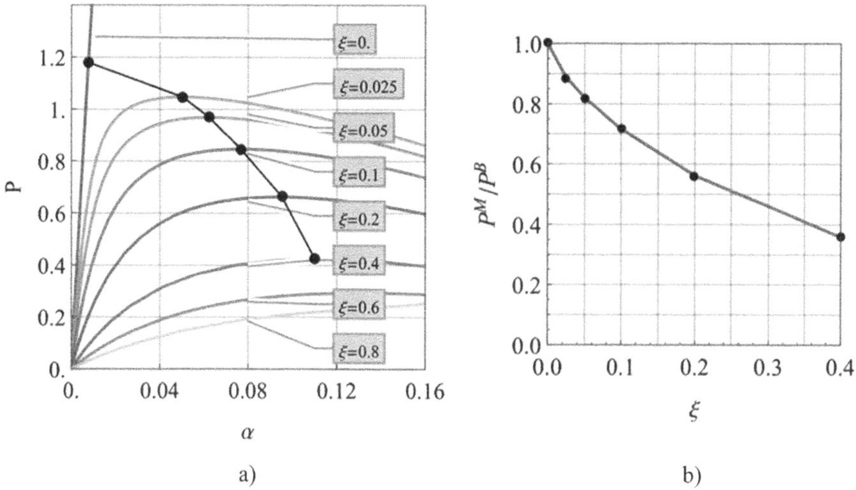

Figure 8.2. *Geometrically nonlinear static response of the two DOF system shown in Figure 8.1. (a) Equilibrium paths for perfect and imperfect configurations. (b) Imperfection-sensitivity curve. For a color version of this figure, see www.iste.co.uk/challamel/mechanics1.zip*

Information about the potential dynamic behavior may be obtained by exploration of the strain energy at the unloaded state P=0, as it would occur in a system subjected to an impulse (Simitses 1990). The potential energy surface for P=0 is shown in Figure 8.3(a) in a three-dimensional view and in Figure 8.3(b) by means of contour curves. The plots show a minimum in the energy at points O and D, at which stable behavior occurs, and equilibrium states at A, B and C exhibiting a maximum (unstable behavior). Point O occurs at $\alpha=\beta=0$, with V_0=0, whereas point D occurs at $\alpha=\beta=0.4$ rad, with V_0=0.16 MNm, which is an energy level much higher than that required to reach P^B.

A local maximum occurs at α=0.2 rad, β=0.2 rad (V_0=0.84 MNm) for point A; α=0.4 rad and β=0.2 rad (V_0=0.18 MNm) for point B; and α=0.2 rad, β=0.4 rad (V_0=0.18 MNm) for point C. For sudden loads and considering the impulse–moment theorem, it is possible to associate load levels with configuration points with known energy. Thus, providing an energy input V_0=0.84 MNm in the form of an impulse should be sufficient to move the system from O to A causing an escape of the oscillation in the vicinity of the origin. This case was identified by Simitses (1990) as a minimum guaranteed critical load (MGCL) and the associated load in this case is P^{MGCL}=231.9 MN. Lower energy levels are required to move the system from O to B (or from O to C); such states were identified by Simitses as minimum possible critical load (MPCL) and in this case they occur at P^{MPCL}=107.3 MN. Such load values are much larger than those found for the static P^B.

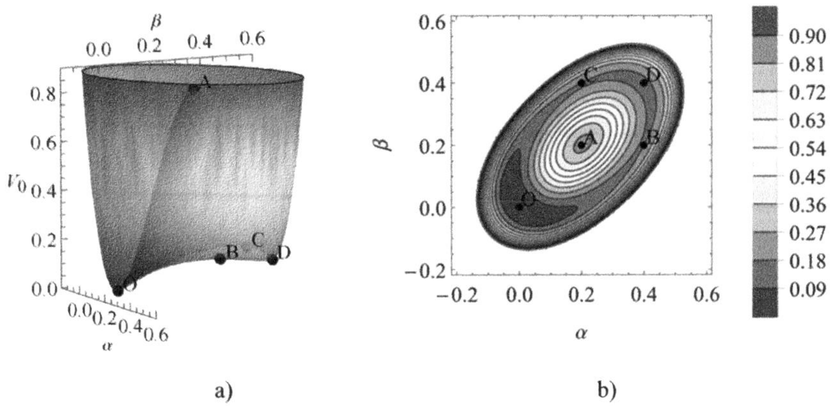

Figure 8.3. *Potential energy under P=0. (a) 3D view. (b) Iso lines. For a color version of this figure, see www.iste.co.uk/challamel/mechanics1.zip*

Next, the dynamic response under an impulsive pressure load is explored. The response of the perfect system (ξ=0) for a load level $P<P^{MPCL}$ (associated with energy points B and C in Figure 8.3) shows a transient response leading to a stable motion about the initial position of the system, with the two DOF oscillating not necessarily in phase. This is also the situation if imperfections are included in the analysis at the same load level, with the difference that the two DOF show oscillations with different phases.

The situation changes if P approaches or exceeds P^{MPCL}, as shown in Figure 8.4 for the perfect system. The oscillations shown in the phase plane (Figure 8.4(b)) move between two stable centers, which are coincident with points O and D in Figure 8.3. The DOF in motion in Figure 8.4(a) are not in phase so the oscillation does not occur along the fundamental mode but away from it. Consideration of an imperfection such as ξ = -0.05 induces an earlier out-of-phase motion of the system. In any case, the introduction of viscous damping causes the system to move to an attractor in the phase plane. In all cases, the escape from the fundamental dynamic path occurs after the load ceases to act on the system.

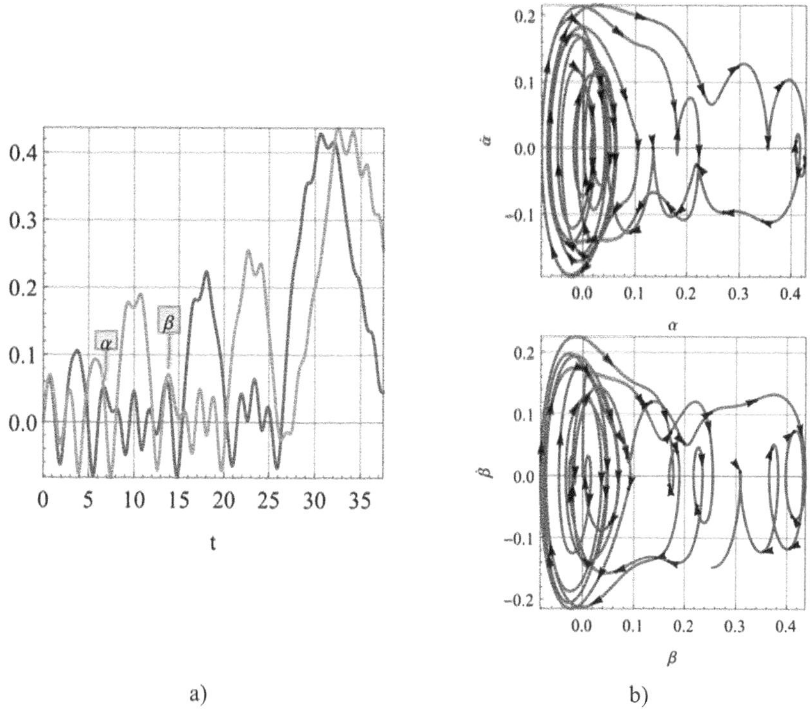

Figure 8.4. *Dynamic analysis of the two DOF system under P_0=128 MN>P^{MPCL} assuming ξ=0. (a) Transit response. (b) Phase plane for α. (c) Phase plane for β. For a color version of this figure, see www.iste.co.uk/challamel/mechanics1.zip*

The visualization through the phase plane is a useful tool for a few DOF; however, generalization to multiple DOF becomes problematic. Not just the early transient response needs to be computed, sometimes it requires the computation of a few hundred seconds in order to visualize reaching a new attractor.

8.4. The dynamic stability criterion due to Lee

Lawrence H.N. Lee (1923–2007), who was a professor at the University of Notre Dame in Indiana, USA, postulated in 1977 that if a perturbation $\delta\zeta$ is imposed onto a system, the dynamic response reaches an unstable configuration whenever the projection of the perturbed acceleration vector on the perturbed displacement vector is positive, i.e. both acceleration and displacement are related in a positive way (Lee 1977). However, the configuration is stable if the projection is negative. Thus, if a positive change in displacement causes a positive change in acceleration, then the system will increase the amplitude of the motion.

According to Lee (1981, p. 81, equation 15), assuming a perturbed motion given by ($\zeta_p = \zeta + \delta\zeta$), the following condition applies for instability:

$$C_L = \delta\ddot{\zeta}_r(t)\,\delta\zeta_r(t) > 0 \quad \text{(sum in } r\text{)}, \quad t > t_{cr} \qquad [8.7]$$

where summation extends to the number of DOF of the system. The result in equation [8.7] is a scalar quantity, which is here called Lee's coefficient C_L. According to Lee (1981, p. 81), when $C_L>0$ after a certain time t_{cr}, for any possible deviated motion, the response ζ grows without bounds and the known motion is unstable. In this way, the system is stable up to time t_{cr} or, in other words, the fundamental motion is bounded or stable when the perturbed force field remains a restoring or retarding one.

The variables $\alpha(t)$ and $\beta(t)$ describe the motion from the initial position α_0 and β_0, i.e. $\alpha(0)=0$ and $\beta(0)=0$, and α_p and β_p designate the motion with a perturbed initial condition $\alpha(0)=\Delta\alpha$ and $\beta(0)=\Delta\beta$. This reference can be associated with equation [8.7] as follows:

$$\begin{aligned} &\alpha = \zeta_1, \quad \alpha_p - \alpha = \delta\alpha = \delta\zeta_1 \\ &\beta = \zeta_2, \quad \beta_p - \beta = \delta\beta = \delta\zeta_2 \end{aligned} \qquad [8.8]$$

To illustrate the use of the criterion proposed by Lee, consider the two DOF system under study, at a load level P_0=110 MN>P^{MPCL}. The transient response of a perfect system without damping is shown in Figure 8.5(a) considering both unperturbed (α, β) and perturbed (α_p, β_p) motions. The two DOF initially oscillate together but at t=3 s the mode changes and each DOF oscillates following a different pattern. The product C_L due to Lee, shown in Figure 8.5(b), is initially negative (stable) and becomes positive at $t\approx 1$ s and again at $t\approx 3$ s, thus indicating an instability. It may be seen in Figure 8.5 that the two DOF depart in their motion in the perturbed system. The conclusion is that Lee's criterion was capable of identifying that the system is unstable at this load level and that it has already gone through a quasi-bifurcation.

Next, consider a lower load level, say P_0=70 MN<P^{MPCL}, and the response of a perfect system without damping is plotted in Figure 8.6. The two DOF oscillate in-phase and they remain bounded; furthermore, it is difficult to note differences between them in these early stages. However, the C_L product becomes positive even before t=3 s, and remains positive for a short time, until it reverses its sign. Thus, we reach the problematic conclusion that the criterion due to Lee would indicate that the system is unstable at this load level, whereas observation of the motion showed the opposite conclusion because the motion is bounded.

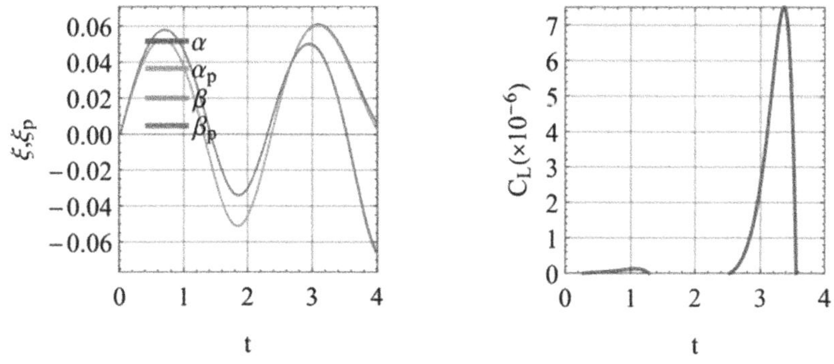

a) b)

Figure 8.5. *Dynamic stability criterion due to Lee for P_0=110 MN≈P^D. Initial perturbation $\Delta\alpha$=0.0001, $\Delta\beta$=0. (a) Transient response. (b) C_L coefficient (equation [8.7]); only positive projection is plotted. For a color version of this figure, see www.iste.co.uk/challamel/mechanics1.zip*

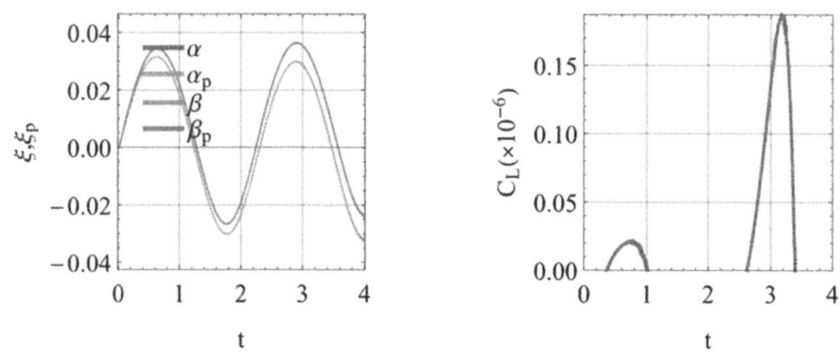

a) b)

Figure 8.6. *Dynamic stability criterion due to Lee for P_0=70 MN<P^{MPCL}. Initial perturbation $\Delta\alpha$=0.0001, $\Delta\beta$=0. (a) Transient response showing α, α_p, β, β_p. (b) C_L coefficient (equation [8.7]; only positive projection is plotted). For a color version of this figure, see www.iste.co.uk/challamel/mechanics1.zip*

This paradox was noted by Kleiber *et al.* (1987, p. 50), who observed that there are situations that satisfy the instability criterion due to Lee and yet the perturbed motion remains within bounds, thus indicating that the motion is stable. The overall conclusion of this section is that Lee's criterion seems to be a necessary but not sufficient condition for instability.

8.5. New stability bounds following Lee's approach

Observation of the plot in Figure 8.6(b) indicates that a short excursion of Lee's coefficient into the positive range with small amplitude is not sufficient to conclude that we are in the presence of an unstable motion. However, Figure 8.5(b) shows that high sustained levels of the coefficient proposed by Lee provide an adequate indication of instability. The question of how we can identify that a positive value of C_L is sufficient to cause instability still remains open.

To overcome this limitation, we can explore the accumulated value of positive C_L up to a given time, and for that, a new stability indicator is defined in this work as the time integral of C_L in the form:

$$C_S = \int_0^{t_f} C'_L(t)dt \qquad [8.9]$$

where

$$C'_L(t) = \begin{cases} C_L(t) & \text{for} \quad C_L(t) > 0 \\ 0 & \text{Otherwise} \end{cases} \qquad [8.10]$$

Since C_L presents lobes of positive values, the integration in equation [8.9] should extend to the time encompassing the lobe. In the present case, the limit of integration was taken as t_f=4 s based on that shown in Figures 8.5(b) and 8.6(b).

To illustrate the procedure, results were obtained considering P_0 in the range 85 MN≤P_0≤115 MN. The analyses were done with and without perturbations, for which $\Delta\alpha$=0.001 rad, $\Delta\beta$=0.001 rad were initially assumed. In each case, the differences between the perturbed and unperturbed motions were computed for both displacements and accelerations, $\delta\alpha, \delta\beta, \delta\ddot{\alpha}$ and $\delta\ddot{\beta}$. The transient responses are shown in Figure 8.7(a) and (b) for 30 s in the motion. For the lower values of P_0, the response remains bounded, and for P_0=115 MN, there is a large amplitude response which is clearly visible after 10 s.

The coefficient C_L in Figure 8.8 becomes positive on reaching 3 s, but a smaller lobe appears as early as 1 s. As stated before, some difficulties arise with the original criterion due to Lee because positive values can be present for low values of P_0 for which the transient response is bounded.

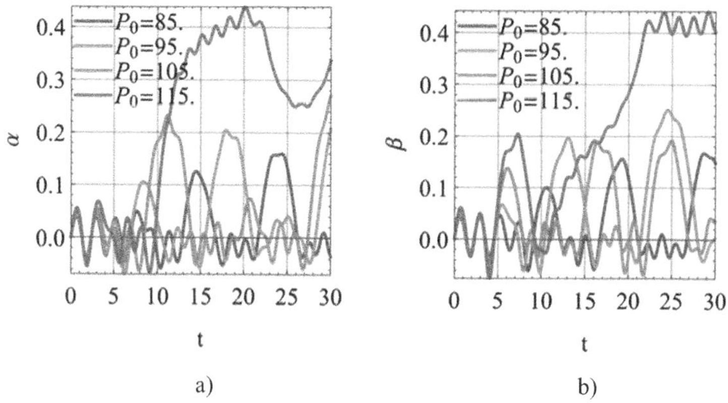

Figure 8.7. *Dynamic response for various load levels. Data: ξ=0, $\Delta\alpha$=0.001, $\Delta\beta$=0.001. (a) Motion in terms of α. (b) Motion in terms of β. For a color version of this figure, see www.iste.co.uk/challamel/mechanics1.zip*

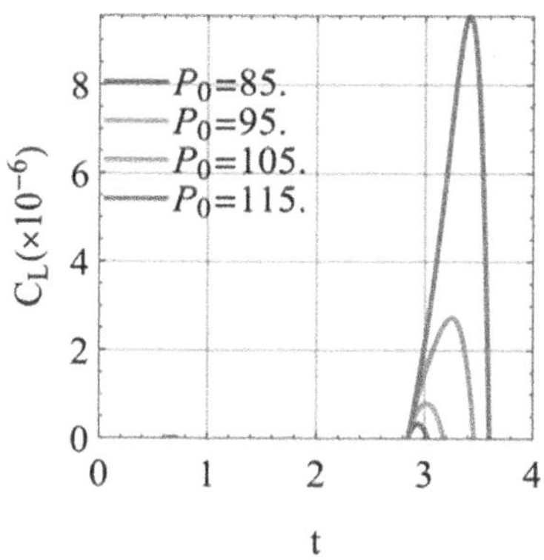

Figure 8.8. *Positive projection of Lee's coefficient C_L for several values of P_0 approaching P^{MPCL}. Data: ξ=0, $\Delta\alpha$=0.001, $\Delta\beta$=0.001. For a color version of this figure, see www.iste.co.uk/challamel/mechanics1.zip*

The plot of C_S is shown in Figure 8.9 for increasing load levels P_0, where C_S has been normalized with respect to a reference value, in this case by adopting C_S^{ref}=C_S(P_0=10 MN). This plot represents the accumulated positive value of C_L computed up to t_f=4 s and for increasing values of impulsive load P_0.

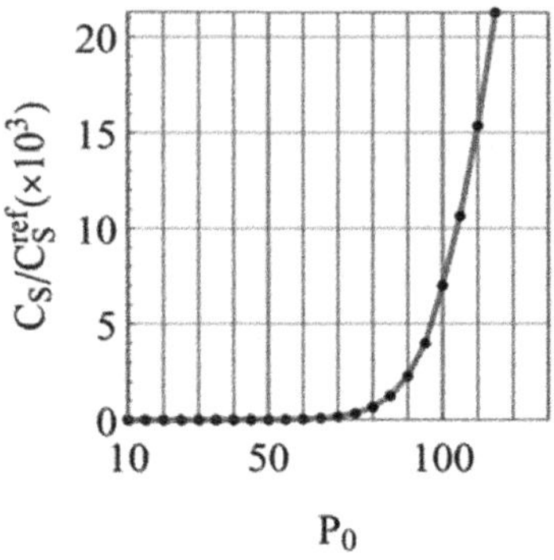

Figure 8.9. *Dynamic response in terms C_S/C^{ref}. Data: ξ=0, t_f=4 s, Δα=0.001, Δβ=0.001. The present normalized stability coefficient C_S for various load levels. For a color version of this figure, see www.iste.co.uk/challamel/mechanics1.zip*

The value of C_S itself does not provide an indication of stability, but what matters is the change in C_S computed for different load levels. The normalized change in C_S may be computed as

$$\Delta C_S = \frac{C_S(P_0 + \delta P_0) - C_S(P_0)}{C_S^{ref}} \qquad [8.11]$$

For low values of P_0, say $P_0 \leq 70$ MN, the plot remains horizontal and no significant increase in C_S is detected. In summary, for $P_0 \leq 70$ MN, no appreciable changes are detected in ΔC_S.

A clear increase is seen in Figure 8.9 for larger values of P_0, for which $\Delta C_S \neq 0$. The rate at which C_S increases follows an almost linear relation with P_0, except for a small transition zone. Along this straight line, we can estimate a constant increase, ΔC_S=constant. We can state with confidence that there is a qualitative change in behavior from $\Delta C_S \approx 0$ to $\Delta C_S \neq 0$, as reflected by the accumulated coefficient of Lee.

This change in behavior does not follow a sudden change occurring at a specific value P_0, as it would be the case of having a vertical line in the variation of C_S versus P_0 in Figure 8.9, with a discontinuity in the plotted variable. The latter situation was found to occur in some cases by Budiansky and Roth (1962), but this is not the case here. What occurs in Figure 8.9 is not a sudden change in the variable itself but in the slope.

Such changes in slope in dynamic buckling have been found by researchers associated with dynamic buckling under seismic excitation (see, for example, Virella *et al.* (2006); Buratti and Tavano (2014)) or in the dynamic buckling of plates (Kubiak 2013).

To better visualize the response, C_S has been plotted for a larger range of values of P_0, and the resulting curve is shown in Figure 8.10. Here, the straight line is seen to extend up to a value of approximately 120 MN, at which a maximum is reached. This suggests that the range of values of C_S for which dynamic buckling may occur (which is at some point along the ΔC_S=constant segment) may be established as bounds.

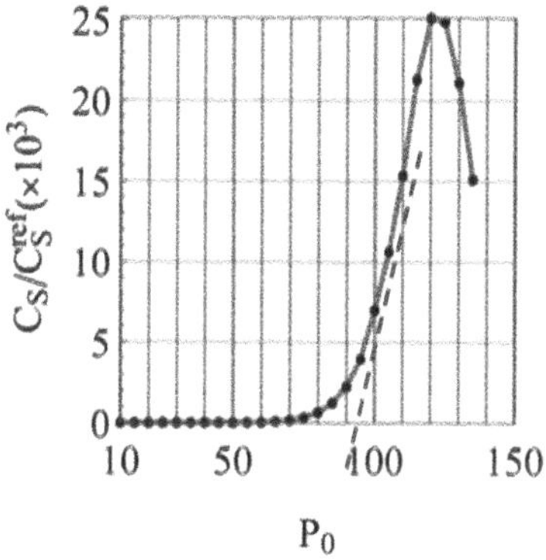

Figure 8.10. *Variation of C_S with the impulsive load P_0 in the range 10 MN<P_0<135 MN, showing lower and upper bounds. For a color version of this figure, see www.iste.co.uk/challamel/mechanics1.zip*

A lower bound may be estimated by drawing a straight line with ΔC_S=constant and finding the intersection with the horizontal axis. This is shown by the dotted line in Figure 8.10, and the intersection occurs for a value of P_0 close to 90 MN. Lower values of P_0 have not accumulated enough positive values of C_L, as reflected by C_S. Such estimates have been used in problems of stable symmetric bifurcations under static loads to identify the critical load.

At the other end of the curve in Figure 8.10, we can see that the constant increase in C_S with rate ΔC_S does not occur indefinitely, but there is a maximum, and for increasing values of P_0 beyond the maximum, the values of C_S decrease. This limit in the growth of the accumulated C_L may be considered as an upper bound to dynamic buckling.

The case studied in this paper serves as an illustration of how this unstable dynamic behavior can be located within bounds. In the present case, it was found that dynamic buckling should occur between 90 MN<P_0<120 MN. This case was also investigated using the present strategy, with various initial perturbations, but they make minor changes in the plots presented here. One of the advantages of this procedure is that the dynamic response needs to be computed for a short time, independently of when the DOF reach a divergent response, which could be much later in the motion.

8.6. Conclusion

A simple, two DOF system has been investigated in this work in order to highlight procedures that can be used to evaluate stability in the dynamic motion of systems under impulsive loads. Like shell structures, the mechanical system considered has both membrane and bending action to equilibrate external loads, and its geometrically nonlinear response displays unstable post-buckling behavior under static loads with high imperfection sensitivity. The original contribution of this work is the analysis of the system under impulsive loads together with the application of the stability criterion due to Lee, and the proposal of a new approach based on Lee's framework of analysis.

The criterion due to Lee, based on a coefficient C_L computed as the projection of the perturbation of the motion on the perturbation of the acceleration of the system, is capable of identifying the stability of the system by means of the observation of sign changes, i.e. a positive sign reveals an unstable state, whereas a negative sign indicates that we are in the presence of a stable state along the motion. However, it is shown that this sign change may also occur at lower load levels, for which the motion is clearly bounded, and instability does not physically occur. The overall conclusion is that the original approach due to Lee is a necessary but not sufficient condition for instability. The passage from negative to positive values of C_L can be a temporary behavior and with short time duration, after which the coefficient becomes negative again.

In this work, the authors propose dealing with the accumulated value of the coefficient due to Lee, by means of a time integral leading to a coefficient named C_S. For increasing impulsive load levels, this coefficient shows a change in the behavior which can be simplified as a bilinear relation. It is postulated that dynamic buckling should occur along the second range in this bilinear relation, and for that, upper and lower bounds were identified.

The behavior discussed in this chapter has been found to occur in other simple systems. The present criterion can be extended to deal with multiple DOF systems discretized with finite elements, such as it would be necessary to model problems in shells fabricated for the oil industry.

The case investigated in this chapter has elastic material behavior, but an extension to impulsive loading leading to plasticity can also be considered in this way. Although derived within the context of nonlinear dynamics, Lee's necessary condition and the present bounds are not limited to impulsive loading and could be used in a number of other stability problems, including thermal and seismic loads, systems with static pre-loading and other non-classical stability problems.

8.7. Acknowledgments

We thank Prof. Isaac Elishakoff for his multiple teachings made throughout the years on the subject of buckling of structures. We have benefited much from his contributions in papers and books; and the second author also acknowledges the many teachings received from Prof. Elishakoff on a personal level, thanks to meeting him on many occasions over the years. This work is part of a research on "vulnerability of infrastructure and physical media associated with fuel storage and transportation". Support for this research was provided by the Science and Technology Research Council of Argentina, CONICET, by a grant awarded to IDIT.

8.8. References

Amabili, M. (2008). *Nonlinear Vibrations and Stability of Plates and Shells*. Cambridge University Press, UK.

Ameijeiras, M.P. (2020). Analysis of oil storage tanks under pressures due to explosions [in Spanish]. PhD Thesis, National University of Cordoba, Argentina.

Baruh, H. (1999). *Analytical Dynamics*. McGraw-Hill, Singapore.

Budiansky, B. and Roth, R.S. (1962). Axisymmetric dynamic buckling of clamped shallow spherical shells. In *Collected Papers on Instability of Shell Structures*, NASA-TN-D-1510 (ed.). Washington, DC.

Buratti, N. and Tavano, M. (2014). Dynamic buckling and seismic fragility of anchored steel tanks by the added mass method. *Earthquake Engineering and Structural Dynamics*, 43, 1–21.

Burmeister, A. and Ramm, E. (1990). Dynamic stability analysis of shell structures. In *Computational Mechanics of Nonlinear Response of Shells*, Kratzig, W.B. and Oñate, E. (eds). Springer, Berlin.

Croll, J.G.A. and Walker, A.C. (1972). *Elements of Structural Stability*. Macmillan, London.

Elishakoff, I., Li Y., Starnes Jr., J.H. (2001). *Non-Classical Problems in the Theory of Elastic Stability*. Cambridge University Press, Cambridge, UK.

Hindmarsh, A.C., Brown, P.N., Grant, K.E., Lee, S.L., Serban, R., Shumaker, D.E., Woodward, C.S. (2005). Sundials: Suite of nonlinear and differential/algebraic equation solvers. *ACM Transactions on Mathematical Software*, 31(3), 363–396.

Kang, K.Y., Heo, Y., Rogstadkjernet, L., Choi, K.H., Lee, J.M. (2016). Structural response of blast wall to gas explosion on semi-confined offshore plant topside. *International Journal of Structural Stability and Dynamics*, 16(10), 1750021.

Kleiber, M., Kotula, W., Saran, M. (1987). Numerical analysis of dynamic quasi-bifurcation. *Engineering Computations*, 4, 48–52.

Kounadis, A.N., Raftoyiannis, J., Mallis, J. (1989). Dynamic buckling of an arch model under impact loading. *Journal of Sound and Vibration*, 134(2), 193–202.

Kratzig, W.B. and Eller, C. (1992). Numerical algorithms for nonlinear unstable dynamic shell responses. *Computers and Structures*, 44(1/2), 263–271.

Kroplin, B. and Dinkler, D. (1986). Dynamic versus static buckling analysis of thin walled shell structures. In *Finite Element Methods for Plate and Shell Structures 2*, Hughes, T. and Hinton, E. (eds). Pineridge Press, Swansea, UK.

Kubiak, T. (2013). *Static and Dynamic Buckling of Thin-Walled Plate Structures*. Springer, New York.

Lee, L.H.N. (1977). Quasi-bifurcation in dynamics of elastic-plastic continua. *Journal of Applied Mechanics*, 44(3), 413–418.

Lee, L.H.N. (1981). Dynamic buckling of an inelastic column. *International Journal of Solids Structures*, 17, 271–279.

Nayfeh, A.H. and Balachandran, B. (2004). *Applied Nonlinear Dynamics*. Wiley-VCH, Weinheim.

Nayfeh, A.H. and Mook, D.T. (1979). *Nonlinear Oscillations*. John Wiley & Sons, New York.

Simitses, G.J. (1990). *Dynamic Stability of Suddenly Loaded Structures*. Springer-Verlag, New York.

Virella, J.C., Godoy, L.A., Suárez L.E. (2006). Dynamic buckling of anchored steel tanks subjected to horizontal earthquake excitation. *Journal of Constructional Steel Research*, 62(6), 521–531.

Wolfram Mathematica (2019). Version 12.1.0.0, Wolfram Research, Inc., Champaign, IL.

9

Singularly Perturbed Problems of Drill String Buckling in Deep Curvilinear Borehole Channels

9.1. Introduction

Over time, the "shale revolution" has become the most influential stimulus for the change in global energy market players and the redistribution of global energy flows. Now, to obtain oil and gas fuels, deep vertical and curvilinear offshore and ground-based boreholes are drilled in complex heterogeneous tectonic mining rocks, under severe climate conditions.

The creation and development of directional drilling techniques and fracturing operations have made it possible to effectively extract shale gas and oil. However, as international practice has shown, the techniques for extracting and exploiting petroleum resources are developing so fast that scientific understanding and substantiation have no time to update them. Hence, various unexpected extraordinary situations and outright breakdowns appear during borehole drivage. Because of these reasons, there is a strong necessity to intensify scientific works in both the field research and the computer modeling of all the critical aspects accompanying drilling operations of extended oil and gas curvilinear wells.

The most severe of them are the loss of bending stability of the drill strings (DSs) and their buckling; their longitudinal and bending vibrations; their sticking during drilling and twisting; the self-triggering of torsional relaxation oscillations and self-excitation of forward, backward and superfast bit whirling (Gulyayev *et al.* 2019).

Chapter written by Valery I. GULYAYEV and Natalya V. SHLYUN.

Amidst the most prevalent emergency situations arising from curvilinear drilling is the bifurcational buckling of a drill string. Usually, it is provoked by the loss of the ability of the drill string to carry an axial compressive force and is connected with a number of effects that negatively affect the entire technological process. Among them are:

– the appearance of intensive distributed contact forces generated between the drill string and the borehole wall;

– the generation of distributed frictional forces between the drill string and the borehole wall, caused by axial and rotational motions of the string;

– the growth of intensity of the bending stresses in the drill string pipes;

– the increase of wear intensity of the drill string pipes;

– the worsening of the penetrability of axial force and torque from the actuating mechanism to the bit;

– the growth of total energy consumption for the well drilling.

Inasmuch, as the buckling prediction exactness is rather low, the stability loss effect (especially in its initial stage) often proceeds unnoticed, and the drilling procedure goes on with abnormal duty when the drill string has partially lost its carrying ability, as a result of its bending. In doing so, the achieved contact and friction forces can lead to the loss of the drill string mobility and its sticking.

In this connection, during the analysis of elastic buckling of a drill string in the well channel, the most attention should be paid to the calculation of the critical loads and bifurcation modes. The first theoretic works in this direction were connected with buckling of DSs in vertical rectilinear boreholes, without paying particular attention to large lengths of the DSs and using simplifying suppositions (Brett *et al.* 1989). Particular questions about this problem were analyzed, with allowances made for the constraining effects of a borehole wall (Sun and Lukasiewicz 2006; Krauberger *et al.* 2007), friction forces (Wang and Yuan 2012) and borehole inclination (Dawson and Paslay 1984). If the drill string is in a curvilinear borehole, such analysis should be performed on the basis of computational mathematic approaches. Through their use, not only can bifurcation state be ascertained, but also the pitch of the bifurcation harmonic, the wave package width and the place of its localization (Gulyayev and Shlyun 2016, 2017; Shlyun and Gulyayev 2020).

When the emergency character is revealed, the measures for its elimination are projected. They can be connected with a change of the string structure and its outfit, readjustment of the operation mode as well as redrilling the entire well or a certain section.

However, until now, there are no effective and reliable tools for the theoretic modeling of buckling effects in curvilinear channels (see papers by Cunha (2004), Mitchell (2008) and Gao and Huang (2015)). This feature is associated with the essential complexity of their nature. Indeed, during operation, the drill string is stretched in its upper zone and compressed below; it undergoes the action of distributed gravity, contact and friction forces; it rotates; it is twisted by a torque and a mud flow is moving inside it with high speed. Moreover, a large number of additional factors complicate the attempts to comprehend this phenomenon (Challamel 2000).

The first one is connected with the specific nature of frictional forces generated in elongated curvilinear wells. The friction force between solids with flat contact surfaces depends on the compressive force clasping them. But in the curvilinear sectors of a borehole, the force of clasping the string and the hole surface is determined by the axial tension of the string. For this reason, in tribology, it is considered that these friction forces have a multiplicative nature (i.e. they are multiplied) and not additive (i.e. they are not added), as this occurs in the body sliding over a rough plane.

The second factor is specified by the length of the string. By conditions of geometric similitude, it corresponds to a human hair and so, a phenomenon occurring at one end of the string can affect, have small influence or be imperceptible on the processes at the other end. In applied mathematics, the equation systems fixing these sorts of effects are called singularly perturbed equations. Usually, their resolving functions have irregularities and singularities in the shapes of boundary effects, breaks, peaks, discontinuities and internal harmonic wavelets. It is obvious that their solutions are poorly converging and their construction is associated with the additional difficulties. For these reasons, in mathematics, the questions associated with the understanding of the specific properties of singularly perturbed systems attract essential attention. Note that Elishakoff *et al.* considered similar problems in 2001. Consider these properties more particularly.

9.2. Singular perturbation theory: elements and history

As pointed out above, the differential correlations of the elastic deformation of drill strings in deep boreholes are singularly perturbed. In the general case, in applied mathematics, this feature of differential equations is connected with the existence of a small multiplier in front of the highest derivative (Chang and Howes 1984; Smith 1985; Johnson 2005). Then, in comparison with derivatives of a lower order, the role of this term falls, but it cannot be neglected, since as a result of its rejection, not only the type of the differential equation, but also the structure of its solution, will change. As analysis shows, the presence of a small coefficient at the highest derivative leads to various violations of regularity in the structure of solutions, since singularities appear in them in the form of discontinuous boundary effects and localized bursts.

Therefore, the question of the behavior of the solutions of the differential equation, with a small change in the value of the parameter included in the equation, is of fundamental interest. Starting from the classical works of Poincare and Lyapunov, the so-called regular case was studied very thoroughly – when the right-hand side of the equation, say, of the second order

$$x'' = F(t,x,x',\varepsilon) \quad (0 \le t \le 1) \qquad [9.1]$$

regularly (continuously, smoothly, analytically) depends on the parameter ε in a neighborhood of the value $\varepsilon = 0$, and solutions of the equation are considered on a finite segment $0 \le t \le 1$ of the change in the independent variable t. In this case, to solve the stated equation for any small ε, there is a solution close to it, corresponding to value $\varepsilon = 0$.

The singular case is much more multifarious and complicated when the solutions depend on a small parameter before the second derivative and when the assumption that the parameter is regular in the equation is not realized. Apparently, the last of these points arises, for example, for the equation

$$\varepsilon x'' = f(t,x,x',\varepsilon) \quad (0 \le t \le 1) \qquad [9.2]$$

where ε is the parameter that takes undefined small positive values (they often write: $0<\varepsilon<<1$). Note that the attempt to perform substitutions $t = \sqrt{\varepsilon}\,\tau$ and $d^2x/dt^2 = d^2x/\varepsilon\, d\tau^2$, may lead equation [9.2] to a "regular form"

$$\varepsilon\frac{d^2x}{dt^2} = \varepsilon\frac{d^2x}{\varepsilon\, d\tau^2} = \frac{d^2x}{d\tau^2} = f(\tau,x,x'\varepsilon)\,, \qquad [9.3]$$

but does not solve the problem. This is because, in this case, only a very small segment (along with ε) $0 \le t \le \varepsilon \cdot 1$ of the old variable t change corresponds to the final segment of the new variable τ change.

"The small multiplier before the highest derivative gave rise to a large theory" – this phrase from mathematical folklore colorfully characterizes the vast branch of the theory of differential equations. It is paradoxical, but this branch, which has gained wide popularity among theoreticians and has great importance for practitioners, has not yet received a single, commonly used name. Some authors call the object of study "an equation with a small parameter in the highest derivatives", others speak of a "singular perturbation", while others use a variety of terminologies generated by numerous applications: the problem "with boundary layer", "with boundary effect", "with shock wave". The term "stiff differential equations" is quite widespread in literature on applied mathematics. This term refers to the case under consideration. Perhaps the first of all the given names is the most accurate, although it is somewhat long and therefore not very convenient. Regarding the term "singular perturbation", despite being generally broader, but vague in meaning, it characterizes the physical process under consideration much more accurately, which is associated with the formation of singularities in the marginal zones of the desired function. The term singular perturbation was contrived in the 1940s by Friedrichs and Wasow. Further, we use this term in reference to problems of singularities in both time and space. A great deal of the earliest motivation in this area arose from Poincare's works on time scales for periodic phenomena in celestial mechanics.

Prandtl drew attention to the applied importance of studying singularly perturbed problems (and to the "unavoidability" of their appearance in mathematical models of natural science) in connection with the theory of the boundary layer in hydrodynamics, which he developed in 1904. Later, singularly perturbed problems have repeatedly arisen (though not always deliberately) in mechanics, physics and technology (we recall, for example, the study of relaxation oscillations in radio engineering by Van der Pol, which began in the 1920s). Further, different aspects of biological systems and chemical reactions were studied, stemming from the ideas typical to these phenomena. Each of these areas yields problems whose solutions have features that vary on disparate length or time scales. The philosophy behind singular perturbation theory is to take advantage of this separation of scales, in order to obtain reduced problems that are simpler than the original full problem. The construction of the full (global) solution of a singular problem, in terms of the solutions of the reduced problems, is a key element of work in this field.

However, unitized mathematical approaches to such problems did not exist for a long time, so each of them were considered in isolation before being solved – with varying degrees of strictness and perfection – by using specific techniques.

These peculiarities also take place in the tasks of bending drill strings. However, they occur in a slightly different form. The fact is that, for example, in the equation

$$EI\frac{d^4w}{dx^4}=q \qquad [9.4]$$

of a rectilinear beam bending, the highest (fourth and only) derivative usually has a coefficient EI, where w is the beam sagging function, E is the modulus of elasticity of the beam material, I is the moment of inertia of its cross-section, x is the axial coordinate and q is the distributed load intensity, and, as a rule, this coefficient is not small. Moreover, if the beam is short, then no features arise in the statement of the problem and in its solution. However, if this beam is installed on an elastic foundation, then an additional summand appears in the equation of its bending

$$EI\frac{d^4w}{dx^4}+kw=q \qquad [9.5]$$

in the form of the desired function (zero derivative) multiplied by the bed coefficient k. In this case, the role of the fourth derivative in the equation decreases and the transverse load q, acting on the beam, begins to be sustained by the elastic foundation, the influence of which increases with the increase of its bed coefficient. Eventually, the effect under consideration increases even more with the beam length enlargement, since in doing so, it becomes flexible, although its coefficient of bending stiffness EI has not changed.

The noted effects are manifested even more clearly in the tasks of drill string bending in the channels of curved wells, as in this case, both the bed coefficient (determined by the stiffness of the rock) and the length of the beam (string) are disproportionately large in comparison with cases typical for conventional equipment. Therefore, it can be argued that the tasks set for buckling curvilinear drill strings are substantially singularly perturbed. Our results of modeling the phenomena of their critical buckling made it possible, for the first time, to build bifurcation modes characteristic for them, which have the shapes of harmonic wavelets localized in the internal or boundary zones of the column. In doing so, it also became possible to establish the main characteristics of the wavelets, such as their pitches (the distance between two adjacent zeros) and widths (a conditional value equal to the distance between its beginning and end).

Using the concept of singularly perturbed systems, it was also possible, for the first time, to answer a question of interest to specialists on the buckling problem, namely, how the boundary conditions affect critical loads and modes. It proved to be that if a buckling wavelet is formed in the inner zone of the borehole, then the boundary conditions do not influence this process, and if buckling takes place in the marginal zone, then the influence of the boundary support is rather weak, since it extends to only one extreme wavelet burst.

It is also important to note one more circumstance. The singular nature of the differential equations does not have to be associated with the presence of a small parameter ε with the highest derivative. There are quite common cases when the singularity of the perturbation of differential equations is hidden (implicit), and is due to the long segment length L of their integration. Indeed, for such problems from the equation

$$\frac{d^2x}{dt^2} = f(t,x,x'), \quad 0 \le t \le L, \quad L >> 1 \qquad [9.6]$$

using replacement

$$t = L\tau, \qquad \frac{d^2x}{dt^2} = \frac{1}{L^2}\frac{d^2x}{d\tau^2} \qquad [9.7]$$

we can go to the equation

$$\frac{1}{L^2}\frac{d^2x}{d\tau^2} = f(\tau,x,x'), \quad 0 \le \tau \le 1 . \qquad [9.8]$$

Because $L>>1$, $1/L^2 = \varepsilon$ and equation [9.8] becomes equivalent to a singularly perturbed equation [9.3].

The singularly perturbed problem is regular if its solution varies smoothly as the perturbation parameter ε approaches zero. The problem is irregular if its solution changes abruptly in some way as ε tends to zero, and it either ceases to exist or may become infinite or degenerate.

This type of equation includes the system of differential equations from the Sturm–Liouville problems considered in the book by Gulyayev *et al.* (2019) for vertical and curved drill columns of super-deep drilling, as well as relaxation torsion self-oscillations. Moreover, the difficulty of solving these equations significantly exceeds the complexity of singularly perturbed equations known in the scientific literature. First, this is due to the fact that in deep drilling problems, the mechanics of the elastic bending of drill strings are described by equations of not second, but fourth order, and the integration interval L reaches a value of 10,000 m. Therefore, the parameter ε with the highest (fourth) derivative, when scaling the integration interval L up to one, takes on value $\varepsilon = 1/L^4 = 10^{-16}\,\mathrm{m}^{-4}$, from which it is not far from "machine zero" in modern computers.

The second complexity of the problems posed in the monograph by Gulyayev *et al.* (2019) is due to the formulation of multipoint boundary value problems, in contrast to the two-point boundary value problems for ordinary differential equations that are usually considered in well-known scientific literature. As the studies conducted in the book showed, these features led, on one hand, to extremely complex problems of constructing solutions of differential equations, and, on the other hand, to beautiful and unexpected solutions describing the studied mechanical phenomena. To date, a huge number of works by many authors who have studied initial and boundary value problems, properties, behavior, representations and estimates of various types of solutions, diverse applications, as well as various modifications of the problem, its generalizations and related questions, have been devoted to the "small factor in the highest derivative" (Miller *et al.* 1996; Roos *et al.* 1996; Farrell *et al.* 2000; Johnson 2005). Methods of the perturbed systems analysis are reviewed by Holmes (1995); their applications to the study of mechanical objects are touched on by Verhulst (2005), Acosta and Lopez-Martinez (2008) and Ouland Ben Zarouala *et al.* (2012).

One-dimensional singularly perturbed equations are widely found in the theory of axisymmetric shells. To solve them, the joint application of the initial parameter method and the Godunov orthogonalization procedure (Gulyayev *et al.* 2019) is effective. This approach is also applied in this work.

9.3. Posing the problem of a drill string buckling in the curvilinear borehole

Consider the case when a drill string lies on the bottom of a long planar curvilinear borehole. If it is immovable, then the distributed contact and static friction forces acting on it equalize the gravity forces, its equilibrium is stable and the probability of its buckling is small. But when the drill string is going down (during drilling or tripping in operation performing), its length enlarges, the friction forces achieve their ultimate values and, together with compressive force acting on the bit, they can lead to the loss of the string quasi-static (inertia forces are neglected) equilibrium stability and provoke its buckling. For this reason, the possibility of the DS buckling and the critical values of load parameters should be analyzed at every stage of the DS descending. To achieve this goal, assume that the borehole trajectory T is prescribed and the DS length is enlarging as the DS is going down (Figure 9.1).

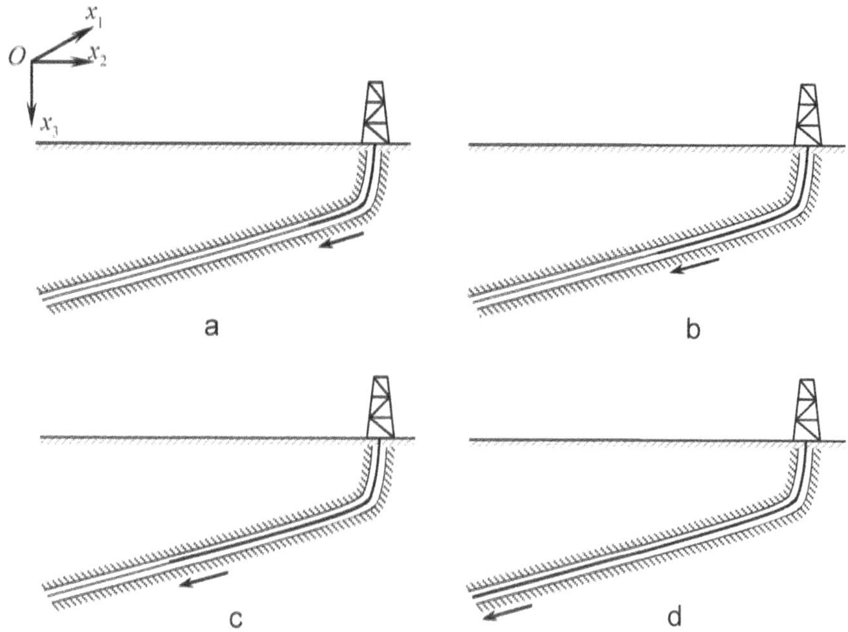

Figure 9.1. *Schematics of the drill string descending in a curvilinear borehole*

Therewith, the distributed external gravity ($\mathbf{f}^{gr}(l)$), contact ($\mathbf{f}^{cont}(l)$) and friction ($\mathbf{f}^{fr}(l)$) forces, as well as internal elastic forces $\mathbf{F}(l)$ and moments $\mathbf{M}(l)$ are acting on every selected element of length Δl (Figure 9.2). Here, l is the natural parameter measured by the axis line length, from some initial point until the considered one.

At the precritical state, the drill string geometry coincides with the geometry of the borehole axis. At the buckling, the DS displaces laterally in the borehole clearance but saves its contact interaction. To deduce the constitutive equations of the DS buckling, first, formulate the nonlinear equations of its equilibrium.

We take into account that the surface of the wall of the well is channel and its cross-section is circular. Then, during deformation, the axial line S of the column will lie on the channel surface Σ of the radius a, equal to the value of the annular gap (clearance).

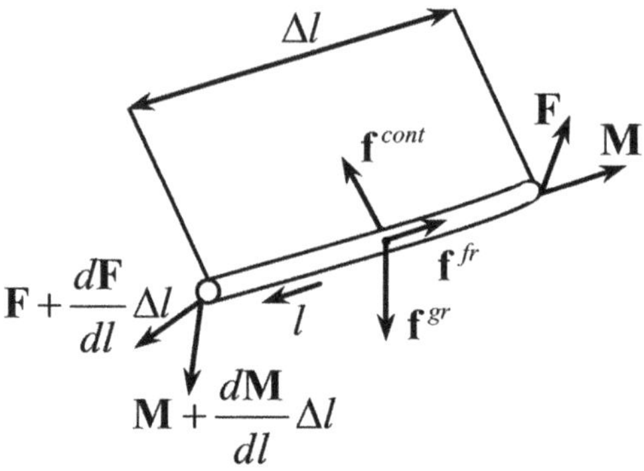

Figure 9.2. *System of forces acting on the drill string*

Parameterize this surface with parameters u, v, w (Figure 9.3), where $0 \leq v \leq 2\pi$, u and w are the natural parameters.

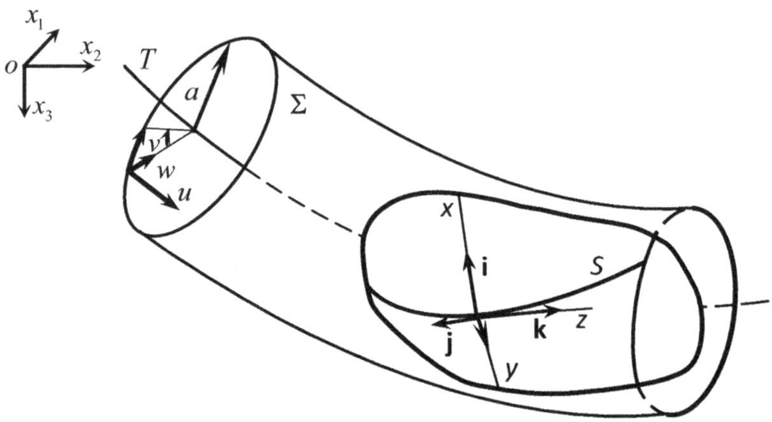

Figure 9.3. *Parameterization of curve S and surface Σ*

Use fixed coordinate system $Ox_1x_2x_3$ to define axial line T

$$\mathbf{r}(u) = x_1(u)\mathbf{i}_1 + x_2(u)\mathbf{i}_2 + x_3(u)\mathbf{i}_3 \qquad [9.9]$$

where $\mathbf{i}_1$, $\mathbf{i}_2$ and $\mathbf{i}_3$ are the unit vectors of coordinate axes.

In the general case, line T can be both three- and two-dimensional. For the sake of simplicity, assume that it is two-dimensional and lies in plane $x_1 = 0$. Then, the surface Σ equation can be represented in the form:

$$\begin{aligned} x_1(u,v) &= a \sin v, \\ x_2(u,v) &= x_2(u) - a\cos v \cdot \sin\left[\varphi(u)\right], \\ x_3(u,v) &= x_3(u) + a\cos v \cdot \cos\left[\varphi(u)\right]. \end{aligned} \qquad [9.10]$$

Here, $\varphi = \operatorname{arctg}(\partial x_2/\partial x_3)$ is the angle of the tangent inclination.

Then, lines $u = const$, $v = const$ make up the curvilinear coordinate system on the surface Σ. In doing so, lines $u = const$ represent circles of radius a and line $v = 0$ lies on the bottom of the channel Σ.

We state the basic correlations defining the geometry of an elastic rod axis S. Its trajectory in the coordinate system $Ox_1x_2x_3$ is described by equation [9.9]. We write the principal differential equalities used for constructing the constitutive equations. We introduce the unit vectors of Frenet:

$$\boldsymbol{\tau} = \frac{d\mathbf{r}}{dl}, \quad \mathbf{n} = R\frac{d^2\mathbf{r}}{dl^2}, \quad \mathbf{b} = \boldsymbol{\tau} \times \mathbf{n} \qquad [9.11]$$

where $\boldsymbol{\tau}$, $\mathbf{n}$ and $\mathbf{b}$ are, respectively, the unit vectors of the tangent, normal and binormal to curve S, and R is the curvature radius.

Proceeding from the condition of the element equilibrium under external distributed forces $\mathbf{f}(l)$, internal elastic forces $\mathbf{F}(l)$ and moments $\mathbf{M}(l)$, we can obtain differential equations

$$\frac{d\mathbf{F}}{dl} = -\mathbf{f}, \quad \frac{d\mathbf{M}}{dl} = -\boldsymbol{\tau} \times \mathbf{F}. \qquad [9.12]$$

Here,

$$\mathbf{f} = \mathbf{f}^{gr} + \mathbf{f}^{cont} + \mathbf{f}^{fr} \qquad [9.13]$$

A characteristic feature of the problem of further transformation and solution of system [9.12] is the definition of vector $\mathbf{f}(l)$. The fact is that the column is pressed to the surface Σ of the well wall by gravity forces $\mathbf{f}^{gr}$ and cannot separate from it.

Therefore, during deformation, the geometry of the column does not change arbitrarily and is determined by the surface of the wall on which it lies. So, in the analysis of stability, methods of differential geometry and the theory of surfaces should also be used. Assume that at buckling, the elastic elongations of the DS can be neglected and its lateral displacements are small. Then, its geometry during buckling can be written as follows:

$$\delta u(l) \approx 0, \;\; v(l) = \delta v(l), \;\; \delta w(l) = 0 \qquad [9.14]$$

When analyzing the 3D bending of curvilinear rods, usually, equations [9.12] are formulated in a movable axis system (the Frenet trihedron with Darboux vectors (Gulyayev *et al.* 2019)). But in our event, taking into account that additional constraints are imposed on the rod, it is more useful to use the generalized vector of Darboux

$$\boldsymbol{\omega} = k_x \mathbf{i} + k_y \mathbf{j} + k_z \mathbf{k}. \qquad [9.15]$$

where the unit vector $\mathbf{i}$ is directed along the internal normal to the surface Σ, the unit vector $\mathbf{j}$ is tangent to the curve S, the unit vector $\mathbf{k}$ adds the system to the right-hand trihedron, k_x and k_y are the appropriate components of the curvature vector $\mathbf{K} = \mathbf{b}/R$, and k_z determines the rotation of system $oxyz$ around the vector $\mathbf{k}$, when it moves along the line S with unit velocity.

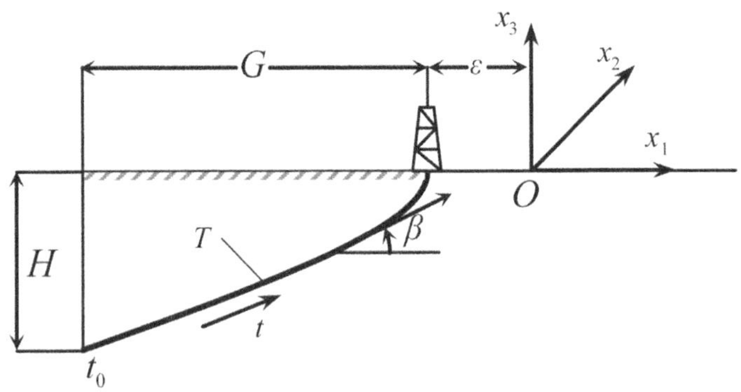

Figure 9.4. *Geometric schematic of an inclined borehole*

Further, to formulate the problem, consider the case when the borehole trajectory T represents a hyperbolic curve (Figure 9.4) characterizing its most representative properties of its geometry. Indeed, it has a vertical tangent at the upper point and it approaches a rectilinear line approximating the hyperbola asymptote.

In the fixed coordinate system $Ox_1x_2x_3$, axial line T is assigned by equalities

$$x_1 = -\varepsilon \operatorname{ch} t\,,\ x_2 = 0\,,\ x_3 = \lambda \operatorname{sh} t. \qquad [9.16]$$

Here, t is the dimensionless parameter, ε is the hyperbola semi-axis and λ is the value characterizing the borehole depth.

To find initial value t_0, we use equalities

$$x_1(t_0) = -\varepsilon \operatorname{ch} t_0 = -(G+\varepsilon),\ \ x_2(t_0) = 0,\ \ x_3(t_0) = \lambda \operatorname{sh} t_0 = -H \qquad [9.17]$$

from Figure 9.4 and equation [9.16].

Therefore,

$$t_0 = \operatorname{arcch}(\varepsilon + G) < 0,\ \ \lambda = -H/\operatorname{sh} t_0 > 0 \qquad [9.18]$$

As,

$$dx_1 = \frac{dx_1}{dt}dt, \qquad [9.19]$$

we have

$$\beta = \operatorname{arctg}\left(\frac{\dot{x}_3}{\dot{x}_1}\right) = \operatorname{arctg}\left(-\frac{\lambda}{\varepsilon}\operatorname{cth} t\right) \qquad [9.20]$$

where the dot over a symbol denotes differentiation with respect to t.

Now, it becomes convenient to describe the borehole geometry with the use of parameter t, but solve equations [9.12] of the DS equilibrium based on parameter l, switching from one to another using the correlation (Gulyayev *et al.* 2019)

$$dl = Ddt = \sqrt{(\varepsilon \operatorname{sh} t)^2 + (\lambda \operatorname{ch} t)^2}\,dt, \qquad [9.21]$$

where D is the scaling multiplicator.

Taking into account the adduced consideration, equations [9.12] can be reduced to the basis $\boldsymbol{\omega}$, where

$$\frac{d\mathbf{F}}{dl}=\frac{\tilde{d}\mathbf{F}}{dl}+\boldsymbol{\omega}\times\mathbf{F},\quad \frac{d\mathbf{M}}{dl}=\frac{\tilde{d}\mathbf{M}}{dl}+\boldsymbol{\omega}\times\mathbf{M}, \qquad [9.22]$$

and the symbol $\sim$ denotes the operation of local differentiation.

Then, in a movable trihedron, equations [9.12] are converted to the form:

$$\frac{\tilde{d}\mathbf{F}}{dl}=-\boldsymbol{\omega}\times\mathbf{F}-\mathbf{f},\quad \frac{\tilde{d}\mathbf{M}}{dl}=-\boldsymbol{\omega}\times\mathbf{M}-\boldsymbol{\tau}\times\mathbf{M}. \qquad [9.23]$$

In this case, the scalar nonlinear equations of the DS element equilibrium are represented as follows:

equations of force equilibrium

$$dF_x/dl=-k_yF_z+k_zF_y-f_x,$$

$$dF_y/dl=-k_zF_x+k_xF_z-f_y, \qquad [9.24]$$

$$dF_z/dl=-k_xF_y+k_yF_x-f_z,$$

equations of moment equilibrium

$$dM_x/dl=-k_yM_z+k_zM_y+F_y,$$

$$dM_y/dl=-k_zM_x+k_xM_z-F_x, \qquad [9.25]$$

$$dM_z/dl=-k_xM_y+k_yM_x.$$

These equations can be applied to the analysis of DS buckling using a two-stage approach. In the first stage, they are integrated in the vicinity of undeformed stage $u=0$, under the action of external forces $\mathbf{f}^{gr}(l)$, $\mathbf{f}^{cont}(l)$, $\mathbf{f}^{fr}(l)$ and internal forces $F_x(l)$, $F_y(l)$, $F_z(l)$. In the second stage, equations [9.24] and [9.25] are linearized in the vicinity of the identified preloaded stage, a system of homogeneous linearized equations is constructed and its degeneracy is verified. If it is degenerated, then it has a non-trivial (non-nil) solution, equal to the trivial one. This state is critical, and this solution determines the mode of buckling. Nevertheless, if the system is not degenerated, then it only has a trivial solution, the state is not critical and the analysis should be continued after the variation of external forces.

Suppose that the DS is tripping in under the action of gravity forces (Figures 9.1–9.3)

$$\mathbf{f}^{gr}(l) = -f^{gr}(l)\mathbf{i}_3 = f_x^{gr}(l)\mathbf{i} + f_y^{gr}(l)\mathbf{j} + f_z^{gr}(l)\mathbf{k} \qquad [9.26]$$

Together with boundary axial forces $F_k(0)\mathbf{k}$ and $F_k(L)\mathbf{k}$, the gravity force generates the contact and friction forces

$$\mathbf{f}^{cont}(l) = f_x^{cont}(l)\mathbf{i}, \quad \mathbf{f}^{fr}(l) = -f_z^{fr}(l)\mathbf{k}, \qquad [9.27]$$

The friction force f^{fr} is determined via the Coulomb formula (Berger 2002)

$$f^{fr} = \mu\left|f^{cont}\right| \qquad [9.28]$$

As a result, the components of the external distributed forces are represented as follows:

$$f_x = f_x^{gr} + f_x^{cont}, \qquad f_y = f_y^{gr}, \qquad f_z = f_z^{gr} - \mu\left|f^{cont}\right|. \qquad [9.29]$$

If we take into account that at the initial (pre-critical) state, the DS lies on the line $v = 0$ of the bottom surface Σ, then $k_x(l) = 0$, $k_z(l) = 0$, $M_x(l) = 0$, $M_z(l) = 0$, $F_y(l) = 0$. Therefore, from equations [9.22] and [9.23], it follows:

$$\begin{aligned} &dF_x / dl = -k_y F_z - f_x^{gr} - f_x^{cont}, \\ &dF_z / dl = k_y F_x - f_z^{gr} + \mu\left|f^{cont}\right|, \\ &F_x = -EIdk_y / dl. \end{aligned} \qquad [9.30]$$

These correlations allow us to deduce the equation

$$dF_z / dl = -EIk_y dk_x / dl - f_z^{gr} + \mu\left|EId^2k_y / dl^2 - k_y F_z - f_z^{gr}\right| \qquad [9.31]$$

Here, only one unknown function $F_z(l)$ is present, so this equation can be easily integrated by the Runge–Kutta method.

In order to investigate the DS stability at the stage $v = 0$, preloaded by the axial force $F_z(l)$ found using equation [9.31], we linearize equations [9.24] and [9.25] in this state vicinity

$$\begin{aligned} d\delta F_x / dl &= -\delta k_y F_z - k_y \delta F_z + \delta k_z F_y + k_z \delta F_y - \delta f_x, \\ d\delta F_y / dl &= -\delta k_z F_x - k_z \delta F_x + \delta k_x F_z + k_x \delta F_z - \delta f_y, \\ d\delta F_z / dl &= -\delta k_x F_y - k_x \delta F_y + \delta k_y F_x + k_y \delta F_x - \delta f_z, \end{aligned} \qquad [9.32]$$

$$\begin{aligned} d\delta M_x / dl &= -\delta k_y M_z - k_y \delta M_z + \delta k_z M_y + k_z \delta M_y + \delta F_y, \\ d\delta M_y / dl &= -\delta k_z M_x - k_z \delta M_x + \delta k_x M_z + k_x \delta M_z - \delta F_x, \\ d\delta M_z / dl &= 0. \end{aligned} \qquad [9.33]$$

Here, the symbol $\delta ...$ indicates a small variation in the appropriate value.

The variables $\delta F_x(l)$ and $\delta F_y(l)$ can be represented in the form:

$$\begin{aligned} \delta F_x &= -EId\delta k_y / dl = 0, \\ \delta F_y &= EId\delta k_x / dl - EIk_y \delta k_z. \end{aligned} \qquad [9.34]$$

Here, E is the DS material elasticity module and I is the inertia moment of the DS cross-section.

It is assumed above that in buckling, the DS does not detach from the borehole wall. Owing to this, we may express DS curvatures k_x and k_y through the geodesic (k^{geod}) and normal (k^{norm}) curvatures of the surface Σ

$$k_x = k^{geod}, \qquad k_y = k^{norm} \qquad [9.35]$$

It follows from the surface theory (Gulyayev *et al.* 2019):

$$k_x = k^{geod} = \sqrt{a_{11}a_{22}} \left[a_{11}(u')^2 + a_{22}(v')^2 \right]^{-3/2} (u''v' - v''u' + Av' - Bu') \qquad [9.36]$$

Here, A and B are expressed via the Christoffel symbols of the Σ surface and the parameters of the first quadratic form are calculated as follows:

$$a_{11} = \varepsilon^2 \operatorname{sh}^2 t + \lambda^2 \operatorname{ch}^2 t, \;\; a_{22} = a^2. \qquad [9.37]$$

After some transformations, we have

$$k_x = k^{geod} = av'' - \frac{\dot{\beta} \sin v}{a_{11}} \sqrt{\varepsilon^2 \operatorname{sh}^2 t + \lambda^2 \operatorname{ch}^2 t}\,. \qquad [9.38]$$

Its variation during buckling in the vicinity of initial state $v = 0$ is

$$\delta k_x = \delta k^{geod} = a\delta v'' - \frac{\dot{\beta}}{a_{11}} \sqrt{\varepsilon^2 \operatorname{sh}^2 t + \lambda^2 \operatorname{ch}^2 t}\,\delta v\,. \qquad [9.39]$$

Analogously, functions k_y and δk_y are expressed using equalities

$$k_y = \frac{\varepsilon\lambda}{(\varepsilon^2 \operatorname{sh}^2 t + \lambda \operatorname{ch}^2 t)^{3/2}}, \qquad [9.40]$$
$$\delta k_y = (-k_1 + k_2)\sin 2\varsigma\delta\varsigma = 0$$

At the initial state, the function $\mathbf{f}^{gr}(l)$ is determined by its components

$$f_x^{gr} = -f^{gr} \cos\beta, \;\; f_y^{gr} = 0, \;\; f_z^{gr} = -f^{gr} \sin\beta. \qquad [9.41]$$

Here, $f^{gr} = \pi g(\rho_t - \rho_l)(r_1^2 - r_2^2)$; g is the gravity acceleration; ρ_t, ρ_l are the densities of the string tube material; and r_1, r_2 are the external and internal radii of the tube, respectively.

At the perturbed state, its components acquire variations

$$\delta f_x^{gr} = 0\,, \;\; \delta f_y^{gr} = f^{gr}(-a\sin\beta d\delta v / ds + \cos\beta\delta v)\,, \;\; \delta f_z^{gr} = 0 \qquad [9.42]$$

Taking into account correlations [9.34] and [9.38]–[9.41] and performing some additional transformations, systems [9.32] and [9.33] can be reduced to one fourth-order differential equation:

$$EI\frac{d^4\delta v}{dl^4}+\frac{EI}{a}\frac{d^2}{dl^2}\left\{\frac{\dot{\beta}}{a_{11}}\left(a\dot{\beta}-\sqrt{\varepsilon^2\,\mathrm{sh}^2\,t+\lambda^2\,\mathrm{ch}^2\,t}\right)\delta v\right\}-$$
$$EI(\varepsilon^2+\lambda^2)\frac{d^2}{dl^2}\left\{\frac{\mathrm{sh}\,t\,\mathrm{ch}\,t}{(a_{11})^{3/2}}\frac{d\delta v}{dl}\right\}-F_z\frac{d^2\delta v}{dl^2}+\frac{F_z(\varepsilon^2+\lambda^2)\mathrm{sh}\,t\,\mathrm{ch}\,t}{(a_{11})^{3/2}}\frac{d\delta v}{dl}-$$
$$-F_z\frac{\dot{\beta}}{aa_{11}}\left(a\dot{\beta}-\sqrt{\varepsilon^2\,\mathrm{sh}^2\,t+\lambda^2\,\mathrm{ch}^2\,t}\right)\delta v- \qquad [9.43]$$
$$-f^{gr}\left(-\sin\beta d\delta v/ds+\frac{\cos\beta}{a}\delta v\right)=0 \qquad 0\leq l\leq L.$$

This equation can only be integrated using numerical methods. Its characteristic feature consists of the fact that it includes two "independent" parameters l and t, simultaneously. The first is used for the differentiating operation, while the second is used for the description of the equation coefficients. But they are not absolutely independent, as they are connected by equations [9.21]. Therefore, at integration, first, transition from t to l is performed and next, integration with respect to l is realized. As noted above, coefficients of this equation depend on unknown internal force $F_z(l)$. To determine this force, it is preliminarily necessary to solve equation [9.31], resulting in an additional problem.

The second feature is more important. It stems from the existence of a nonlinear constraint in the form of a channel surface and is associated with the structure of equation [9.43], which contains both unknown functions $\delta v(l)$ and fourth derivative $d^4\delta v/dl^4$. Owing to this property, it becomes, in some sense, similar to equation [9.5], determining the equilibrium of a beam on the elastic foundation considered in section 1.2. If equation [9.43] is integrated in a large length, it also becomes singularly perturbed and its eigenmodes come to be in the shape of localized wavelets. Positions of their localizations are not known beforehand. For this reason, we cannot separate the singularity for independent study and the problem should be solved using the global statement.

Because of this reason, typical computational difficulties connected with bad convergence of numerical methods and the necessity to use large numbers of approximating variables appear in these cases. To overcome these severities, a small increment of load parameter and small steps of finite difference discretization are used.

9.4. Modeling the drill string buckling in lowering operation

Let the borehole have the trajectory represented by equation [9.16] (Figure 9.4). By selecting different values of parameters ε, λ, and t_0, it is possible to modify the outline of its profile. Suppose that a DS is lowering inside this borehole (Figure 9.1). The friction forces $f^{fr}(l)$ counteract this operation and compress the DS. There is a need to analyze the stability and buckling of the DS at different stages of its descending. It is obvious that these tasks cannot be solved by analytical methods; therefore, in our case, the finite difference method was used. Moreover, the stated problem is multi-parametric, as the buckling process depends on the values of the string tube bending stiffness EI, borehole curvature, inclination angle β, its length L, clearance a and friction coefficient μ. In this chapter, the values $E = 2.1 \cdot 10^{11}$ Pa, $I = 2.7 \cdot 10^{-4}$ m^4, $\rho_t = 7.8 \cdot 10^3$ kg/m^3, $\rho_l = 1.3 \cdot 10^3$ kg/m^3, $r_1 = 0.1$ m, $r_2 = 0.09$ m, and $\varepsilon = 2$ km were chosen. Two types of borehole outlines determined by depth $H = 2$ km and 2.5 km and extension $G = 6$ km and 7.5 km (Figure 9.4) were considered.

H (km)	G (km)	DS length	$\mu = 0$		$\mu = 0.3$	
			$F_z^{ext}(l_0)$, kN	F_z^{ext}, kN	$F_z(l_0)$, kN	F_z^{ext}, kN
2	6	$L/4$	228.807	–	225.313	–
		$L/2$	228.337	–	221.909	224.508
		$3L/4$	228.250	–	214.929	224.387
		L	228.318	–	196.417	224.543
2.5	7.5	$L/4$	227.776	–	224.619	–
		$L/2$	227.608	–	221.919	224.077
		$3L/4$	227.634	–	211.899	224.053
		L	227.666	–	198.435	224.072

Table 9.1. *Critical values of the axial force F_z*

The calculated critical axial forces F_z^{ext} are shown in Table 9.1 for two combinations of parameters H and G. They conform to the cases when the DS is partially lowered into the well with its different parts $L/4$, $L/2$, $3L/4$ and L of its full length L (Figure 9.1). The results testify that critical boundary values $F_z^{ext}(l_0)$ do not depend on the DS length (case $\mu = 0$), but if the friction forces are taken into account ($\mu = 0.3$), then the buckling can occur inside the DS length. In doing so, critical values F_z^{ext} essentially do not change. They correspond to the buckling modes shown in Figures 9.5 and 9.6.

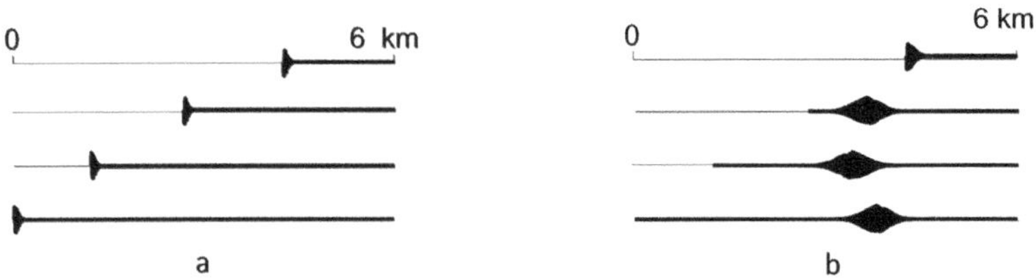

Figure 9.5. *Buckling modes of the DS at different stages of its lowering (H = 2 km, G = 6 km); a - μ = 0, b - μ = 0.3*

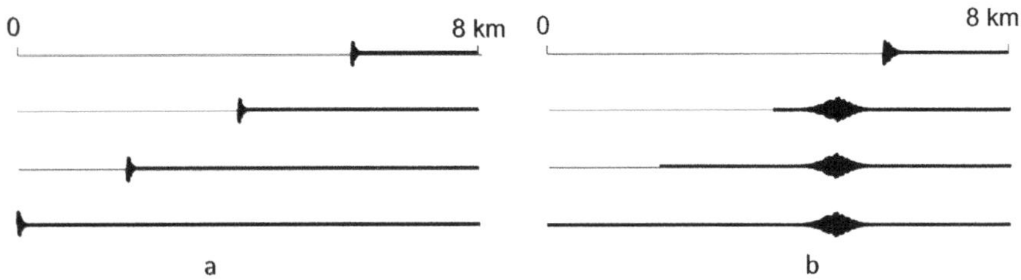

Figure 9.6. *Buckling modes of the DS at different stages of its lowering (h = 2.5 km, G = 7.5 km); a - μ = 0, b - μ = 0.3*

The research results indicate that if there is no frictional interaction between the string and the bore $(\mu = 0)$, then the greatest value of the axial force $(F(l_0))$ is realized at the end $l = l_0 = 0$ of the string, and it is equal to the force acting on the bit (Figures 9.5(a) and 9.6(a)). In this case, the singular perturbation takes place in the form of a boundary effect. If the DS descent regime is such that frictional effects ($\mu = 0.3$) are generated, then, depending on the angle β of the column inclination, the axial friction force f_z^{fr} can exceed the axial component of the gravitational force f_z^{gy}. Then, the friction forces lead to the redistribution of not only external, but also internal forces, and the extreme value F_z^{ext} of the internal axial compressive force is generated in the inner zone of the column. As shown by global analysis, the bifurcation wavelet of the buckle is localized in the vicinity of this point (Figures 9.5(b) and 9.6(b)). In this case, the extreme value F_z^{ext} of the axial force exceeds the axial force $F_z(l_0)$ acting on the bit (Table 9.1) and its position changes with the length of the column lowering into the well (Figure 9.5(a)), also resulting in the shifting of location of the bifurcation wavelet (Figures 9.5(b) and 9.6(b)).

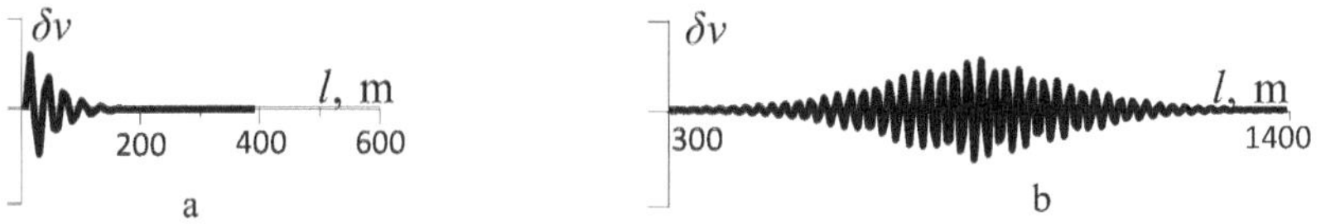

Figure 9.7. *Modes of buckling in the shapes of boundary effect (a) and internal wavelet (b)*

In Figure 9.7, fragments of stability loss realized in the form of a boundary effect (a) and an internal wavelet (b) are presented. We emphasize that since the zone where the extreme values of the force F_z^{ext} are unknown in advance, the position of the buckling zone is also unknown. Therefore, despite the fact that the bifurcation displacements δv outside of this zone are almost equal to zero, the solution of the problem cannot be limited to the analysis of any small section of the column and it is necessary to integrate the resolving equations on the full length of the column lowered into the well. Such an approach corresponds to a global analysis of a singularly perturbed problem.

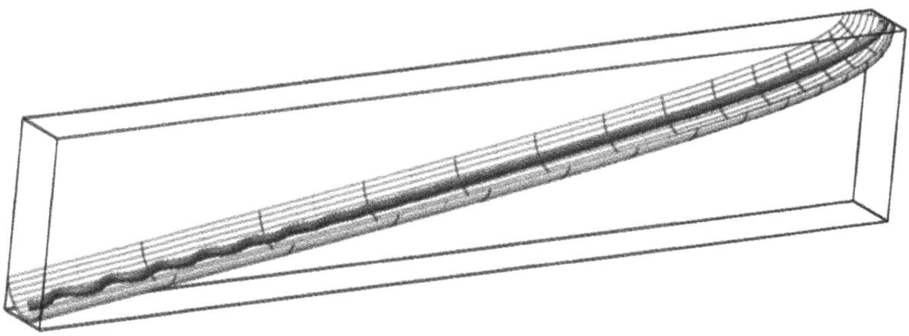

Figure 9.8. *Typical 3D schematic fragment of the drill string buckling wavelet in the inclined low boundary zone (frictionless model)*

Inferring these results, it can be noted that if the lower section of the hole is inclined and the friction effects are removed, the maximal values of the compressive axial forces occur at the bottom zone of the DS. In these cases, the buckling mode represents a boundary effect in the shape of a short wavelet localized in the bottom zone of the drill string, where the hole trajectory is near-rectilinear. The appropriate mode is schematically shown in Figure 9.8. This case is not very difficult for analysis. If during tripping in operation, the friction forces are generated, then they act opposite to the motion direction and the maximal axial force shifts upward. Therefore, buckling wavelets also shift upward. This situation becomes more serious. In this case, the maximal compressive force $F_z(l)$ is generated in the knee zone and therefore, the buckling wavelets are born here (Figure 9.9).

The contact interaction of the DS with the borehole channel wall surface is characterized by essential obstacles conditioned by the geometrical complexity of the interface surface. In this case, the DS stability analysis should especially be performed.

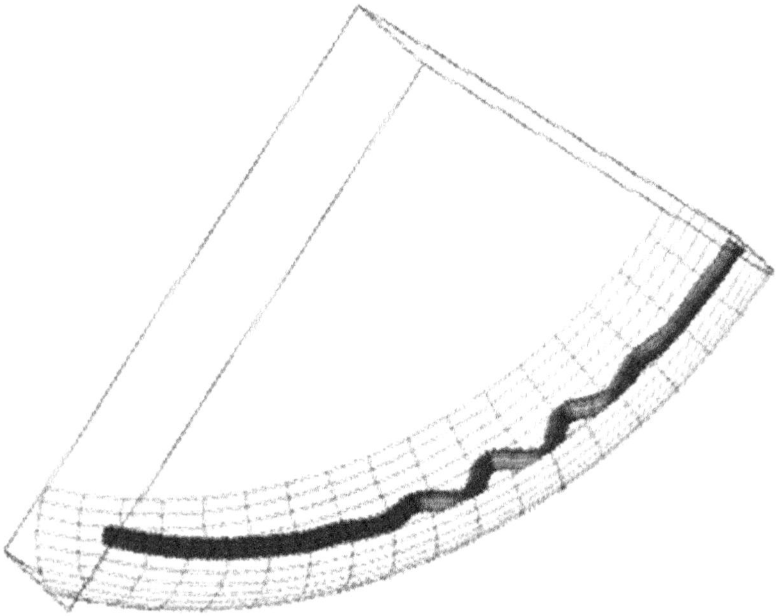

Figure 9.9. *Typical 3D schematic fragment of the drill string buckling wavelet in the knee zone (frictional model)*

Thus, the elaborated techniques for global computer analysis of DS buckling in the curvilinear boreholes allow us to calculate the critical values of mechanical parameters, as well as to find the modes of stability loss and the places of their localizations.

9.5. References

Acosta, J.A. and Lopez-Martinez, M. (2008). On singular perturbation of unstable underactuated mechanical systems with underactuation degree ≥ 1. *IFAC Proceedings Volumes*, 41(2), 5998–6003.

Berger, E.J. (2002). Friction modeling for dynamic system simulation. *Appl. Mech. Rev.*, 55(6), 535–576.

Brett, J.F., Beckett, A.D., Holt, C.A., Smith, D.L. (1989). Uses and limitations of drillstring tension and torque models for monitoring hole conditions. *SPE Drill. Eng.*, 4, 223–229.

Challamel, N. (2000). Rock destruction effect on the stability of a drilling structure. *J. Sound Vib.*, 233(2), 235–254.

Chang, K.W. and Howes, F.A. (1984). *Nonlinear Singular Perturbation Phenomena*. Springer-Verlag, New York.

Cunha, J.C. (2004). Buckling of tubulars inside wellbore: A review on recent theoretical and experimental works. *SPE Drill. Complet.*, 19(1), 13–19.

Dawson, R. and Paslay, R.R. (1984). Drill pipe buckling in inclined holes. *J. Pet. Technol.* 36(10), 1734–1738.

Elishakoff, I., Li, Y., Starnes Jr., J.H. (2001). *Non-Classical Problems in the Theory of Elastic Stability*. Cambridge University Press, Cambridge.

Farrell, P.A., Hegarty, A.F., Miller, J.J.H., O'Riordan, E., Shishkin, G.I. (2000). *Robust Computational Techniques for Boundary Layers*. Chapman and Hall, CRC Press, London, Boca Raton.

Gao, D.L. and Huang, W.J. (2015). A review of down-hole tubular string buckling in well engineering. *Pet. Sci.*, 12(3), 443–457.

Gulyayev, V.I. and Shlyun, N.V. (2016). Influence of friction on buckling of a drill string in the circular channel of a bore hole. *Pet. Sci.*, 13, 698–711.

Gulyayev, V.I. and Shlyun, N.V. (2017). Global analysis of drill-string buckling in the channel of a curvilinear bore-hole. *J. Nat. Gas Sci. Eng.*, 40, 168–178.

Gulyayev, V.I., Glazunov, S., Glushakova, O., Vashchilina, E., Shevchuk, L., Shlyun, N.V., Andrusenko, E. (2019). *Modelling Emergency Situations in the Drilling of Deep Boreholes*. Cambridge Scholars Publishing, Newcastle upon Tyne.

Holmes, M.H. (1995). *Introduction to Perturbation Methods*. Springer-Verlag, New York.

Johnson, R.S. (2005). *Singular Perturbation Theory: Mathematical and Analytical Techniques with Applications to Engineering*. Springer-Verlag, New York.

Krauberger, N., Saje, M., Planinc, I., Bratina, S. (2007). Exact buckling load of a restrained RC column. *Struct. Eng. Mech.*, 27(3), 293–310.

Miller, J.J.H., O'Riordan, E., Shishkin, G.I. (1996). *Fitted Numerical Methods for Singular Perturbation Problems*. World Scientific, Singapore.

Mitchell, R.F. (2008). Tubing buckling. The state of art. *SPE Drill. Complet.*, 25(4), 361–370.

Ouland Ben Zarouala, R., Vivas, C., Acosta, J.Á., and El Bakkali, L. (2012). On singular perturbations of flexible and variable-speed wind turbines. *International Journal of Aerospace Engineering*, 1687–5966.

Roos, H.G., Stynes, M., Tobiska, L. (1996). *Numerical Methods for Singularly Perturbed Differential Equations*. Springer-Verlag, New York.

Shlyun, N.V. and Gulyayev, V.I. (2020). Buckling of a drill-string in two-sectional bore-holes. *Int. J. Mech. Sci.*, 172, 105427.

Smith, D.R. (1985). *Singular-Perturbation Theory*. Cambridge University Press, Cambridge.

Sun, C. and Lukasiewicz, S. (2006). A new model on the buckling of a rod in tubing. *J. Petr. Sci. Eng.*, 50, 78–82.

Verhulst, F. (2005). *Methods and Applications of Singular Perturbation: Boundary Layers and Multiple Timescale Dynamics*. Springer-Verlag, New York.

Wang, X. and Yuan, Z. (2012). Investigation of frictional effects on the nonlinear buckling behavior of a circular rod laterally constrained in a horizontal rigid cylinder. *J. Petr. Sci. Eng.*, 90–91, 70–78.

10

Shape-optimized Cantilevered Columns under a Rocket-based Follower Force

This chapter discusses the dynamic stability of shape-optimized cantilevered columns subjected to a follower force. The stability analysis and shape optimum design of the columns are described for giving theoretical predictions of the critical (flutter) loads of the test columns. Optimum designs are found by both maximization of the critical load under the constraint of a constant volume and by minimization of volume under the constraint of critical load. The main objective of this chapter is concerned with an experimental verification of optimum shape design of these columns. Referring to the approach of earlier experimental works by Sugiyama *et al.* (1995a, 1995b, 2000, 2012), the procedure of test runs for the present experiment is formulated. The follower force is realized by the thrust of a small solid-propellant rocket motor. Some rather wide, plate-like columns were used, in order to amplify the significance of designed optimum shape. Optimally shaped test columns with smoothly varying cross-sections were specially manufactured for the present experiment. With these columns, it turned out that the experimental flutter bound was significantly lower than the predicted value by theory. However, a comparison of the dynamic responses of the uniform/initial column and the optimally shaped column with same volume shows that the optimal tapering clearly stabilizes the vibrations. Three extra test runs were conducted to find the actual critical bound of a uniform plate-like column. This chapter discusses a possible cause of the apparent gap between theory and experiment.

10.1. Background

The dynamics of an elastic cantilevered beam subjected to a tangential follower force, the so-called Beck's column, has been one of the most speculative models in

---

Chapter written by Yoshihiko SUGIYAMA, Mikael A. LANGTHJEM and Kazuo KATAYAMA.

the field of applied mechanics. Beck (1952), with a cantilevered column under a follower force, and Ziegler (1952) with a double pendulum model under the follower force launched the study on dynamics of simple structures subjected to a follower force. Bolotin (1963) published the first monograph on the topic entitled *Nonconservative Problems of the Theory of Elastic Stability*, which encouraged many structural engineers and scientists to work on the subject. Since then, a large number of journal papers dealing with the effect of a follower force, as well as the effect of small damping, have been published. Some review articles have been published (Langthjem and Sugiyama 2000a; Venkateswara Rao and Singh 2001; Elishakoff 2005), and books (Kirillov 2013; Bigoni and Kirillov 2019; Sugiyama *et al.* 2019) on recent advances of the study of simple structures under follower loads.

A considerable amount of work has been done on determining the optimum shape design of Beck's column since initial studies by Claudon (1975), Odeh and Tajdbakhsh (1975) and Plaut (1975). Figure 10.1 shows a simple image of the shape optimization of Beck's column. Formulated as a problem of maximizing the critical (flutter) load under a constant volume, the objective function (critical load) is not smooth, and most optimization algorithms tend to strand on not very high local optima. It was already noted by Plaut (1975) that load optimization under a constant volume and volume minimization under a constant critical load are dual problems, i.e. the solutions are identical except for a scaling factor. And in the volume minimization problem, the objective function is smooth. Gutskowski *et al.* (1993) were the first to use this in an optimization algorithm. At each design step, the volume was minimized followed by a rescaling to the initial volume. However, the magnitude of the critical load of their strongest column was limited by interaction of the eigenvalue branches, causing a discontinuous drop in the critical load.

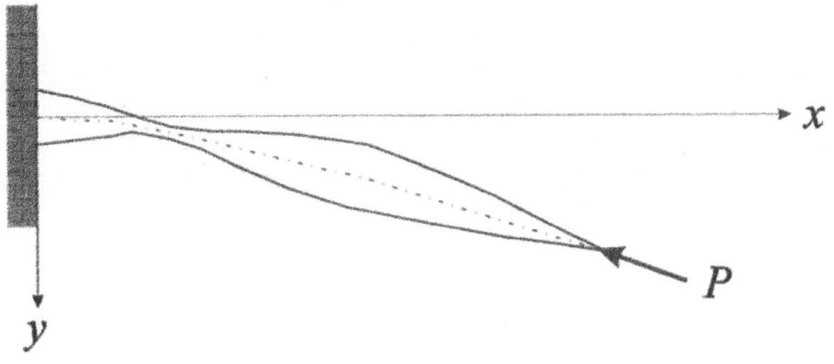

Figure 10.1. *Image of the optimally shaped Beck's column*

Hanaoka and Washizu (1980), in an earlier paper, realized the importance of keeping the eigenvalue branches apart, by imposing constraints on the distance between neighboring branches. The technique was further developed by Tada *et al.* (1988) and used by Ringertz (1994) in connection with the volume minimization formulation; while Hanaoka and Washizu (1980) and Tada *et al.* (1988) considered the critical load formulation. By scaling up the minimum volume design to the initial volume, Ringertz (1994) obtained the strongest, to date and to the authors' knowledge, optimized Beck's column with a critical (flutter) load of 188.07 EI_0/L^2, where EI_0 is the bending stiffness and L is the column length. Recalling the critical load for Beck's column, $20.05EI_0/L^2$, Ringertz's striking value is 9.4 times as high as the load for Beck's column, the non-optimized/uniform column. Recalling that Beck's value is 8.1 times as high as the load for Euler's column under a conservative loading and Ringertz's value is 76 times higher than the critical load for Euler's column. Optimum design of cantilevered columns under the combined action of conservative and non-conservative forces has been discussed by Langthjem and Sugiyama (1999a, 2000b, 2000c).

A striking feature of Ringertz's solution is that the cross-sectional area is vanishing at the tip, implying that the curvature there becomes very large indeed. However, in reality, the follower load has to be applied through some finite-size mechanism, such as a small solid-propellant rocket motor (see Figure 10.2). An optimum design with a needle-thin tip is unrealistic.

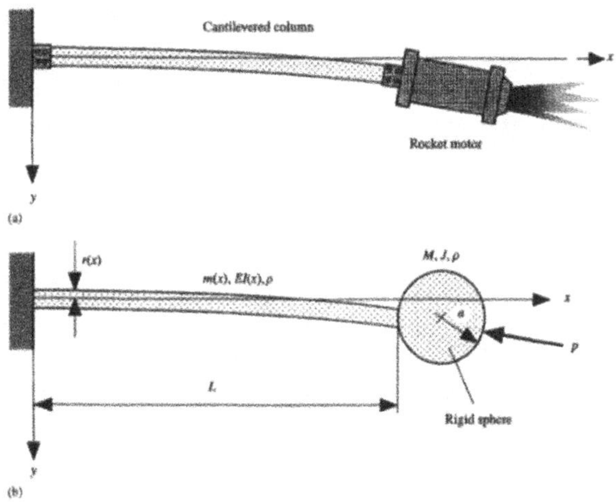

Figure 10.2. *Column subjected to a rocket-based follower force and having a rigid body at the tip (to be shape-optimized) (Langthjem and Sugiyama 1999b): (a) sketch of a realistic column and (b) mathematical model of the column with a rigid sphere at the tip*

This led to the study by Langthjem and Sugiyama (1999b) on the optimum design of Beck's column with a rigid sphere, as sketched in Figure 10.2, where the column is assumed to have circular cross-sections, to be optimized. For practical value of the ratio

$$\mu = \frac{\text{tip mass (mass of rocket motor)}}{\text{mass of column}}, \qquad [10.1]$$

with realistic rocket motors, it has been found that the critical load for the optimized columns can only get a profit by a factor 1.3–1.4. It is interesting to note that the optimum load is similar to what can be obtained in the case of columns under a conservative load (Tadjbakhsh and Keller 1962; Ishida and Sugiyama 1995), and certainly very different from the factor 9.4 obtained in the speculative case of zero tip mass (Ringertz 1994).

10.2. Aims

The primary aim of this chapter is to describe an experiment with a series of test runs planned to verify the reality of improved performance by shape optimization of cantilevered columns under a rocket-based follower force. The columns are designed optimally by both maximization of the critical load under the constraint of a constant volume (mass of the column), and minimization of the volume under the constraint of a constant critical load. The follower force was provided in the form of thrust by a small solid-propellant rocket motor mounted to the tip of the test columns, as shown in Figure 10.3. Sugiyama *et al.*, in collaboration with Daicel Corporation (Harima Plant, Tatsuno, Japan), have had significant experience with these physical systems, and three successful experiments have been conducted (Sugiyama *et al.* 1995a, 1995b, 2000, 2019).

The aim of the first experiment with a rocket-based follower force was to show that the cantilevered column under a follower force loses stability by flutter, to investigate the much-discussed effect of (structural) damping on flutter, and to verify the theoretical prediction of the critical load. A relaxed stability criterion was introduced, which takes the finite (short) duration of the follower load into account (Sugiyama *et al.* 1995a). The predicted stability limit with the proposed criterion is almost identical to the limit obtained from the usual Lyapunov criterion, if damping is neglected. One of the important conclusions of Sugiyama *et al.* (1995a) is thus that the destabilizing effect of small damping is not of any significance when the load duration is short, while the asymptotic/mathematical stability criteria defines the stability in an infinite time interval.

The aim of the second experiment with a rocket-based follower force was to discuss the effect of a lumped mass mounted on the column (Sugiyama *et al.* 1995b). It was found that a lumped mass mounted on the column at its mid span destabilizes the column.

In the third experiment with a rocket-based follower force, the test runs demonstrated that a follower load may dynamically re-stabilize a standing column buckled under a dead load due to self-weight of the motor at the tip, although the total compressive load (the buckling load + the follower thrust) is significantly increased (Sugiyama *et al.* 2000).

Through these three experiments, some interesting features of the effect of a follower force on the dynamics of modified versions of Beck's column have been demonstrated experimentally. The three experiments have been described in detail in the recent book by Sugiyama *et al.* (2019), and two recordings of the dynamic responses observed in the series of test runs have been released as online supplemental materials of the book[1].

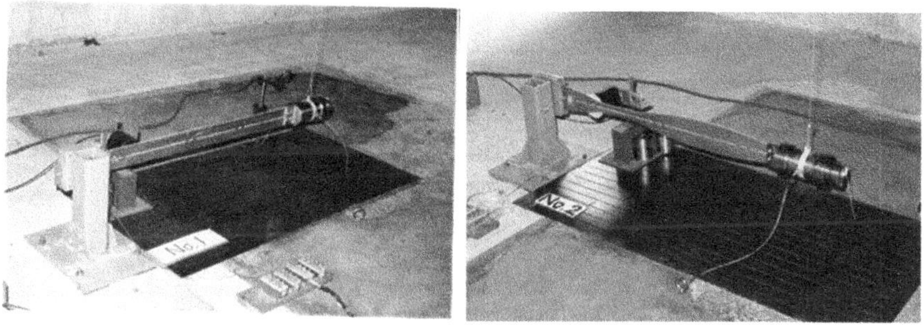

a) b)

Figure 10.3. *Photographs of the experimental setups for the present experiment: (a) setup with a uniform column and (b) setup with a shape-optimized column. For a color version of this figure, see www.iste.co.uk/challamel/mechanics1.zip*

In continuation of these experiments, the aim of the present experiment is to give experimental validation of the stabilizing effect of the shape optimum design of columns under a rocket-based follower force. The setups in the experiment are shown in Figure 10.3.

1 Springer Extra Materials Archive: https://extras.springer.com/

10.3. Numerical analysis

10.3.1. *Stability analysis*

This section is concerned with the stability of small motion about the equilibrium position. Rotatory inertia and shear effect of the column are ignored in the mathematical model, as shown in Figure 10.4. Damping is also ignored. As already touched upon in sections 10.1 and 10.2, this might be questioned in light of the well-known "destabilizing effect of small damping" (Ziegler 1952; Herrmann and Jong 1965; Bolotin and Zhinzher 1969). However, through previous experimental and theoretical studies (Sugiyama *et al.* 1995a, 1995b, 2000), it has been verified that the destabilizing effect of small damping is not manifested when the load duration is short, just like in the case of the present rocket thrust applied test, and that a mathematical model without damping gives the correct stability limit (see also the discussions in Langthjem and Sugiyama (1999b) and (2000a)).

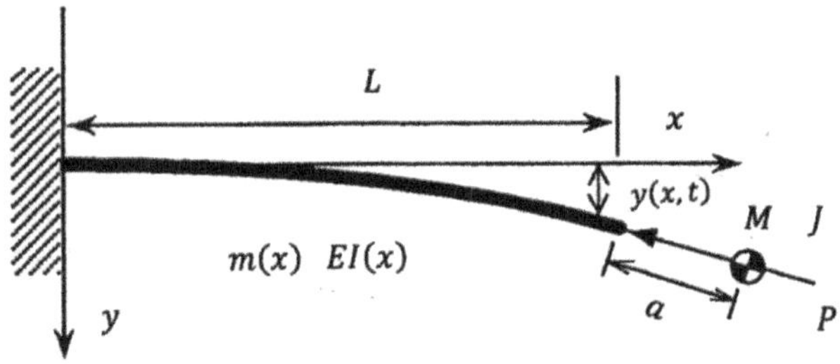

Figure 10.4. *Mathematical model*

The equation of motion can then be written as

$$m\frac{\partial^2 y}{\partial t^2} + \frac{\partial^2}{\partial x^2}\left\{EI\frac{\partial^2 y}{\partial x^2}\right\} + P\frac{\partial^2 y}{\partial x^2} = 0, \qquad [10.2]$$

where $y(x,t)$ is the lateral displacement as a function of station x and time t. $m = m(x)$ is the mass distribution function, $EI = EI(x)$ is the variable bending stiffness, and P is the applied rocket thrust.

The rotatory inertia of the rocket motor is significant, even though the flutter frequency is low, as will be seen later. Taking this into account, the boundary conditions are

$$y = 0, \qquad \frac{\partial y}{\partial x} = 0, \qquad \text{at } x = 0,$$

$$EI\frac{\partial^2 y}{\partial x^2} + M\frac{\partial^2}{\partial t^2}\left(y + a\frac{\partial y}{\partial x}\right)a + J\frac{\partial^2}{\partial t^2}\left(\frac{\partial y}{\partial x}\right) = 0,$$

and

$$\frac{\partial}{\partial x}\left\{EI\frac{\partial^2 y}{\partial x^2}\right\} - M\frac{\partial^2}{\partial t^2}\left(y + a\frac{\partial y}{\partial x}\right) = 0, \text{ at } x = L, \qquad [10.3]$$

where M is the nominal (average during burning) mass of the rocket motor, J is the rotatory inertia of the motor about an axis through its center, and a is the distance from the tip end of the column to the center of gravity of the motor.

The variables x and t are separated according to

$$y(x,t) = \tilde{y}(x)\exp(\lambda t), \quad \lambda = \alpha + i\omega. \qquad [10.4]$$

With the non-dimensional variables

$$\bar{x} = \frac{x}{L}, \quad \bar{y} = \frac{\tilde{y}}{L}, \quad \bar{p} = \frac{PL^2}{EI_0}, \quad \mu = \frac{M}{m_0 L}, \quad \bar{m} = \frac{m}{m_0},$$

$$\bar{s} = \frac{I}{I_0}, \quad \bar{a} = \frac{a}{L}, \quad \bar{J} = \frac{J}{m_0 L^3}, \quad \bar{t} = \frac{t}{L^2}\sqrt{\frac{EI_0}{m_0}}, \quad \bar{\lambda} = \frac{\lambda t}{\bar{t}}, \qquad [10.5]$$

equations [10.2] and [10.3] are converted into the following non-dimensional boundary value problem:

$$By \equiv \lambda^2 my + \{sy''\}'' + py'' = 0,$$

$$y = 0 \text{ and } y' = 0 \quad \text{at } x = 0,$$

$$sy'' + \lambda^2(J + a^2\mu)y' + \lambda^2 a\mu y = 0$$

and

$$\{sy''\}' - \lambda^2\mu\,(ay' + y) = 0 \text{ at } x = 1, \qquad [10.6]$$

wherein the differential operator B has been defined. All the "overbars" have been dropped, for the sake of simple notation. By making use of the adjoint eigenvalue problem, obtained from the identity

$$\int_0^1 [vBy - yB^*v]\,dx = 0, \qquad [10.7]$$

we can set up the functional

$$L(y,v) = \lambda^2 \left[\int_0^1 myvdx + \mu\{yv + a(yv' + y'v) + (J/\mu + a^2)y'v'\}_{x=1} \right]_M$$

$$+ \left[\int_0^1 sy''dx \right]_S + p\left[\{y'v\}_{x=1} - \int_0^1 y'v'dx \right]_Q, \qquad [10.8]$$

where v is the eigenvector of the adjoint system. The functional [10.8] is stationary with respect to variations $\delta y, \delta v$ just satisfying the kinematic boundary conditions $y(0) = 0$ and $y'(0) = 0$ (and $v(0) = 0,\ v'(0) = 0$). The functional is discretized via the finite element method. The square bracket with index M gives the symmetric mass matrix $\mathbf{M}$, the one with index S gives the symmetric stiffness matrix $\mathbf{S}$, and finally, the one with index Q gives the non-symmetric load matrix $\mathbf{Q}$. The asymmetry of $\mathbf{Q}$ is due to the first term. The discretized equation of motion can be written as

$$\mathbf{Ly} = [\lambda^2\mathbf{M}(\mathbf{m}) + \mathbf{S}(\mathbf{m}) + p\mathbf{Q}]\mathbf{y} = \mathbf{0}, \qquad [10.9]$$

where the vector

$$\mathbf{m} = \{m_1, m_2, \cdots, m_{N_e+1}\}^T \qquad [10.10]$$

represents the design variables which are the value of the mass distribution function at $N_e + 1$ nodal points. N_e specifies the number of finite elements. The mass distribution varies linearly between each element. Given the load p, the eigenvalues λ are determined numerically by using the QR method.

Flutter is the only possible type of instability for the present system. In terms of $\lambda = \alpha + i\omega$, the vibrations are stable, if all $\alpha < 0$, and unstable, if some $\alpha > 0$. In terms of λ^2, the vibrations are stable if all $\mathrm{Im}(\lambda^2) < 0$, and unstable if some $\mathrm{Im}(\lambda^2) > 0$ (where $\mathrm{Im}(z)$ denotes the imaginary part of z). However, as damping is absent in the present mathematical model, all $\alpha \equiv 0$ (and so $\mathrm{Im}(\lambda^2) \equiv 0$) for load values below the critical, p_{cr}, and the vibrations are only neutrally stable. Flutter is initiated when two eigensolutions (eigenvalues and eigenvectors) coalesce. Instability is initiated already at the coalescence point because the differential equation [10.2] here has a solution in the form

$$y(x,t) = \{y(x) + ty_a(x)\}\exp(i\omega t), \qquad [10.11]$$

where $y_a(x)$ is the so-called associated eigenfunction. After this point, i.e. at a slightly higher load value than the critical one, the before double eigenvalue splits into two complex ones in the form $\lambda_{a,b} = \pm\alpha + i\omega, \alpha \neq 0$. The oscillations thus grow exponentially in time.

Let **v** be the eigenvector of the adjoint system

$$\mathbf{L}^T\mathbf{v} = \mathbf{0}. \qquad [10.12]$$

The flutter criterion can then be written (Plaut 1972; Langthjem and Sugiyama 2000) in the form

$$\mathbf{v}_f^T\mathbf{M}\mathbf{y}_f = 0, \qquad [10.13]$$

where the index f indicates the onset of flutter. This also implies that

$$\mathbf{v}_f^T\left[\mathbf{S} + p_f\mathbf{Q}\right]\mathbf{y}_f = 0. \qquad [10.14]$$

In the numerical computations, the flutter points were determined by using a bisection method. The criterion [10.13] was used as a check.

10.3.2. *Optimum design*

The intended aim of the experiments was to verify both (I) critical load maximization, keeping the volume constant and (II) volume minimization, keeping the critical load constant. Without end mass and non-zero geometrical constraints, these two formulations are equivalent, meaning they give solutions which are identical, except for a scaling factor. This is not so for the present problem.

10.3.2.1. *Optimization problem I: maximization of critical load under a constant volume*

Let $p_1, p_2, \cdots, p_{N_e}$ be the "flutter loads" at which pairs of eigensolutions coincide, and let p_{cr} be the smallest of these values. For the uniform column, it is assumed that $p_1 < p_2 < \cdots < p_{N_e}$. Figure 10.5 defines the multiple critical loads $p_1, p_2, \cdots$. In the case shown, the critical load p_{cr} equals the minimum critical load p_1.

Optimization problem I is to determine the design vector **m** and can be posed as follows:

Maximize critical load $p_{cr} = p_{cr}(\mathbf{m})$

subject to

$$\text{i) } p_{cr} \leq \{p_1, p_2, \cdots, p_{N_e}\},$$

$$\text{ii) } m_j^l \leq m_j \leq m_j^u, \ j = 1, 2, \cdots, N_e + 1,$$

$$\text{iii) Volume} = \mathbf{l}^T \mathbf{m} = \text{constant},$$

$$\text{iv) } \omega_{n+1} - \omega_n > \begin{cases} 0 & \text{for } n = 1, 3, \cdots, 2N_e - 1, \\ c_+ & \text{for } n = 2, 4, \cdots, 2N_e - 2, \end{cases} \quad \text{for all } p <, p_{cr}. \qquad [10.15]$$

where m_j^l and m_j^u are the lower and upper bounds, respectively, on the design variables. $\mathbf{l}$ is the vector whose elements are the linear combination of the non-dimensional element length l_e:

$$\mathbf{l} = \frac{1}{2}\{l_1, l_1 + l_2, l_2 + l_3, \cdots, l_{N_e-1} + l_{N_e}, l_{N_e}\}^T \qquad [10.16]$$

and c_+ is a positive constant. With N_e elements of equal length, $l_e \equiv 1/N_e$. Figure 10.5 shows typical load–frequency curves (the frequencies $\omega_j = \text{Im}(\lambda_j)$, $j = 1, 2, \cdots, 2N_e$, as a function of the load level p), and the meaning of the minimum allowable curves distance c_+.

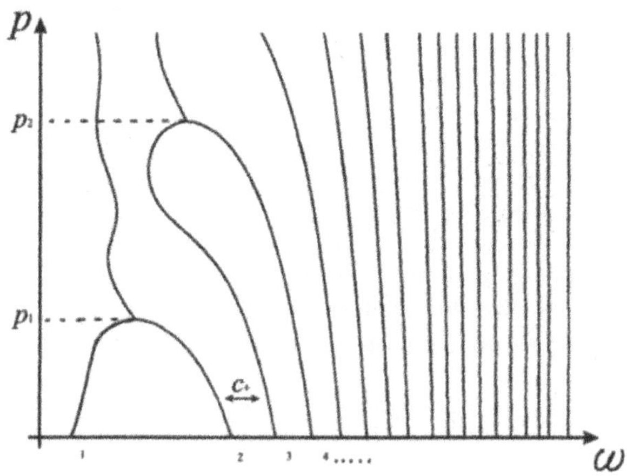

Figure 10.5. *Typical load–frequency curves (ω values are plotted on a logarithmic scale)*

10.3.2.2. *Optimization problem II: minimization of volume under a constant critical load*

Again let p_{cr} be the smallest load at which two eigensolutions coincide, and let p_{cr}^0 be the value for the initial/uniform design. Optimization problem II is to determine the design vector $\mathbf{m}$ and is posed as follows:

$$\text{Minimize volume } = \min(\mathbf{l}^T \mathbf{m})$$

subject to

$$\text{i) } p_{cr} \geq p_{cr}^0,$$

$$\text{ii) } m_j^l \leq m_j \leq m_j^u,\ j = 1, 2, \cdots, N_e + 1,$$

$$\text{iii) } \omega_{n+1} - \omega_n > \begin{cases} 0 \text{ for } n = 1, 3, \cdots, 2N_e - 1 \\ c_+ \text{ for } n = 2, 4, \cdots, 2N_e - 2 \end{cases} \text{ for all } p < p_{cr}. \quad [10.17]$$

The inclusion of the constraint on the load–frequency curves, [10.15 (iv)] and [10.17 (iii)], is necessary to ensure a robust optimization algorithm which is able to provide good optimization design for any set of system parameters (Langthjem and Sugiyama 1999b, 2000b). However, in the present case, where the mass ratio $\mu =$ (mass of rocket motor)/(mass of column) ≈ 4.6, this constraint did not become active (see also section 10.3.2.4).

10.3.2.3. *Solution by sequential linear optimization*

The nonlinear optimization problems [10.15] and [10.17] are linearized and solved iteratively by using sequential linear optimization (linear programming). After k iterations, the design is specified by vector $\mathbf{m}^k$. The $(k + 1)$th optimal redesign problem, for determining the design vector $\mathbf{m}^{k+1} = \mathbf{m}^k + \Delta\mathbf{m}$, is written as follows:

$$\text{Maximize critical load } p_{cr} = p_{cr}(\mathbf{m})$$

subject to

$$\text{i) } p_{cr} \leq p_n^{k+1} \approx p_n^k + \sum_{j=1}^{N_e+1} \frac{\partial p_n^k}{\partial m_j} \Delta m_j,\ n = 1, 2, \cdots, N_e,$$

$$\text{ii) } m_j^l \leq m_j^k + \Delta m_j \leq m_j^u,\ j = 1, 2, \cdots, N_e + 1,$$

$$\text{iii) } \mathbf{l}^T \Delta\mathbf{m} = \mathbf{0},$$

$$\text{iv) } \omega_{n+1}^{k+1} - \omega_n^{k+1} \approx \omega_{n+1}^k - \omega_n^k$$

$$+ \sum_{j=1}^{N_e} \frac{\partial(\omega_{n+1}^k - \omega_n^k)}{\partial m_j} \Delta m_j > \begin{cases} 0 & \text{for } n = 1, 3, \cdots, \\ c_+ & \text{for } n = 2, 4, \cdots, \end{cases}$$

for

$$p = p_1^*,\ p_2^*, \cdots, p_{N_c}^*, \qquad [10.18]$$

for problem [10.15], and

$$\text{Minimize volume} = \min(\mathbf{l}^T \mathbf{m}^{k+1}) = \min(\mathbf{l}^T\{\mathbf{m}^k + \Delta\mathbf{m}\}) = \min(\mathbf{l}^T \Delta\mathbf{m})$$

subject to

$$\text{i) } p_{cr}^{k+1} \approx p_{cr}^k + \sum_{j=1}^{N_e} \frac{\partial p_{cr}^k}{\partial m_j} \Delta m_j \geq p_{cr}^0,$$

$$\text{ii) } m_j^l \leq m_j^k + \Delta m_j \leq m_j^u, \quad j = 1, 2, \cdots, N_e + 1,$$

$$\text{iii) } \omega_{n+1}^{k+1} - \omega_n^{k+1} \approx \omega_{n+1}^k - \omega_n^k$$

$$+ \sum_{j=1}^{N_e} \frac{\partial(\omega_{n+1}^k - \omega_n^k)}{\partial m_j} \Delta m_j > \begin{cases} 0 & \text{for } n = 1, 3, \cdots, \\ c_+ & \text{for } n = 2, 4, \cdots, \end{cases}$$

for

$$p = p_1^*,\ p_2^*, \cdots, p_{N_c}^*, \qquad [10.19]$$

for problem [10.17]. The N_c discrete load values $p = p_1^*,\ p_2^*,\ \cdots,\ p_{N_c}^*$ in the constraints [10.18 (iv)] and [10.19 (iii)] are in the range $0 < p < p_{cr}$. The derivative of the flutter loads with respect to the design variables, $\partial p_n / \partial m_j$, and the derivative of the frequencies, $\partial \omega / \partial m_j$, are given by :

$$\frac{\partial p_n}{\partial m_j} = -\left[\frac{\mathbf{v}^T(-\omega^2 \partial\mathbf{M}/\partial m_j - \partial\mathbf{S}/\partial m_j)\mathbf{y}}{\mathbf{v}^T\mathbf{Q}\mathbf{y}}\right]_{p=p_n}$$

and

$$\frac{\partial \omega}{\partial m_j} = \left[\frac{\mathbf{v}^T(-\omega^2 \partial\mathbf{M}/\partial m_j + \partial\mathbf{S}/\partial m_j)\mathbf{y}}{2\omega\mathbf{v}^T\mathbf{M}\mathbf{y}}\right]_{p \neq p_n} \qquad [10.20]$$

respectively. The size of the design change is governed by move limits Δm_j^{min}, Δm_j^{max}, $j = 1, 2, \cdots, N_e + 1$, such that

$$0 \geq \Delta m_j^{min} \leq \Delta m_j \leq \Delta m_j^{max} \geq 0. \qquad [10.21]$$

The linearized problems [10.18] and [10.19] are converted into a pure linear programming problem by introduction of the positive variables

$$\Delta m_j^+ = \Delta m_j - \Delta m_j^{min} \geq 0. \quad [10.22]$$

In terms of these variables, the design change limits are given by

$$0 \leq \Delta m_j^+ \leq \Delta m_j^{max} - \Delta m_j^{min}. \quad [10.23]$$

The sequence of linear programming problems [10.18] and [10.19] are solved, for $k = 1, 2, \cdots$, by applying the Simplex method (Press *et al.* 1992). During the iterations, the move limits are decreased automatically according to the size of the gradients (Press *et al.* 1992). Further computational details are given in Langthjem and Sugiyama (2000b, 2000c).

10.3.2.4. *Optimality criterion by a single (simple) flutter load*

It was found in Langthjem and Sugiyama (1999b) that for columns equipped with a rocket motor of a certain size, such that the mass ratio $\mu \gtrsim 1$, convergence of the optimum design is reached by a single (simple) flutter load. It was shown that the optimum design then satisfies the optimality criterion (Pedersen and Seyranian 1983)

$$\frac{\partial p_{cr}}{\partial m_j} \frac{1}{l_{j-1} + l_j} = \begin{cases} c^+ & \text{if } m_j > m_j^l \\ c_j^\pm < c^+ & \text{if } m_j = m_j^l \end{cases}$$

$$\text{for all } j = 1, 2, \cdots, N_e + 1, \ with \ l_0 = l_{N_e+1} = 0. \quad [10.24]$$

where c^+ is a positive constant and $c_j^\pm$ are positive or negative constants (but any one smaller than c^+).

10.4. Experiment

10.4.1. *General description*

The test columns are clamped into a solid holder, fixed to the concrete floor of the test dome at Harima Plant, Daicel Corporation (see Figure 10.3). The dome is licensed by the authorities for tests with rocket motors (pyrotechnic devices). The test columns are placed horizontally, and the vibrations are allowed only in the horizontal plane, such that gravity has no influence on the motions, as long as motions with small amplitude take place. The rocket motor is suspended from the ceiling by a long, thin wire. A strain gauge is mounted 70 mm from the clamped end, on both sides of the column, to measure the axial strain, which is converted to an axial force. The lateral deflections are measured by an eddy current displacement sensor. For this purpose, a target disk is mounted 170 mm from the clamped end. Figure 10.6 shows a schematic of the test setup.

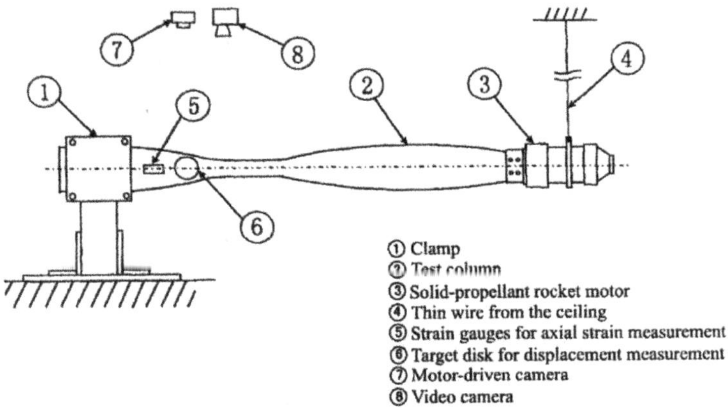

Figure 10.6. *Schematic of the experimental setup (side view)*

In order to protect the equipment from the, possibly very violent, fluttering motion of the test column, a couple of thin but strong and light-weighted wires of good length (harnesses for security) are loosely connected between the rocket motor and two holders on the floor (see Figure 10.3).

10.4.2. *Rocket motor*

In the case of normal experiments with columns, for example under conservative loading, the test column is fixed, while the load is variable, increasing or decreasing. In the present rocket-applied experiment, a batch of identical rocket motors was specially prepared. Thus, the load is fixed, and the column design is variable. This means that the load, the rocket thrust in the present case, is specified first, and then the test columns are designed for test runs.

The rocket motors used are of the solid-propellant type. They were specially designed and manufactured for the present experiment by Daicel Corporation (Harima Plant, Tatsuno, Japan). The casing has a total length (including mounting part) of 280 mm and a diameter of 103 mm, as shown in Figure 10.7. The mass of the empty motor casing is 3.4 kg. The rocket motor is filled with propellant with a mass of 1.1 kg. The total mass of a fresh motor is thus 4.5 kg, and it is assumed that the average mass of the motor is $M = 3.95$ kg ($= 3.4 + 1.1/2$). The average mass moment of inertia around the center of gravity is $J = 3.01 \times 10^{-2}$ kg m^2. The distance from the end of the column to the center of gravity of the motor is $a = 149.9$ mm.

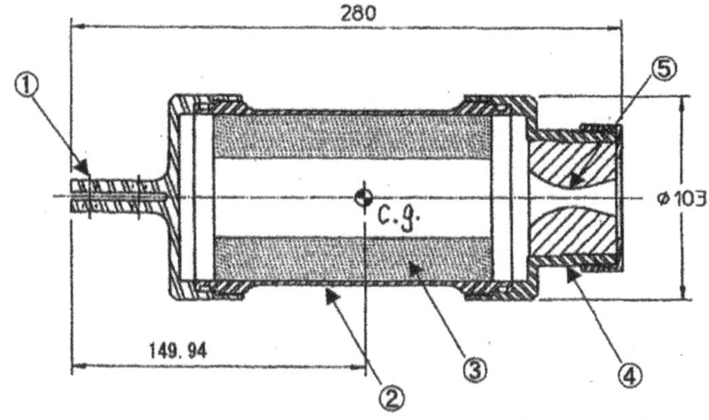

Figure 10.7. *Sketch of the cross-section of the rocket motor: ① forward cap, ② casing, ③ propellant, ④ after cap, ⑤ nozzle*

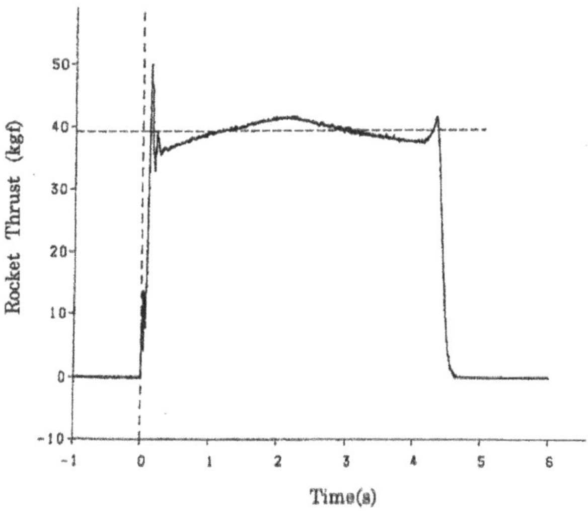

Figure 10.8. *Thrust curve*

The thrust curve is shown in Figure 10.8. The burnout time is approximately 4.5 s. During the first 0.13 s, the thrust increases from 0 to a peak value of 49.9 kgf (489 N). After the peak thrust point, the thrust decreases to a reasonably steady level, with the average value P_{av} = 39.0 kgf (382 N). Based on experience from previous experiments (Sugiyama *et al.* 1995, 2000), the very short initial peak value is ignored, and the motor is assumed to produce a steady/nominal thrust of P = 39.0 kgf (382 N).

The tests described in this chapter were conducted on metric engineering units, where the unit of force is kilogram force (kgf). The unit of kgf can be converted into the SI force unit of Newton (N) by multiplication by the conversion factor $g \approx 9.8\, m\, s^{-2}$.

10.4.3. *Columns*

The test pieces are beams cut out from an aluminum plate of thickness ("height" in the case of test beams) t = 4 mm. The modulus of elasticity (Young's modulus) is E = 68.0 Gpa (obtained by the free vibration test in section 10.4.4). The density is $\rho = 2.71 \text{ x } 10^3 \text{ kg m}^{-3}$.

The aim of the experiment is just to verify that a column designed by maximization of critical load under a constant volume is stable at a critical load higher than that for the initial column, and possibly stabilized by shape optimization with a targeted critical load less than that for the initial column, and with less volume. As only five rocket motors are presented, five tests can run. Under this restriction, the following missions are formulated:

Mission I: using two motors for two test runs, the aim of the mission is to show that a column, designed by maximization of critical load under fixed volume (column no. 2), is stable while the initial column (column no. 1) is unstable.

Mission II: using three motors for three test runs, the aim of the mission is to show that columns designed by minimization of volume under a fixed critical load (column nos. 4 and 5) may be stable, while the designed column (column no. 3) with less volume than that of the initial column is unstable.

In these missions, five test runs are planned with the aim for each test run. And then, five test columns are designed to achieve the aims of the columns. It is noted that, as the applied load is constant and the length of the test columns is fixed, the variable in the experiment is the designed shape of the test columns, and so the design variable in the optimization is breadth b. In the search for the designed shape, the finite element method is applied with 20 equal finite elements, and with 21 design variables. Assuming that the nominal thrust is constant during loading and of magnitude P = 39.0 kgf (382 N), five test columns are designed and machined for five test runs.

Column no. 1: the aim of the first test run is to confirm that the initial column no. 1 becomes unstable by flutter. The critical flutter load for the initial column is designed as a load just less than the nominal thrust. Thus, the length of the initial column is determined as $L = 990$ mm to have the critical load 38.2 kgf(374 N) $<$ $P = 39.0$ kgf (382 N). The length of the other four columns is fixed as $L = 990$ mm.

Column no. 1 is the uniform column having the length $L = 990$ mm, the initial breadth $b_0 = 80$ mm, and the height $h = 4$ mm. The column gives the initial/reference column with the aspect ratio of the cross-section $AR = b/h = 20.0$. This gives a mass per unit length of $m = 0.866$ kg m^{-1}, and a radius of gyration of $r = \sqrt{I/A} = 1.15$ mm, where A is the cross-sectional area and I is the area moment of inertia. The slenderness ratio is given as $s = L/r = 861$.

The design for column nos. 2–5 is conducted under the constraint for the design variable b with minimum allowable breadth, i.e. $b \geq b_{\text{min}} = b_0/2 = 40.0$ mm. The minimum allowable breadth was determined from an engineering point of view. Vanishing minimum breadth seems obviously unrealistic. Most importantly, it is noted that the present optimum design is based on the assumption that the column is elastic during loading. If the cross-section vanishes, then the compressive stress on the cross-section at the vanishing breadth becomes infinite, going beyond the elastic limit of the column, and out of application of theory.

Column no. 2: the aim of the second run is to confirm that the shape-optimized column, having the same volume as the initial column, is stable under the same thrust as that applied to column no. 1, validating the stabilizing effect of shape optimization by maximization of critical load. The design is obtained by application of [10.18]. The designed critical load is 43.9 kgf (430 N) $>$ $P = 39.0$ kgf (382 N), corresponding to an increase of 15 % over the initial design. The column is thus designed and expected to exhibit stable oscillations when being subjected to the average rocket thrust.

The design of column nos. 3, 4 and 5 follows mission II. Thus, the shape optimization problem is defined as follows: mimimize volume (mass) subject to p_{cr} = fixed, by application of [10.19].

Column no. 3: the aim of the test is to show that the shape-optimized column under the constraint of fixed load with less volume is stable. The critical load is designed as 38.2 kgf (374 N), exactly the same as for run no. 1, and the column is thus predicted to be unstable. The volume is 87.4 % of the initial design's volume, V_0, aiming at a volume reduction of about 12 %.

Column no. 4: the aim of the test is to validate the stabilizing effect of shape optimization under fixed load with less volume. The critical load P_{cr}^{o} was targeted to be 39.0 kgf (382 N) in the course of designing a test column with the load value coinciding with the flutter limit for test run no. 3. The resulting designed column has the critical load 39.1 kgf (383 N) and a volume of 89.35% of V_0. The gain of volume reduction is about 10%. The stability of the vibration is expected to be critical under the applied load.

Column no. 5: the aim of the test is to validate experimentally the stabilizing effect of shape optimization of the test column with higher critical load and less volume. The test column is targeted to have the critical load of $1.03P(40.0\ \text{kgf}/392\ \text{N})$ and volume of 0.914 V_0. The critical load is raised 2.6% and the volume is reduced by about 8.6%. The column is expected to remain stable under the applied load.

Five test beams were prepared as shown in Figure 10.9. It is noted that the length of the naked beams, as shown in the figure, is 1138 mm. The length comprises of the clamp part of 100 mm, the column part of 990 mm, and the mounting part of 48 mm. It is interesting to see that there is a narrow and straight "corridor" between the clamped side and the forward portion in the cases of nos. 2–5, which is attributable to the minimum allowable breadth $b_{min} = (1/2)b_0 = 40$ mm. Geometrical data of the four optimized columns, nos. 2–5, are listed in the appendix (section 10.8).

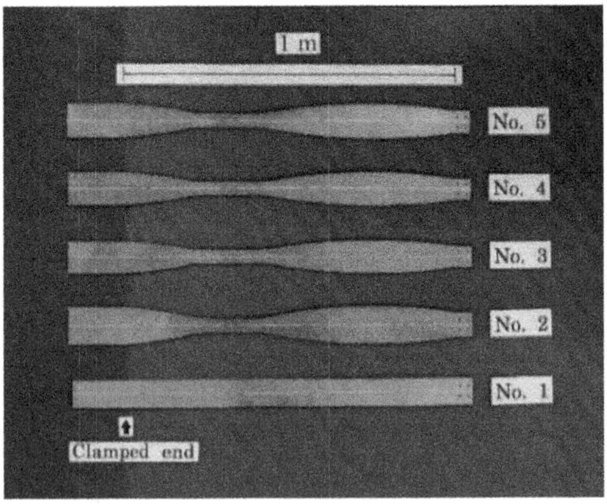

Figure 10.9. *Test beams. For a color version of this figure, see www.iste.co.uk/challamel/mechanics1.zip*

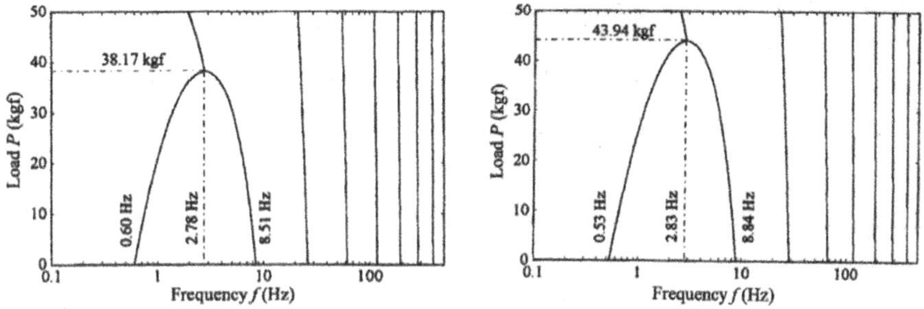

a) b)

Figure 10.10. *Load–frequency curves: (a) for the initial column no. 1 and (b) for the optimized column no. 2 with the maximum critical load*

Figure 10.10(a) shows the load–frequency curves for the initial column no. 1, while Figure 10.10(b) shows the curves for the optimized column no. 2. It can be observed that there are no significant changes in the development of these curves, contrary to cases with small or vanishing end mass (Langthjem and Sugiyama 1999b). The curves are similar for any of the five test columns. In addition, any of the optimized column nos. 2, 3, 4 and 5 satisfy the optimality criterion [10.24]. The designed columns are summarized in Table 10.1.

Column no.	Volume	Designed critical thrust (kgf)	Predicted stability
No. 1	V_0	38.2 (0.98P)	Flutter
No. 2	V_0	43.9 (1.13P)	Stable
No. 3	$0.874V_0$	37.2 (0.95P)	Flutter
No. 4	$0.894V_0$	39.1 (1.00P)	Critical
No. 5	$0.914V_0$	40.0 (1.03P)	Stable

Table 10.1. *Columns for five test runs*

10.4.4. *Free vibration test*

Before the planned flutter test, free vibration tests with the five test beams having a total length of 1038 mm (no rocket motor at tip end) were conducted. The test beams were fixed with the same clamp as was used in the rocket motor tests. The free vibrations, caused by a disturbance given at the free end for the first mode, were recorded to find the first natural eigenfrequency. The eigenfrequencies were measured three times and their average was taken to be the nominal values of the test results. The test results are compared with the numerical predictions by the finite element method with 20 segments, as shown in Table 10.2.

Column no.	No. 1	No. 2	No. 3	No. 4	No. 5
Numerical predictions	3.01	2.50	2.56	2.55	2.54
Test results	3.00	2.53	2.55	2.54	2.57

Table 10.2. *First natural eigenfrequencies (Hz) of the five test beams*

Young's modulus (modulus of elasticity) was determined by the free vibration test of the reference beam. The first eigenfrequency of a uniform cantilevered beam is given by $f_1 \approx (0.560/L^2)\sqrt{EI/\rho A}$, where L is the length, E is Young's modulus, I is the area moment of inertia, ρ is the mass per unit length, and A is the cross-sectional area. The eigenfrequency f_1 was measured by using the laser tracing system. Measurements were done on two different but identical uniform beams (for test beam no. 1), and on three different measure points on each beam. All measurements were repeated three times, giving 18 test cases in total. These tests gave $f_1 = 3.23\ (\pm 0.006)$Hz, and then $E = 68.0$ Gpa was obtained.

10.5. Flutter test

According to mission I, the first two test runs with column nos. 1 and 2 were conducted.

Run no. 1: the vibrations of column no. 1 became unstable, as predicted, but in a much more drastic way than expected in the planned plot. In fact, column no. 1 was twisted up, breaking the harness wires (see Figures 10.3 and 10.11), in several rounds about the clamp, as shown in Figure 10.11 (a). Figure 10.12 (a) shows the dynamic response, i.e. the lateral deflection, observed in run no. 1. The response was measured 170 mm from the clamped end (see Figures 10.3 and 10.4).

Run no. 2: the vibrations of column no. 2 became unstable, although they were predicted to remain stable in the design of the column. Figure 10.11 (b) shows the damaged column of test run no. 2. It is interesting to see that the narrow straight "corridor", the weakest portion of the test column, remains straight, recalling that the weakest portion results in the heaviest bend in the case of conservative loading for buckling. Figure 10.12 (b) shows the response curve in run no. 2. The deflection turned on one side and then on the opposite side showing that the motion was not in divergence, but in flutter.

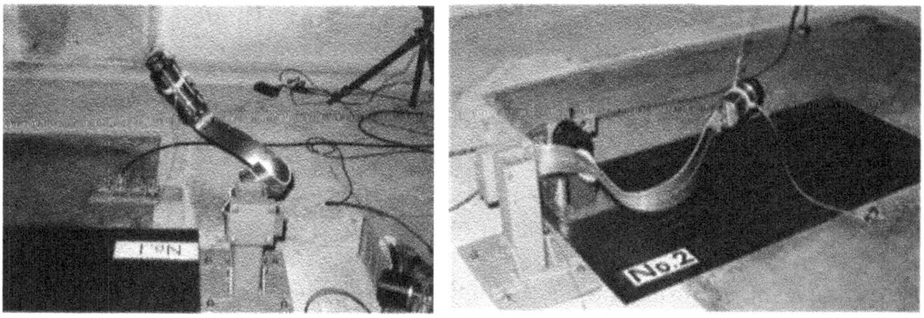

a) b)

Figure 10.11. *Damaged test columns: (a) damaged column no. 1 and (b) damaged column no. 2. For a color version of this figure, see www.iste.co.uk/challamel/mechanics1.zip*

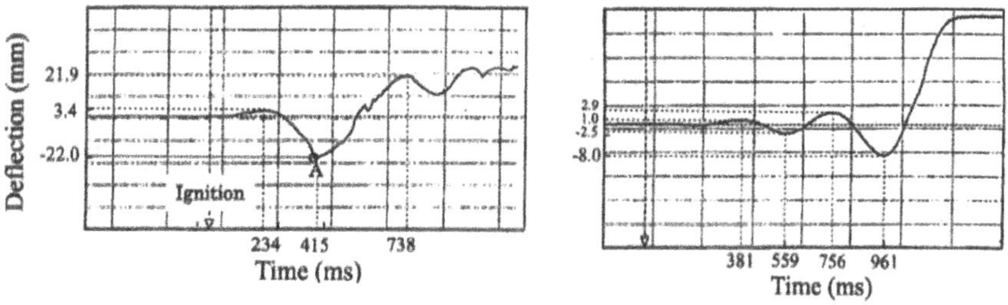

a) b)

Figure 10.12. *Dynamic responses in test runs nos. 1 and 2: (a) response in run no. 1 (the reading is only correct until the point "A") and (b) response in run no. 2*

From careful observation and comparison of the two dynamic responses (Figure 10.12), it may be seen, at least qualitatively, that the vibrations are stabilized by the optimal material distribution. In this sense, the experiment with the two test runs was partially successful in verifying the stabilizing effect of shape optimization; however, it must be said that the experiment was not completely successful with respect to a quantitative verification.

The two test runs showed that the theoretical predictions of the critical length for the uniform column (and likewise for the optimally designed column as well) were not adequate, and that the actual critical length would be shorter than that predicted. Accordingly, planned test runs nos. 3, 4 and 5 were called off. Instead, it was decided to use the remaining three rocket motors to check the critical length of a uniform column, as used in run no. 1.

Another three extra test runs were conducted with uniform columns with decreasing length, shorter than the initial length of 990 mm.

Run no. 6: the uniform column of length 900 mm was used to find that the vibrations still became unstable. The dynamic response is shown in Figure 10.13(a). It is observed by the curve in Figure 10.13(a) that the column loses stability by vibrations with increasing amplitude, i.e. by flutter.

Run no. 7: the uniform column of length 850 mm was used to find the vibrations to remain stable. The dynamic response is shown in Figure 10.13(b).

Run no. 8: the uniform column of length 875 mm, the mean value of the two lengths employed above, was used to find that the vibrations went unstable. The response is shown in Figure 10.13(c).

The result of the extra runs is summarized graphically in Figure 10.14. The theoretical stability boundary is drawn with a full curve. The dots represent the test results. Stable and unstable vibrations are depicted by open and solid marks, respectively. Accordingly, the critical length of the uniform column exists in the range of 850 mm $\leq L_{cr} < 875$ mm. The mean value $L_{cr}^{me} = 863$ mm may be employed as the critical length of the uniform column. If one takes the value of $L_{cr} = 863$ mm as the (estimated) critical length for the uniform column, the length is 13% shorter than the initial length of 990 mm, which was predicted for the first series of test runs.

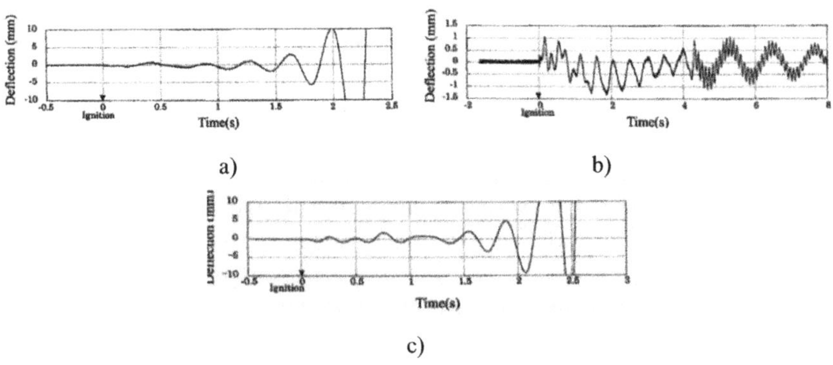

Figure 10.13. *Dynamic responses in the extra test runs with uniform columns: (a) response in run no. 6, (b) response in run no. 7 and (c) response in run no. 8*

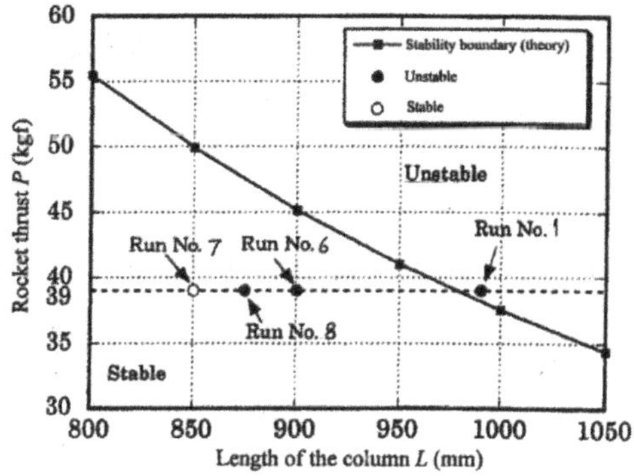

Figure 10.14. *Stability chart for the extra runs*

10.6. Concluding remarks

Optimum design of Beck's column has been studied almost continuously during the last four decades. This chapter is one of the rare attempts to provide experimental validation of the stabilizing effect of shape optimization of slender structures subjected to follower loads (Borglund 1998, 1999). However, this chapter could not give a definite experimental verification of the stabilizing effect of shape optimization, as the theoretical prediction for the critical length failed to match the dynamic responses of the columns under a rocket-based follower force. Then, was the theory wrong? Or was the experiment done inappropriately? It is noted that the experiment has demonstrated some physical realities nonetheless, while the theory could not predict the dynamic response correctly. The stability analysis in this chapter has been developed based on the Euler–Bernoulli beam theory, assuming that the beam is sufficiently slender. However, the present test columns were prepared with plate-like (deep) beams, having a rather wide breadth of 80 mm and a height of 4.0 mm, i.e. the aspect ratio of the cross-section $AR = b/h = 20.0$. It is recalled that the earlier three experiments (Sugiyama *et al.* 1995a, 1995b, 2000, 2019) were done successfully by using slender columns with aspect ratios of the cross-section in the range of $3.3 \leq AR \leq 5.0$. This advised the authors to return to test columns with proper (as beams) sizes of the initial breadth of 30 mm and the height of 7 mm, with the aspect ratio $AR = b/h = 4.3 \ll 20.0$. The authors did so and succeeded in gaining good experimental verification of the stabilizing effect of shape optimization of a generalized version of Beck's column (Sugiyama *et al.* 2012).

The plate-like test columns were applied for the present experiment as a result of the authors' intention to magnify the glamorous optimal shape design. The authors proceeded from the successful free vibration tests, where the test results of the first eigenfrequency were well predicted by numerical analysis, to the somewhat unsuccessful flutter tests. However, it was revealed by the present experiment that the dynamic responses of the plate-like columns under a follower compressive loading were not predicted so as in the case of the free vibration test with deep beams, where the beams were free of an axial compressive loading. The present experiment suggests, in one more lesson, that there may be a gray zone between the beam theory and the plate theory, specifically in the case of deep beams under compressive loading. This chapter might eventually step into this zone. The gray zone remains to be investigated further in the future.

For beams having plate-like form, Bolotin discussed in his book (Bolotin 1963, pp. 66–72, 104–119) the stability of the plane form of bending under the action of a transverse load in its plane of maximum stiffness. He suggested that the dynamics of the plane form can be described by two displacements, the lateral displacement and the angle of torsion (around the longitudinal axis x). The dynamic responses in the present experiment might be a kind of coupled bending–torsion flutter. This suggests to us that we might be better to study the dynamic stability of plate-like columns under a follower loading by considering both lateral and torsional motion.

10.7. Acknowledgments

The authors would like to express their gratitude to Mr. M. Kobayashi (presently at Kawasaki Heavy Industries, Ltd) and Mr. T. Iwama (presently at Toyota Motor Corporation), former master students at the Sugiyama Laboratory, Department of Aerospace Engineering, Osaka Prefecture University; and Mr. H. Yutani at the Harima Plant, Daicel Corporation, who partly joined the present research project.

10.8. Appendix

Designed breadth b (mm) of the optimized column numbers 2, 3, 4 and 5 (height h = 4.0 mm and length L = 990.0 mm)

Station (mm)	No. 2	No. 3	No. 4	No. 5
0.0	109.5	93.4	95.9	98.4
49.5	99.6	85.1	87.3	89.6
99.0	86.1	73.5	75.4	77.4
148.5	71.5	61.3	62.9	64.5
198.0	48.0	40.0	40.8	42.1
247.5	40.0	40.0	40.0	40.0

Station (mm)	No. 2	No. 3	No. 4	No. 5
297.0	40.0	40.0	40.0	40.0
346.5	40.0	40.0	40.0	40.0
396.0	40.0	40.0	40.0	40.0
445.5	58.6	48.6	50.3	51.8
495.0	75.3	65.3	66.9	68.5
544.5	88.3	76.0	78.8	79.0
594.0	98.5	85.0	87.2	89.3
643.5	105.6	91.2	93.5	95.8
893.0	109.9	94.9	97.3	99.7
742.5	111.2	96.0	98.5	100.9
792.0	109.5	94.6	97.0	99.4
841.5	104.8	90.6	92.9	95.1
891.0	97.0	83.8	85.9	88.0
940.5	85.8	74.2	76.1	77.9
990.0	71.3	61.7	63.3	64.8

10.9. References

Beck, M. (1952). Die Knicklast des einseitig eingespannten, tangential gedrückten Stabes. *Z. Angew. Math. Phys.*, 3, 225–228.

Bigoni, D. and Kirillov, O. (2019). *Dynamic Stability and Bifurcation in Nonconservative Mechanics*. Springer, Switzerland.

Bolotin, V.V. (1963). *Nonconservative Problems of the Theory of Elastic Stability*. Pergamon Press, New York.

Bolotin, V.V. and Zhinzher, N.I. (1969). Effects of damping on stability of elastic systems subjected to nonconservative force. *Int. J. Solids Struct.*, 5, 965–989.

Borglund, D. (1998). On the optimal design of pipes conveying fluid. *J. Fluids Struct.*, 12, 353–365.

Borglund, D. (1999). Active nozzle control and integrated design optimization of a beam subjected to fluid-dynamic forces. *J. Fluids Struct.*, 12, 353–365.

Claudon, J.-L. (1975). Characteristic curves and optimum design of two structures subjected to circulatory loads. *J. Méca.*, 14, 531–543.

Elishakoff, I. (2005). Controversy associated with the so-called follower forces. *Appl. Mech. Rev.*, 58, 117–142.

Gutkowski, W., Mahrenholt, O., Pyrz, M. (1993). Minimum weight design of structures under nonconservative forces. In *Optimization of Large Structural Systems*, Rozvany, G.I.N. (ed.). Kluwer, Dordrecht.

Hanaoka, H. and Washizu, K. (1980). Optimum design of Beck's column. *Comp. Struct.*, 11, 473–480.

Herrmann, G. and Jong, I.-C. (1965). On the destabilizing effect of damping in nonconservative elastic systems. *ASME J. Appl. Mech.*, 32, 592–597.

Ishida, R. and Sugiyama, Y. (1995). Proposal of constructive algorithm and discrete shape design of the strongest column. *AIAA J.*, 33(3), 401–406.

Kirillov, O. (2013). *Nonconservative Stability Problems of Modern Physics*. De Gruyter, Berlin.

Langthjem, M.A. and Sugiyama, Y. (1999a). Optimum design of Beck's column with a constraint on the static buckling load. *Struct. Optim.*, 18, 228–235.

Langthjem, M.A. and Sugiyama, Y. (1999b). Optimum shape design against flutter of a cantilevered column with an end-mass of finite size subjected to a non-conservative load. *J. Sound Vib.*, 226, 1–23.

Langthjem, M.A. and Sugiyama, Y. (2000a). Dynamic stability of columns subjected to follower loads: A survey. *J. Sound Vib.*, 238(5), 809–851.

Langthjem, M.A. and Sugiyama, Y. (2000b). Optimum design of cantilevered columns under the combined action of conservative and nonconservative forces, Part I: The undamped case. *Comp. Struct.*, 74, 385–398.

Langthjem, M.A. and Sugiyama, Y. (2000c). Optimum design of cantilevered columns under the combined action of conservative and nonconservative forces, Part II: The damped case. *Comp. Struct.*, 74, 399–408.

Odeh, F. and Tadjbakhsh, I. (1975). The shape of the strongest column with a follower force. *J. Optim. Theo. Applicat.*, 15, 103–108.

Pedersen, P. and Seyranian, A.P. (1983). Sensitivity analysis for problems of dynamic stability. *Int. J. Solids Struct.*, 19, 315–335.

Plaut, R.H. (1972). Determining the nature of instability in nonconservative problems. *AIAA J.*, 10, 967–968.

Plaut, R.H. (1975). Optimal design for stability under dissipative, gyroscopic, or circulatory loads. In *Optimization in Structural Design*, Sawczuk, A. and Mroz, Z. (eds). Springer-Verlag, Berlin.

Press, W.H., Teukolsky, S.A., Vetterling, V.T., Flannery, B.P. (1992). *Numerical Recipes in Fortran*, 2nd edition. Cambridge University Press, Cambridge.

Ringertz, U.T. (1994). On the design of Beck's column. *Struct. Optim.* (*Struct. Multidisci. Optim.*), 8, 120–124.

Ryu, B.-J. and Sugiyama, Y. (2000). Dynamic stability of cantilevered Timoshenko columns subjected to a rocket thrust. *Comp. Struct.*, 51, 331–335.

Sugiyama, Y., Katayama, K., Kinoi, S. (1995a). Flutter of cantilevered column under rocket thrust. *ASCE J. Aeros. Eng.*, 8, 9–15.

Sugiyama, Y., Matsuike, J., Ryu, B.-J., Katayama, K., Kinoi, S., Enomoto, N. (1995b). Effect of concentrated mass on stability of cantilevers under rocket thrust. *AIAA J.*, 33, 499–503.

Sugiyama, Y., Katayama, K., Kiriyama, K., Ryu, B.-J. (2000). Experimental verification of dynamic stability of vertical cantilevered columns subjected to a sub-tangential force. *J. Sound Vib.*, 236, 193–207.

Sugiyama, Y., Langthjem, M.A., Iwama, T., Kobayashi, M., Katayama, K., Yutani, H. (2012). Shape optimization of cantilevered columns subjected to a rocket-based follower force and its experimental verification. *Struct. Multidisc. Optim.*, 46, 826–838.

Sugiyama, Y., Langthjem, M.A., Katayama, K. (2019). *Dynamic Stability of Columns under Nonconservative Forces: Theory and Experiment*. Springer, Switzerland.

Tada, Y., Matsumoto, R., Oku, A. (1988). Shape determination of nonconservative structural systems (determination of optimum shape with stable critical load). In *Computer Aided Optimum Design of Structures: Recent Advances*, Hernández, S., Kassab, A.J., Brebbia, C.J. (eds). WIT Press, Southampton, UK.

Tadjbakhsh, I. and Keller, J.B. (1962). Strongest columns and isoperimetric inequalities for eigenvalues. *ASME J. Appl. Mech.*, 29, 159–164.

Tezak, E.G., Nayfeh, A.H., Mook, D.T. (1982). Parametrically excited nonlinear multi-degree-of-freedom systems with repeated frequencies. *J. Sound Vib.*, 85, 459–472.

Venkateswara Rao, G. and Singh, G. (2001). Revisit to the stability of a uniform cantilever column subjected to Euler and Beck loads – Effect of realistic follower forces. *Indian J. Eng. Mate. Sci.*, 8, 123–128.

Ziegler, H. (1952). Die Stabilitätskrierien der Elastomechnik. *Ing.-Archiv*, 20, 45–56.

11

Hencky Bar-Chain Model for Buckling Analysis and Optimal Design of Trapezoidal Arches

This paper presents a Hencky bar-chain model (HBM) for buckling analysis and optimal design of funicular trapezoidal arches. The HBM comprises rigid beam segments connected by frictionless hinges and elastic rotational springs with stiffness $C = EI/a$, where EI is the member flexural rigidity and a is the segmental length. Based on the energy approach, canonical elastic and geometric stiffness matrices are derived for the HBM trapezoidal arch. The compatibility condition for trapezoidal arch shapes with general supports at different elevations is handled by using Lagrange multipliers. The eigenvalue equation for the buckling of trapezoidal arches is established by applying the principle of minimum potential energy. The lowest positive eigenvalue of the eigenvalue equation furnishes the buckling load. Two trapezoidal arch problems are solved to demonstrate the HBM for buckling analysis. Based on the developed HBM, we optimize the shape of symmetric trapezoidal arches of a prescribed volume of material for maximum buckling capacity. We consider trapezoidal arches under vertical point loads at different positions from the supports, as well as various base support conditions (such as hinged–hinged, fixed–hinged and fixed–fixed). The optimal arch height h and the cross-sectional area ratio A_I/A_H of the inclined members to the horizontal member are obtained, and a sensitivity analysis is performed to examine how the buckling load varies as the arch height changes, as well as how least-weight fully stressed arch solutions fare in comparison to the optimal arch designs against buckling.

Chapter written by Chien Ming WANG, Wen Hao PAN and Hanzhe ZHANG.

11.1. Introduction

Funicular arches are elegant structures that transmit loads to the supports by purely compressive forces; there is no bending moment in the arch. In the design of funicular arches, it is important to check the arches against in-plane and out-of-plane buckling, which may cause premature failure.

This paper focuses on funicular trapezoidal arches under two vertical point loads applied at the joints. In literature, some studies have been conducted on the stability of trapezoidal arches. Salem (1969) obtained the sway buckling load of a trapezoidal frame under axial compression. It was found that the elastic critical load increases as the inclination of columns or beams increases up to a maximum value, and then decreases again. The percentage increase in the critical load also depends on the bending stiffness of the frame members, i.e. the percentage increase is higher for hinged base frames than their fixed base counterparts. Wang (2007) investigated the in-plane buckling and nonlinear post-buckling of a trapezoidal frame with rigid members and flexible joints, where the horizontal member carries a rigid block. He also optimized the shape of the trapezoidal frame against buckling. Note that the deformation of the frame is only caused by the angle changes of four flexible connections whose stiffnesses are the same. Wang (2011) further studied the shape optimization problem of a similar trapezoidal frame against buckling by considering four flexible connections with different stiffnesses. Mohamadain *et al.* (2010) studied the stability behavior of trapezoidal frames exposed to elevated temperatures. They evaluated the effects of the column inclination, material degradation, support conditions and heating regimes of trapezoidal frames on their critical carrying capacities, and predicted the critical temperature of trapezoidal frames under design loads. Wang (2020) examined the stability and shape optimization problem of a trapezoidal vault with semi-rigid joints that has been subjected to hydrostatic pressure. It was found that the vault could have maximal volume and highest critical pressure when the top three rigid members were 39.64% of the base length forming a symmetrical trapezoid. The two upper joints should also be heavily strengthened.

In this paper, we present the Hencky bar-chain model (HBM) for the in-plane elastic buckling analysis of funicular trapezoidal arches under two point loads. The three straight members of the trapezoidal arch may take on different sizes and the end supports may take on any elastically rotationally restrained condition. Based on the HBM, we investigate the optimal design of the funicular trapezoidal arch for maximum in-plane buckling capacity.

11.2. Buckling analysis of trapezoidal arches based on the HBM

Consider the buckling problem of a trapezoidal funicular arch under two vertical concentrated loads P and αP at the member connections, as shown in Figure 11.1.

The three inclined members have flexural rigidities EI_1, EI_2 and EI_3, lengths L_1, L_2 and L_3, and inclination angles ϕ_1, ϕ_2 and ϕ_3 with respect to the horizontal line. The supports at A and B are hinged supports with rotational spring stiffnesses K_{RA} and K_{RB}, respectively. When $K_{RA} = 0$ and $K_{RB} = 0$, we have hinged supports, and when $K_{RA} = \infty$ and $K_{RB} = \infty$, we have fixed supports. From a geometrical consideration, the inclination angle γ of the sloping line passing through the two supporting points (see Figure 11.1) is given by

$$\gamma = \arctan\left(\frac{L_1 \sin\phi_1 + L_2 \sin\phi_2 + L_3 \sin\phi_3}{L}\right) \tag{11.1}$$

where the horizontal span length L is given by

$$L = L_1 \cos\phi_1 + L_2 \cos\phi_2 + L_3 \cos\phi_3 \tag{11.2}$$

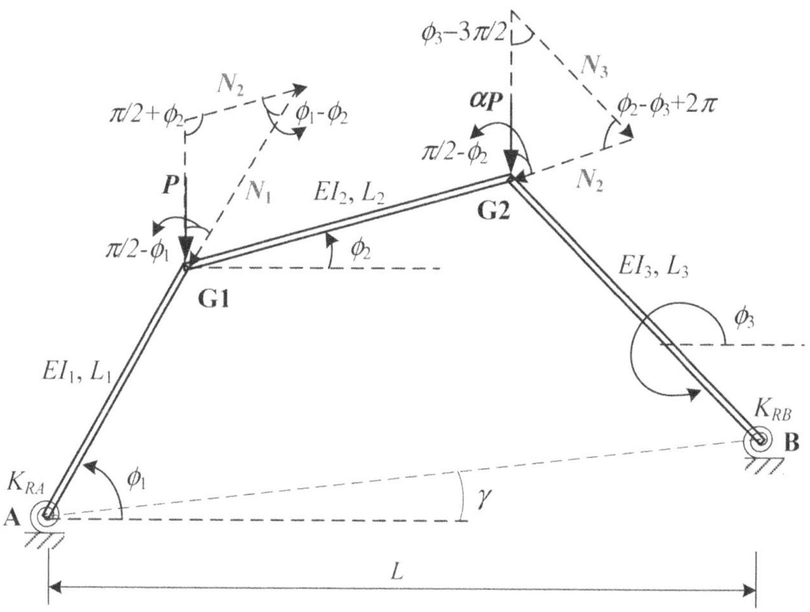

Figure 11.1. *Trapezoidal arch under two vertical concentrated loads. For a color version of this figure, see www.iste.co.uk/challamel/mechanics1.zip*

Based on the equilibrium of forces at the two joints G1 and G2, the axial forces in the three inclined members are given by

$$N_1 = \frac{P\cos\phi_2}{\sin(\phi_1 - \phi_2)} \quad [11.3a]$$

$$N_2 = \frac{P\cos\phi_1}{\sin(\phi_1 - \phi_2)}, \text{ or } N_2 = \frac{\alpha P\cos\phi_3}{\sin(\phi_2 - \phi_3)} \quad [11.3b]$$

$$N_3 = \frac{\alpha P\cos\phi_2}{\sin(\phi_2 - \phi_3)} \quad [11.3c]$$

Equation [11.3b] gives the following condition for the funicular shape of the trapezoidal arch:

$$\frac{P\cos\phi_1}{\sin(\phi_1 - \phi_2)} = \frac{\alpha P\cos\phi_3}{\sin(\phi_2 - \phi_3)} \Rightarrow \alpha = \frac{\sin(\phi_2 - \phi_3)\cos\phi_1}{\sin(\phi_1 - \phi_2)\cos\phi_3} \quad [11.4]$$

Equation [11.4] implies that the load ratio α is fixed when the member inclination angles ϕ_1, ϕ_2 and ϕ_3 are prescribed. Alternatively, for a specified load ratio α and any of the two inclination angles ϕ, the remaining angle ϕ is fixed.

11.2.1. *Description of the HBM*

The Hencky bar-chain model (HBM) comprises a finite number of rigid segments connected by frictionless hinges and rotational springs. The HBM was proposed by Hencky (1920) based on an idea to replace deflection curves with piecewise linear curves. Since then, many researchers have been using this model for the analysis of beams and frames. These researchers have called the HBM by other names, such as the discrete element model (Sun *et al.* 1995), the segmented/articulated column (Wang 2001), the discrete link–spring model (Krishna and Ram 2007) and the microstructured beam model (Zhang *et al.* 2013).

Silverman (1951) highlighted that the HBM is actually equivalent to the first-order central finite difference model (FDM), if the segmental length of the HBM is made equal to the nodal spacing of the FDM. Based on this equivalence, Wang *et al.* (2015) derived the boundary spring stiffnesses of the Hencky bar chain for elastic member-end restraints. Continuing along this line of investigation, Ruocco *et al.* (2016), Wang *et al.* (2016) and Zhang *et al.* (2016) developed the Hencky bar chain for beams with internal elastic restraints, with allowance for self-weight and varying cross-sections along the length. Zhang *et al.* (2017) and Pan *et al.* (2019a, 2019b) developed the Hencky bar chain for the buckling analysis of frames and circular arches, respectively. The HBM for buckling analysis of general funicular arches was further presented in the book by Wang *et al.* (2020b).

The problem at hand is to determine the elastic buckling load P of the trapezoidal arch using the HBM. The HBM of the considered trapezoidal arch has each of its three members divided into n rigid segments of equal length, i.e. $a_1 = L_1/n$ for the first member, $a_2 = L_2/n$ for the second member and $a_3 = L_3/n$ for the third member (see Figure 11.2). The rigid segments are connected by frictionless hinges with elastic rotational stiffnesses $C_1 = EI_1/a_1$, $C_2 = EI_2/a_2$ and $C_3 = EI_3/a_3$ for the first, second and third members, respectively.

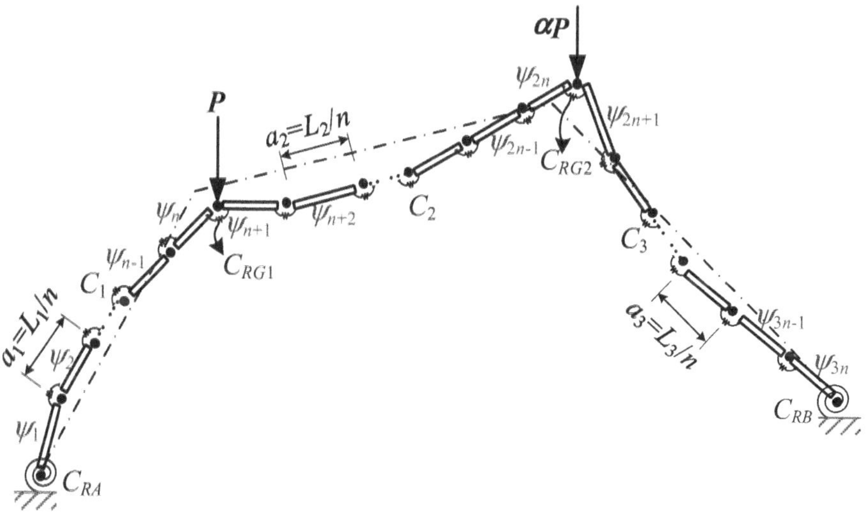

Figure 11.2. *HBM for the trapezoidal arch. For a color version of this figure, see www.iste.co.uk/challamel/mechanics1.zip*

The HBM rotational spring stiffnesses at the supporting points, C_{RA} and C_{RB}, are given by (Wang *et al.* 2015)

$$C_{RA} = \frac{1}{\dfrac{1}{2C_1} + \dfrac{1}{K_{RA}}} \quad C_{RB} = \frac{1}{\dfrac{1}{2C_3} + \dfrac{1}{K_{RB}}} \qquad [11.5]$$

Note that for hinged supports, $C_{RA} = C_{RB} = 0$ since $K_{RA} = K_{RB} = 0$, whereas for fixed supports, $C_{RA} = 2C_1$ and $C_{RB} = 2C_3$ since $K_{RA} = K_{RB} = \infty$.

The HBM rotational spring stiffnesses at the joints are given by (Pan *et al.* 2019a; Wang *et al.* 2020b)

$$C_{RG1} = \frac{2}{\dfrac{1}{C_1} + \dfrac{1}{C_2}} \quad C_{RG2} = \frac{2}{\dfrac{1}{C_2} + \dfrac{1}{C_3}} \qquad [11.6]$$

11.2.2. *HBM stiffness matrix formulation*

For the trapezoidal arch problem, the $3n$-order basis deformation vector of segment rotations is selected as $\Psi = (\psi_1, \cdots, \psi_n, \psi_{n+1}, \cdots, \psi_{2n}, \psi_{2n+1}, \cdots, \psi_{3n})^T$. The elastic strain energy U of the HBM is given by

$$U = \frac{1}{2} \Psi^{\mathrm{T}} [K_{U\Psi}] \Psi \qquad [11.7]$$

where $[K_{U\Psi}]$ is the elastic stiffness matrix associated with the basis deformation vector, which is given by

$$[K_{U\Psi}] = \begin{bmatrix} C_1[K_{U0}] & & \\ & C_2[K_{U0}] & \\ & & C_3[K_{U0}] \end{bmatrix}$$

$$+ \begin{bmatrix} C_{RA} & & & & & & & & & \\ & [0_{n-2}] & & & & & & & & \\ & & C_{RG1} & -C_{RG1} & & & & & & \\ & & -C_{RG1} & C_{RG1} & & & & & & \\ & & & & [0_{n-2}] & & & & & \\ & & & & & C_{RG2} & -C_{RG2} & & & \\ & & & & & -C_{RG2} & C_{RG2} & & & \\ & & & & & & & [0_{n-2}] & & \\ & & & & & & & & C_{RB} \end{bmatrix} \qquad [11.8]$$

where $[K_{U0}]$ is a canonical elastic stiffness matrix for the internal rotational springs in one member:

$$[K_{U0}] = \begin{bmatrix} 1 & -1 & & & \\ -1 & 2 & -1 & & \\ & -1 & \ddots & -1 & \\ & & -1 & 2 & -1 \\ & & & -1 & 1 \end{bmatrix} \qquad [11.9]$$

The potential energy V, due to the axial compressive load P in the HBM, can be expressed as

$$V = \frac{1}{2} P \cdot \Psi^{\mathrm{T}} [K_{V\Psi}] \Psi \qquad [11.10]$$

where the geometric stiffness matrix $[K_{V\Psi}]$ is given by

$$[K_{V\Psi}] = -\begin{bmatrix} \frac{\cos\phi_2}{\sin(\phi_1 - \phi_2)} a_1 [I_n] & & \\ & \frac{\cos\phi_1}{\sin(\phi_1 - \phi_2)} a_2 [I_n] & \\ & & \frac{\alpha \cos\phi_2}{\sin(\phi_2 - \phi_3)} a_3 [I_n] \end{bmatrix} \qquad [11.11]$$

The total potential energy function is given by

$$\Pi = U + V = \frac{1}{2} \Psi^{\mathrm{T}} \left([K_{U\Psi}] + P \cdot [K_{V\Psi}] \right) \Psi \qquad [11.12]$$

11.2.3. *Governing equation considering compatibility conditions*

By considering the deformation at the buckled state and the fact that supports A and B do not move from their positions during buckling, the compatibility condition for the basis deformation vector Ψ can be established from geometrical considerations as

$$[K_\mu] \Psi = \begin{Bmatrix} 0 \\ 0 \end{Bmatrix} \qquad [11.13]$$

where the ($2\times3n$)-order matrix $\left[K_\mu\right]$ for considering the compatibility conditions is given by

$$\left[K_\mu\right]=\begin{bmatrix} a_1\sin\phi_1\cdot[1 & \cdots & 1] & a_2\sin\phi_2\cdot[1 & \cdots & 1] & a_3\sin\phi_3\cdot[1 & \cdots & 1] \\ a_1\cos\phi_1\cdot[1 & \cdots & 1] & a_2\cos\phi_2\cdot[1 & \cdots & 1] & a_3\cos\phi_3\cdot[1 & \cdots & 1] \end{bmatrix} \qquad [11.14]$$

In order to obtain the governing equation for buckling, we minimize the total potential energy subject to the satisfaction of the compatibility conditions (which contains the boundary conditions). By using Lagrange multipliers, we transform the constrained extremum problem to an unconstrained one. So the augmented total potential energy Γ (or Lagrangian) is given by

$$\Gamma(\Psi,\mu_1,\mu_2)=\Pi(\Psi)+\{\mu_1 \quad \mu_2\}\cdot\left[K_\mu\right]\Psi \qquad [11.15]$$

where μ_1 and μ_2 are the Lagrange multipliers.

The stability criterion corresponds to the minimum value of the Lagrangian (equation [11.15]), i.e.

$$\frac{\partial\Gamma(\Psi,\mu_1,\mu_2)}{\partial(\Psi,\mu_1,\mu_2)}=\begin{bmatrix} \left[K_{U\Psi}\right]+P\cdot\left[K_{V\Psi}\right] & \left[K_\mu\right]^{\mathrm{T}} \\ \left[K_\mu\right] & 0 \end{bmatrix}\begin{pmatrix} \Psi \\ \mu_1 \\ \mu_2 \end{pmatrix}=\{0\} \qquad [11.16]$$

In order to determine the buckling load of the arch, equation [11.16] has to take on a non-trivial solution of the vector (Ψ, μ_1, μ_2), which is achieved by setting the determinant of the governing stiffness matrix to zero, i.e.

$$\det\begin{bmatrix} \left[K_{U\Psi}\right]+P\cdot\left[K_{V\Psi}\right] & \left[K_\mu\right]^{\mathrm{T}} \\ \left[K_\mu\right] & 0 \end{bmatrix}=0 \qquad [11.17]$$

The buckling load P_{cr} is obtained by solving this characteristic equation [11.17] for the lowest positive root. The eigenvector (Ψ, μ_1, μ_2) in equation [11.16], associated with P_{cr}, furnishes the buckling mode.

11.2.4. *Verification of the HBM*

Consider funicular trapezoidal arches under two loads P and αP, as shown in Figure 11.1. Table 11.1 presents the prescribed parameters (ϕ_1, ϕ_2, ϕ_3, L_1, L_2, L_3, EI_1,

EI_2, EI_3) for the two trapezoidal arches considered. The span length L_h and the load factor α are calculated from equations [11.2] and [11.4], respectively. Table 11.2 presents the buckling load parameters $PL^2/(EI)$ for n-segment HBM trapezoidal arches for the different shapes. It can be seen that the buckling loads converge monotonically to their corresponding continuum solutions ($n = \infty$) from below with increasing segment numbers n. The buckling modes of the trapezoidal arches 1 and 2, modeled by the HBM with $n = 8$, are shown in Figures 11.3 and 11.4, respectively.

Arch shape	Symmetric trapezoidal arch 1 with $\phi_1 = \pi/3$, $\phi_2 = 0$, $\phi_3 = 5\pi/3$ $EI_1 = EI_2 = EI_3 = EI$ $L_1 = L_2 = L_3 = L_0$	Non-symmetric trapezoidal arch 2 with $\phi_1 = \pi/4$, $\phi_2 = 11\pi/6$, $\phi_3 = 5\pi/3$ $EI_1 = EI_3 = 2EI$, $EI_2 = EI$ $L_1 = 2L_0$, $L_2 = 1.5L_0$, $L_3 = L_0$
Span length L	$\left[\cos\left(\frac{\pi}{3}\right) + \cos(0) + \cos\left(\frac{5\pi}{3}\right)\right] L_0 = 2L_0$	$L_1 \cos\left(\frac{\pi}{4}\right) + L_2 \cos\left(\frac{11\pi}{6}\right) + L_3 \cos\left(\frac{5\pi}{3}\right) = 3.213 L_0$
Load factor α	$\alpha = \dfrac{\sin\left(-\frac{5\pi}{3}\right)\cos\frac{\pi}{3}}{\sin\frac{\pi}{3}\cos\frac{5\pi}{3}} = 1$	$\dfrac{\sin\left(\frac{11\pi}{6} - \frac{5\pi}{3}\right)\cos\frac{\pi}{4}}{\sin\left(\frac{\pi}{4} - \frac{11\pi}{6}\right)\cos\frac{5\pi}{3}} = 0.732$

Table 11.1. *Parameters for non-symmetrical trapezoidal arches 1 and 2*

n	Trapezoidal arch 1	Trapezoidal arch 2
4	38.784	68.647
8	40.314	71.352
16	40.705	72.032
32	40.804	72.202
∞*	40.828	72.260

*The continuum solution is obtained using the matrix stiffness method with stability functions (Yuan 2008).

Table 11.2. *Buckling load parameters $PL^2/(EI)$ of trapezoidal arches – n-segment HBM solutions and continuum solutions*

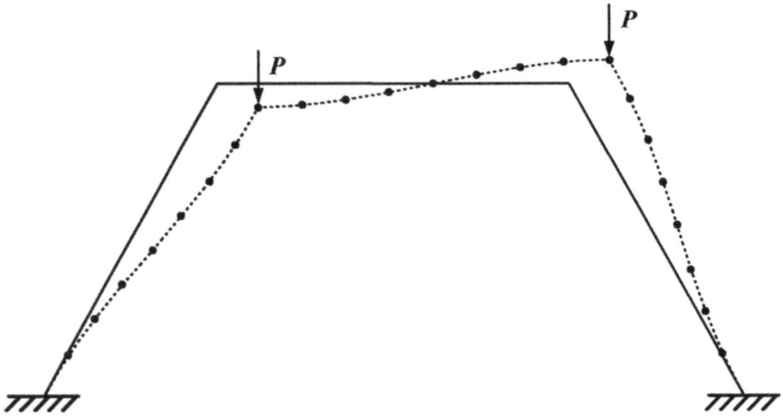

Figure 11.3. *Buckling mode of trapezoidal arch 1 modeled by the HBM*

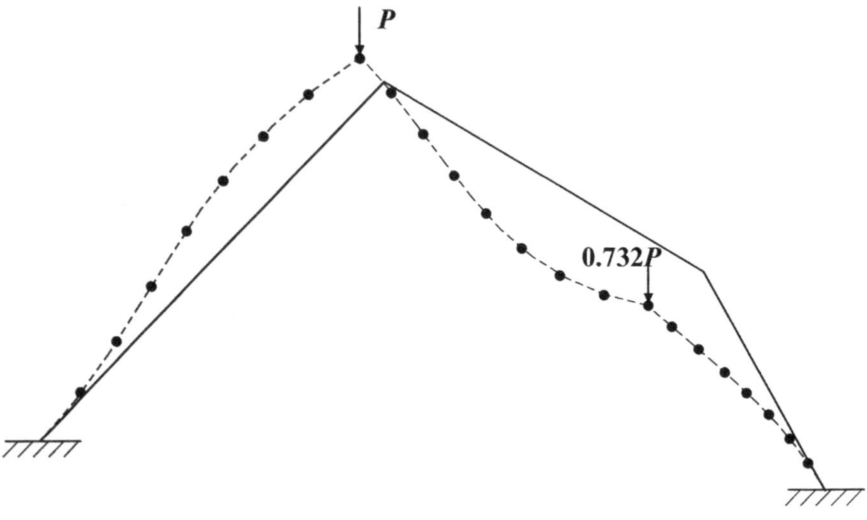

Figure 11.4. *Buckling mode of trapezoidal arch 2 modeled by the HBM*

It can be seen that the HBM furnishes accurate buckling loads for the trapezoidal arches by taking a sufficient number of segments, say 20 segments per member. As the HBM is a discrete structural model, any local changes to the cross-section of the arch due to damage or stiffening can be readily accommodated by changing the rotational spring stiffnesses appropriately at that location. Also, semi-rigid joints and support conditions may be easily taken care of by changing the rotational spring stiffnesses at the joints and supporting points.

11.3. Optimal design of symmetric trapezoidal arches

11.3.1. *Problem definition*

Consider the elastic in-plane buckling of a symmetric trapezoidal arch, carrying two equal vertical point loads P at the joints G1 and G2, as shown in Figure 11.5.

The span length of the arch is L. The horizontal distances of the total span length L divided by P are b, b_{H} and b. The height of the arch is h.

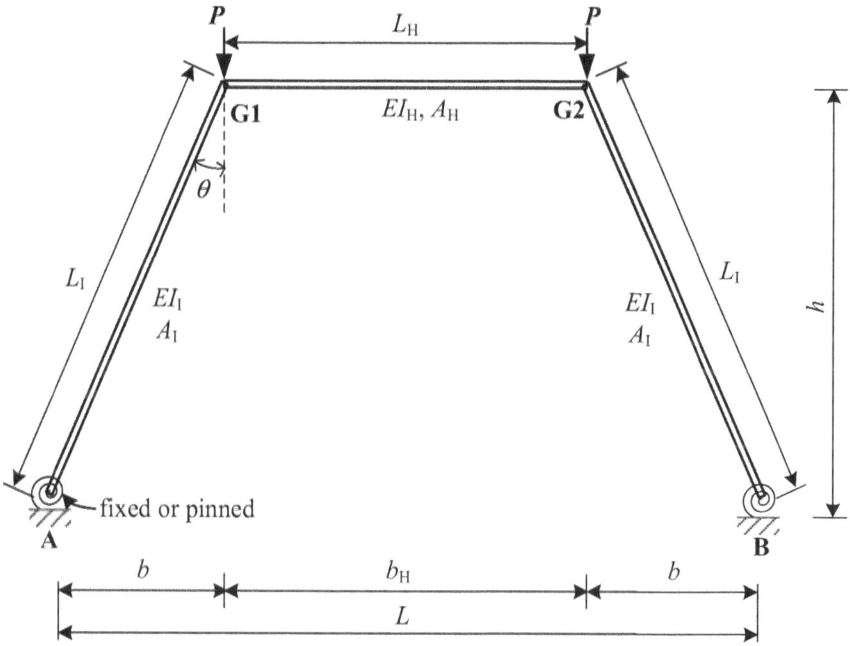

Figure 11.5. *Trapezoidal arch*

The inclined members have a flexural rigidity EI_{I}, cross-sectional area A_{I} and length L_{I}, while the horizontal member has a flexural rigidity EI_{H}, cross-sectional area A_{H} and length L_{H}. Considering the members of a geometrically similar cross-section, the second moment of area I is related to the cross-sectional area A by

$$I = \alpha A^2 \qquad [11.18]$$

where the parameter α depends on the cross-sectional shape. For example, $\alpha = 1/(4\pi)$ for a circular solid cross-section and $\alpha = 1/12$ for a square solid section.

The arch supporting points A and B are either hinged–hinged, fixed–fixed or fixed–hinged.

The optimization problem at hand may be stated as follows:

Maximize buckling load P_{cr} subjected to a prescribed arch volume V constraint, i.e.

$$2A_I L_I + A_H L_H = V \quad [11.19]$$

By using this constraint, the decision variables involved in this optimization problem are the arch height h and the cross-sectional area ratio A_I/A_H of the two members. Both variables are free to take any positive value in the optimization procedure.

11.3.2. *Optimization procedure*

The buckling load (calculated from the HBM with 40 rigid segments per member) has to be maximized by selecting the appropriate arch height h and the cross-sectional area ratio A_I/A_H of the two members. The final unconstrained optimization problem, involving two decision variables (h and A_I/A_H), may be solved by using the Nelder and Mead simplex method (Nelder and Mead 1965), which does not require the evaluations of any gradients.

11.3.3. *Optimal solutions*

Figure 11.6 shows the variations of the optimal height h_{opt}/L of the trapezoidal arch with respect to the horizontal distance b/L of the point load from the support, for various combinations of support conditions. It can be seen that the optimal height h_{opt}/L increases from approximately 0.15 to 0.25, as the horizontal distance b/L varies from 0.15 to 0.4. The optimal arch solution with h/L = 0.15–0.25 is considered to be a high arch (Bažant and Cedolin 2010; Wang *et al.* 2020a) because there is no significant increase in axial forces in members during the deformation of the arch. Therefore, the bifurcation buckling due to the compressive axial forces in the members is the dominant behavior for the considered trapezoidal arch structure.

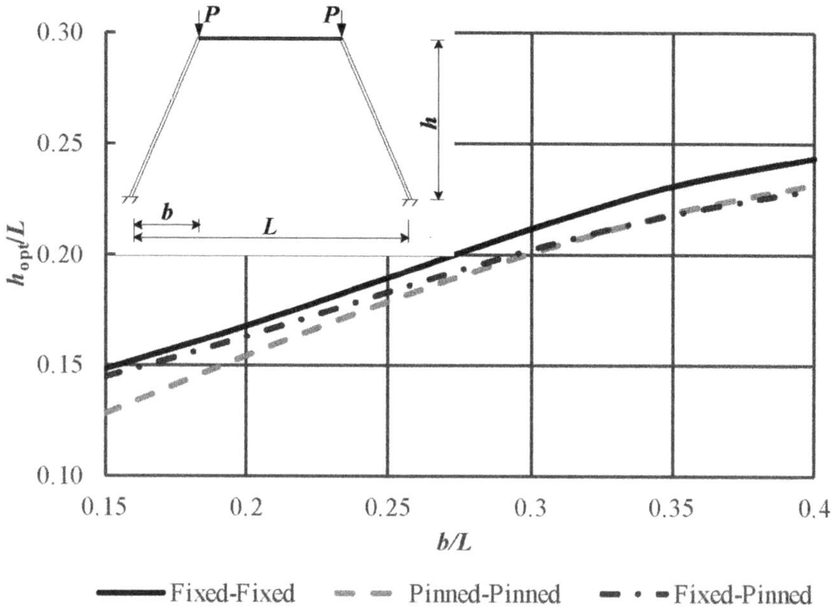

Figure 11.6. *Variations of optimal height of trapezoidal arches for different support conditions with respect to b/L. For a color version of this figure, see www.iste.co.uk/challamel/mechanics1.zip*

Figure 11.7 shows the variations of the optimal cross-sectional area ratio A_I/A_H, with respect to the horizontal distance b/L of the point load from the support for different base support conditions. It can be seen that the optimal cross-sectional area ratio A_I/A_H varies from approximately 0.8 to 1.6, as the horizontal distance b/L varies from 0.15 to 0.4.

Figure 11.8 shows the variations of the non-dimensional maximum buckling load $P_{max}L^4/\left(\alpha EV^2\right)$, with respect to the horizontal position b/L of the load from the support for the various base support conditions. For the fixed–hinged arch, the maximum buckling load lies within the buckling loads of the fixed–fixed and hinged–hinged arches.

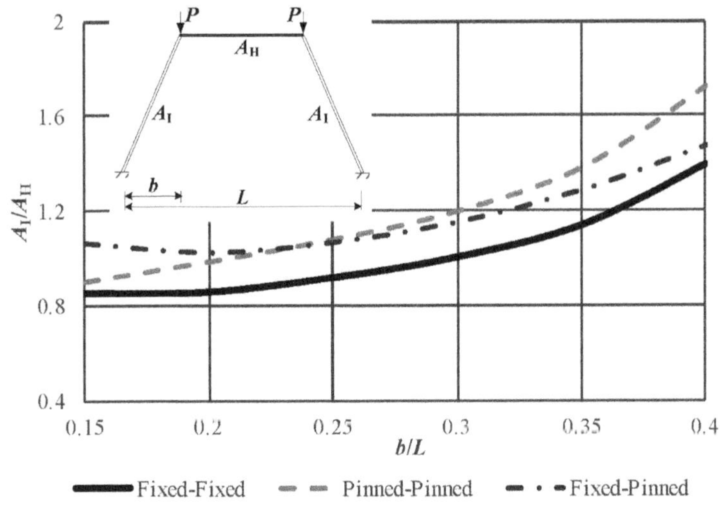

Figure 11.7. *Variations of optimal cross-sectional area ratio of trapezoidal arches for different base support conditions with respect to b/L. For a color version of this figure, see www.iste.co.uk/challamel/mechanics1.zip*

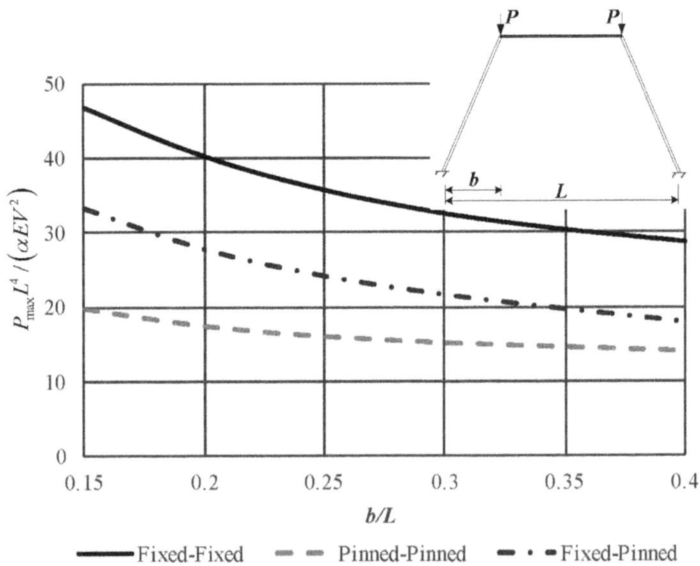

Figure 11.8. *Variations of the maximum buckling load of trapezoidal arches for different base support conditions with respect to b/L. For a color version of this figure, see www.iste.co.uk/challamel/mechanics1.zip*

11.3.4. *Sensitivity analysis of optimal solutions*

Consider a trapezoidal arch of a given volume of material and subject to two vertical point loads, whose horizontal position *b/L* from the support is prescribed. Figure 11.9(a)–(c) shows the sensitivity of the maximum buckling load for this arch with respect to the vertical height *h/L*. It can also be seen from Figure 11.9 that the optimal height for the maximum buckling load *h/L* varies from approximately 0.15 to 0.25. Based on this sensitivity analysis, it can be noted that the trapezoidal arch with a vertical height *h/L* that is non-optimal may have significantly smaller buckling loads than the corresponding optimal arch designed against buckling. Take fixed–fixed trapezoidal arches as an example. When $b/L = 0.2$, the trapezoidal arch with a height $h/L = 0.4$ has a buckling load that is only 47.3% of that associated with the optimal arch design against buckling (where $h/L = 0.1678$).

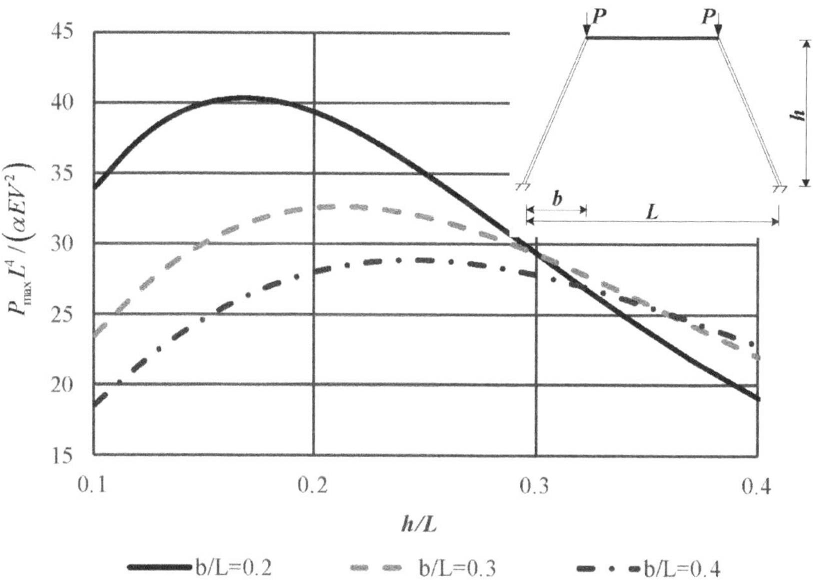

(a) Fixed–fixed arches

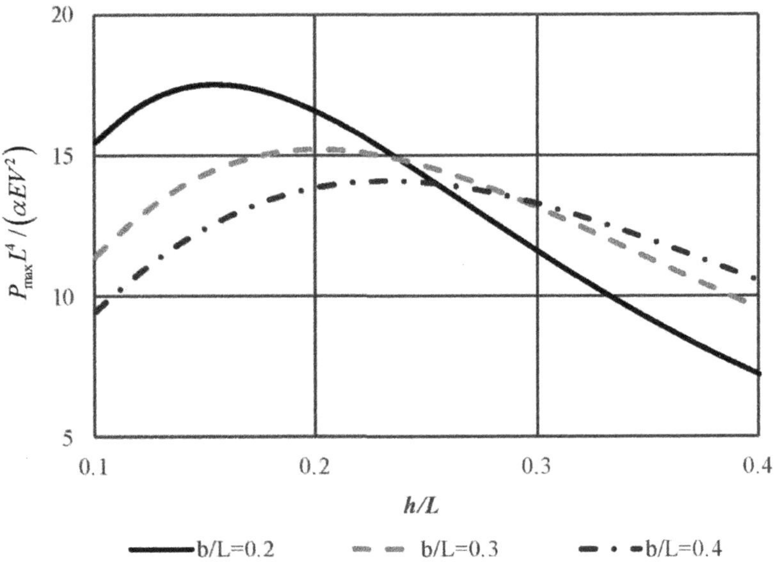

(b) Hinged–hinged arches

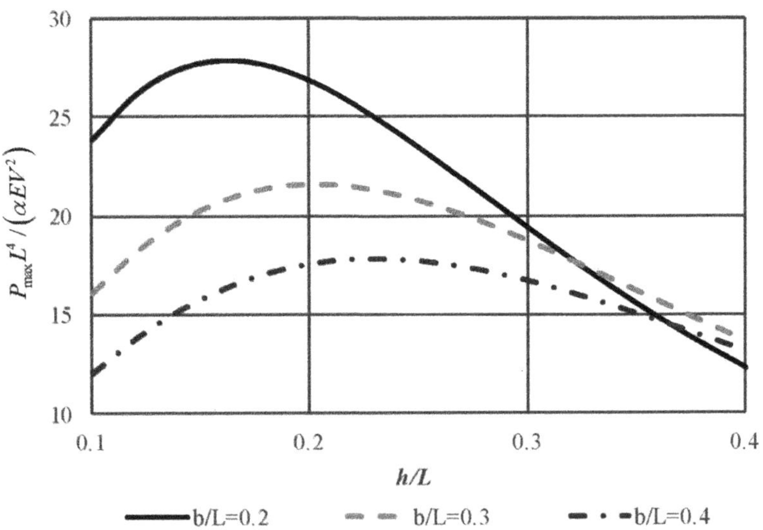

(c) Fixed–hinged arches

Figure 11.9. *Sensitivity of the maximum buckling load of trapezoidal arches with respect to h/L, for three base support conditions. For a color version of this figure, see www.iste.co.uk/challamel/mechanics1.zip*

11.3.5. *Comparison with the buckling load of optimal fully stressed trapezoidal arches*

The optimal design of fully stressed trapezoidal arches for least weight was investigated by Rozvany and Wang (1983). The shape of the optimal arch with base supports at the same level has to satisfy the unit mean square slope optimality condition, i.e.

$$\frac{\int_0^L \left(\frac{dy}{dx}\right)^2 dx}{L} = 1.0 \tag{11.20}$$

where y is the elevation of the member above the horizontal supporting line.

For a symmetrical trapezoidal arch under two equal vertical point loads, the above optimality condition yields

$$h_{\text{opt}} = \frac{\sqrt{bL}}{\sqrt{2}} \tag{11.21}$$

$$\frac{A_{\text{I}}}{A_{\text{II}}} = \sqrt{\frac{1+2b/L}{2b/L}} \tag{11.22}$$

Such optimal fully stressed trapezoidal arches, whose shapes are described by equation [11.21], may have significantly smaller buckling loads than the corresponding optimal arch designed against buckling. For example, consider a hinged–hinged trapezoidal arch with loading at $b/L = 0.3$. According to equation [11.21], the optimal fully stressed arch has an optimal height $h/L \approx 0.39$, and its buckling load is only 65% of that associated with the optimal arch that is designed against buckling (where $h/L = 0.201$).

11.4. Concluding remarks

In this chapter, we have presented the Hencky bar-chain model (HBM) for the elastic in-plane buckling analysis of a funicular trapezoidal frame, under two point loads at the joints between members. The supports may be hinged (defined by an HBM rotational spring stiffness of zero) or fixed (defined by an HBM rotational spring stiffness that is twice the HBM internal rotational spring stiffness of the member). The HBM furnishes accurate buckling solutions when a sufficient number of bar segments is adopted (say 20 segments) for each member.

Based on the developed HBM, optimal designs of trapezoidal arches against buckling were obtained for various load positions and base support conditions. As the horizontal distance b/L varies from 0.15 to 0.4, the optimal height h_{opt}/L increases from approximately 0.15 to 0.25, and the optimal cross-sectional area ratio A_I/A_H varies from approximately 0.8 to 1.6. A sensitivity analysis of the maximum buckling load with respect to h/L indicates that the trapezoidal arch with a vertical height h/L that is non-optimal may have significantly smaller buckling loads than the corresponding optimal arch designed against buckling. Therefore, for an economical design of trapezoidal arches considering both strength and stability, the optimal arch solution against buckling should also be used in addition to the condition for the optimal fully stressed trapezoidal arches.

11.5. References

Hencky, H. (1920). Über die angenäherte Lösung von Stabilitätsproblemen im Raum mittels der elastischen Gelenkkette. *Der Eisenbau*, 11, 437–452.

Krishna, S.G. and Ram, Y.M. (2007). Discrete model analysis of optimal columns. *International Journal of Solids and Structures*, 44(22–23), 7307–7322.

Mohamadain, M.A., El-Arabi, I.A., Nassef, W.M. (2010). Inelastic stability of steel trapezoidal frames at elevated temperatures. *Ain Shams Engineering Journal*, 1(1), 21–30.

Nelder, J.A. and Mead, R. (1965). A simplex method for function minimization. *The Computer Journal*, 7(4), 308–313.

Pan, W.H., Wang, C.M., Zhang, H. (2019a). Hencky bar-chain model for buckling analysis of non-symmetric portal frames. *Engineering Structures*, 182, 391–402.

Pan, W.H., Wang, C.M., Zhang, H. (2019b). Matrix method for buckling analysis of frames based on Hencky bar-chain model. *International Journal of Structural Stability and Dynamics*, 19(8), 1950093.

Rozvany, G.I.N. and Wang, C.M. (1983). On plane Prager-structures I. *International Journal of Mechanical Sciences*, 25(7), 519–527.

Ruocco, E., Zhang, H., Wang, C.M. (2016). Hencky bar-chain model for buckling analysis of non-uniform columns. *Structures*, 6, 73–84.

Salem, A.H. (1969). Buckling of trapezoidal frames permitted to sway. *Journal of the Structural Division*, 95(12), 2621–2640.

Silverman, I.K. (1951). Discussion on the paper of "Salvadori M.G., Numerical computation of buckling loads by finite differences, Transactions of the ASCE, 116, 590–636, 1951". *Transactions of the ASCE*, 116(1), 625–626.

Sun, C., Shaw, W.J.D., Vinogradov, A.M. (1995). Discrete-element model for buckling analysis of thin ring confined within rigid boundary. *Journal of Engineering Mechanics*, 121(1), 71–79.

Wang, C.Y. (2001). Stability of a heavy segmented column. *Mechanics Research Communications*, 28(5), 493–497.

Wang, C.Y. (2007). Optimum trapezoidal frame with rigid members and flexible joints. *International Journal of Non-linear Mechanics*, 42(5), 760–764.

Wang, C.Y. (2011). Stability and optimization of the portal frame with rigid members and flexible joints. *Journal of Engineering Mechanics*, 137(5), 366–370.

Wang, C.Y. (2020). Stability and optimization of a pressure-loaded trapezoidal vault with semi-rigid joints. *International Journal of Structural Stability and Dynamics*, 20(1), 2071001.

Wang, C.M., Zhang, H., Gao, R.P., Duan, W.H., Challamel, N. (2015). Hencky bar-chain model for buckling and vibration of beams with elastic end restraints. *International Journal of Structural Stability and Dynamics*, 15(7), 1540007.

Wang, C.M., Zhang, H., Challamel, N., Xiang, Y. (2016). Buckling of nonlocal columns with allowance for selfweight. *Journal of Engineering Mechanics*, 142(7), 04016037.

Wang, C.M., Pan, W.H., Zhang, J.Q. (2020a). Optimal design of triangular arches against buckling. *Journal of Engineering Mechanics*, 146(7), 04020059.

Wang, C.M., Zhang, H., Challamel, N., Pan, W.H. (2020b). *Hencky Bar-Chain/Net for Structural Analysis*. World Scientific, Singapore.

Yuan, S. (2008). *Programming Structural Mechanics*, 2nd edition. Higher Education Press, Beijing.

Zhang, Z., Challamel, N., Wang, C.M. (2013). Eringen's small length scale coefficient for buckling of nonlocal Timoshenko beam based on microstructured beam model. *Journal of Applied Physics*, 114(11), 114902.

Zhang, H., Wang, C.M., Challamel, N. (2016). Buckling and vibration of Hencky bar-chain with internal elastic springs. *International Journal of Mechanical Sciences*, 119, 383–395.

Zhang, H., Wang, C.M., Challamel, N. (2017). Small length scale coefficient for Eringen's and lattice-based continualized nonlocal circular arches in buckling and vibration. *Composite Structures*, 165, 148–159.

List of Authors

Jacob ABOUDI
Tel Aviv University
Israel

Mariano P. AMEIJEIRAS
CEFyN-UNC
Argentina

Romesh C. BATRA
Virginia Polytechnic Institute
and State University
USA

Atul BHASKAR
University of Southampton
UK

Emanuela BOLOGNA
Univeristy of Palermo
Italy

Devin BURNS
Virginia Polytechnic Institute
and State University
USA

Noël CHALLAMEL
University of Southern Brittany
France

Srinandan DASMAHAPATRA
University of Southampton
UK

Aharon DEUTSCH
Technion – Israel Institute
of Technology
Israel

Mario DI PAOLA
University of Palermo
Italy

Moshe EISENBERGER
Technion – Israel Institute
of Technology
Israel

Uğurcan EROĞLU
Izmir University of Economics
Turkey

Rivka GILAT
Ariel University
Israel

Luis A. GODOY
CEFyN-UNC
Argentina

Valery I. GULYAYEV
National Transport University
Kyiv
Ukraine

Andrii IAKOVLIEV
University of Southampton
UK

Julius KAPLUNOV
Keele University
UK

Kazuo KATAYAMA
Daicel Corporation
Japan

Mikael A. LANGTHJEM
Aarhus University
Denmark

Wen Hao PAN
Zhejiang University
China

Achille PAOLONE
Sapienza University of Rome
Italy

Vincent PICANDET
University of Southern Brittany
France

Giuseppe RUTA
Sapienza University of Rome
and
National Group for Mathematical Physics
Italy

Natalya V. SHLYUN
National Transport University
Kyiv
Ukraine

Yoshihiko SUGIYAMA
Osaka Prefecture University
Japan

Izuru TAKEWAKI
Kyoto University
Japan

Joscph TENENBAUM
Technion – Israel Institute of Technology
Isreal

Ekrem TÜFEKCI
Istanbul Technical University
Turkey

Chien Ming WANG
Queensland University
Australia

Hanzhe ZHANG
Beijing Institute of Technology
China

Massimiliano ZINGALES
University of Palermo
Italy

Index

G, H, I, J

L, M, N, O

P, R

S, T

U, V, W

Summary of Volume 2

Chapter 3. A Softening–Hardening Nanomechanics Theory for the Static and Dynamic Analyses of Nanorods and Nanobeams: Doublet Mechanics

Ufuk GUL and Metin AYDOGDU

Chapter 4. Free Vibration of Micro-Beams and Frameworks Using the Dynamic Stiffness Method and Modified Couple Stress Theory

J.R. BANERJEE

Chapter 10. Dynamic Soil Stiffness of Foundations Supported by Layered Half-Space

Yang ZHOU and Wei-Chau XIE

Summary of Volume 3

Chapter 3. On the Applicability of First-Order Approximations for Design Optimization under Uncertainty

Benedikt KRIEGESMANN

Chapter 4. Understanding Uncertainty

Maurice LEMAIRE

Chapter 5. New Approach to the Reliability Verification of Aerospace Structures

Giora MAYMON

Chapter 6. A Review of Interval Field Approaches for Uncertainty Quantification in Numerical Models

Matthias FAES, Maurice IMHOLZ, Dirk VANDEPITTE and David MOENS

Chapter 7. Convex Polytopic Models for the Static Response of Structures with Uncertain-but-bounded Parameters

Zhiping QIU and Nan JIANG

Chapter 8. On the Interval Frequency Response of Cracked Beams with Uncertain Damage

Roberta SANTORO

Chapter 9. Quantum-Inspired Topology Optimization

Xiaojun WANG, Bowen NI and Lei WANG

Chapter 10. Time Delay Vibrations and Almost Sure Stability in Vehicle Dynamics

Walter V. WEDIG

Chapter 11. Order Statistics Approach to Structural Optimization Considering Robustness and Confidence of Responses

Makoto YAMAKAWA and Makoto OHSAKI

Other titles from

ISTE

in

Mechanical Engineering and Solid Mechanics

2021

DAHOO Pierre Richard, POUGNET Philippe, EL HAMI Abdelkhalak
Applications and Metrology at Nanometer Scale 1: Smart Materials, Electromagnetic Waves and Uncertainties
(Reliability of Multiphysical Systems Set – Volume 9)

2020

LEDOUX Michel, EL HAMI ABDELKHALAK
Heat Transfer 1: Conduction

SALENÇON Jean
Elastoplastic Modeling

2019

BAYLE Franck
Reliability of Maintained Systems Subjected to Wear Failure Mechanisms: Theory and Applications
(Reliability of Multiphysical Systems Set – Volume 8)

BEN KAHLA Rabeb, BARKAOUI Abdelwahed, MERZOUKI Tarek
Finite Element Method and Medical Imaging Techniques in Bone Biomechanics
(Mathematical and Mechanical Engineering Set – Volume 8)

IONESCU Ioan R., QUEYREAU Sylvain, PICU Catalin R., SALMAN Oguz Umut
Mechanics and Physics of Solids at Micro- and Nano-Scales

LE VAN Anh, BOUZIDI Rabah
Lagrangian Mechanics: An Advanced Analytical Approach

MICHELITSCH Thomas, PÉREZ RIASCOS Alejandro, COLLET Bernard, NOWAKOWSKI Andrzej, NICOLLEAU Franck
Fractional Dynamics on Networks and Lattices

SALENÇON Jean
Viscoelastic Modeling for Structural Analysis

VÉNIZÉLOS Georges, EL HAMI Abdelkhalak
Movement Equations 5: Dynamics of a Set of Solids
(Non-deformable Solid Mechanics Set – Volume 5)

2018

BOREL Michel, VÉNIZÉLOS Georges
Movement Equations 4: Equilibriums and Small Movements
(Non-deformable Solid Mechanics Set – Volume 4)

FROSSARD Etienne
Granular Geomaterials Dissipative Mechanics: Theory and Applications in Civil Engineering

RADI Bouchaib, EL HAMI Abdelkhalak
Advanced Numerical Methods with Matlab® 1: Function Approximation and System Resolution
(Mathematical and Mechanical Engineering SET – Volume 6)
Advanced Numerical Methods with Matlab® 2: Resolution of Nonlinear, Differential and Partial Differential Equations
(Mathematical and Mechanical Engineering SET – Volume 7)

SALENÇON Jean
Virtual Work Approach to Mechanical Modeling

2017

BOREL Michel, VÉNIZÉLOS Georges
Movement Equations 2: Mathematical and Methodological Supplements
(Non-deformable Solid Mechanics Set – Volume 2)
Movement Equations 3: Dynamics and Fundamental Principle
(Non-deformable Solid Mechanics Set – Volume 3)

BOUVET Christophe
Mechanics of Aeronautical Solids, Materials and Structures
Mechanics of Aeronautical Composite Materials

BRANCHERIE Delphine, FEISSEL Pierre, BOUVIER Salima, IBRAHIMBEGOVIC Adnan
From Microstructure Investigations to Multiscale Modeling: Bridging the Gap

CHEBEL-MORELLO Brigitte, NICOD Jean-Marc, VARNIER Christophe
From Prognostics and Health Systems Management to Predictive Maintenance 2: Knowledge, Traceability and Decision
(Reliability of Multiphysical Systems Set – Volume 7)

EL HAMI Abdelkhalak, RADI Bouchaib
Dynamics of Large Structures and Inverse Problems
(*Mathematical and Mechanical Engineering Set – Volume 5*)
Fluid-Structure Interactions and Uncertainties: Ansys and Fluent Tools
(Reliability of Multiphysical Systems Set – Volume 6)

KHARMANDA Ghias, EL HAMI Abdelkhalak
Biomechanics: Optimization, Uncertainties and Reliability
(Reliability of Multiphysical Systems Set – Volume 5)

LEDOUX Michel, EL HAMI Abdelkhalak
Compressible Flow Propulsion and Digital Approaches in Fluid Mechanics (Mathematical and Mechanical Engineering Set – Volume 4)
Fluid Mechanics: Analytical Methods (Mathematical and Mechanical Engineering Set – Volume 3)

MORI Yvon
Mechanical Vibrations: Applications to Equipment

2016

BOREL Michel, VÉNIZÉLOS Georges
Movement Equations 1: Location, Kinematics and Kinetics (Non-deformable Solid Mechanics Set – Volume 1)

BOYARD Nicolas
Heat Transfer in Polymer Composite Materials

CARDON Alain, ITMI Mhamed
New Autonomous Systems (Reliability of Multiphysical Systems Set – Volume 1)

DAHOO Pierre Richard, POUGNET Philippe, EL HAMI Abdelkhalak
Nanometer-scale Defect Detection Using Polarized Light (Reliability of Multiphysical Systems Set – Volume 2)

DE SAXCÉ Géry, VALLÉE Claude
Galilean Mechanics and Thermodynamics of Continua

DORMIEUX Luc, KONDO Djimédo
Micromechanics of Fracture and Damage (Micromechanics Set – Volume 1)

EL HAMI Abdelkhalak, RADI Bouchaib
Stochastic Dynamics of Structures (Mathematical and Mechanical Engineering Set – Volume 2)

GOURIVEAU Rafael, MEDJAHER Kamal, ZERHOUNI Noureddine
From Prognostics and Health Systems Management to Predictive Maintenance 1: Monitoring and Prognostics (Reliability of Multiphysical Systems Set – Volume 4)

KHARMANDA Ghias, EL HAMI Abdelkhalak
Reliability in Biomechanics
(Reliability of Multiphysical Systems Set –Volume 3)

MOLIMARD Jérôme
Experimental Mechanics of Solids and Structures

RADI Bouchaib, EL HAMI Abdelkhalak
Material Forming Processes: Simulation, Drawing, Hydroforming and Additive Manufacturing
(Mathematical and Mechanical Engineering Set – Volume 1)

2015

KARLIČIĆ Danilo, MURMU Tony, ADHIKARI Sondipon, MCCARTHY Michael
Non-local Structural Mechanics

SAB Karam, LEBÉE Arthur
Homogenization of Heterogeneous Thin and Thick Plates

2014

ATANACKOVIC M. Teodor, PILIPOVIC Stevan, STANKOVIC Bogoljub, ZORICA Dusan
Fractional Calculus with Applications in Mechanics: Vibrations and Diffusion Processes
Fractional Calculus with Applications in Mechanics: Wave Propagation, Impact and Variational Principles

CIBLAC Thierry, MOREL Jean-Claude
Sustainable Masonry: Stability and Behavior of Structures

ILANKO Sinniah, MONTERRUBIO Luis E., MOCHIDA Yusuke
The Rayleigh−Ritz Method for Structural Analysis

LALANNE Christian
Mechanical Vibration and Shock Analysis – 5-volume series – 3rd edition
Sinusoidal Vibration – Volume 1
Mechanical Shock – Volume 2
Random Vibration – Volume 3
Fatigue Damage – Volume 4
Specification Development – Volume 5

LEMAIRE Maurice
Uncertainty and Mechanics

2013

ADHIKARI Sondipon
Structural Dynamic Analysis with Generalized Damping Models: Analysis

ADHIKARI Sondipon
Structural Dynamic Analysis with Generalized Damping Models: Identification

BAILLY Patrice
Materials and Structures under Shock and Impact

BASTIEN Jérôme, BERNARDIN Frédéric, LAMARQUE Claude-Henri
Non-smooth Deterministic or Stochastic Discrete Dynamical Systems: Applications to Models with Friction or Impact

EL HAMI Abdelkhalak, RADI Bouchaib
Uncertainty and Optimization in Structural Mechanics

KIRILLOV Oleg N., PELINOVSKY Dmitry E.
Nonlinear Physical Systems: Spectral Analysis, Stability and Bifurcations

LUONGO Angelo, ZULLI Daniele
Mathematical Models of Beams and Cables

SALENÇON Jean
Yield Design

2012

DAVIM J. Paulo
Mechanical Engineering Education

DUPEUX Michel, BRACCINI Muriel
Mechanics of Solid Interfaces

ELISHAKOFF Isaac *et al.*
Carbon Nanotubes and Nanosensors: Vibration, Buckling and Ballistic Impact

GRÉDIAC Michel, HILD François
Full-Field Measurements and Identification in Solid Mechanics

GROUS Ammar
Fracture Mechanics – 3-volume series
Analysis of Reliability and Quality Control – Volume 1
Applied Reliability – Volume 2
Applied Quality Control – Volume 3

RECHO Naman
Fracture Mechanics and Crack Growth

2011

KRYSINSKI Tomasz, MALBURET François
Mechanical Instability

SOUSTELLE Michel
An Introduction to Chemical Kinetics

2010

BREITKOPF Piotr, FILOMENO COELHO Rajan
Multidisciplinary Design Optimization in Computational Mechanics

DAVIM J. Paulo
Biotribolgy

PAULTRE Patrick
Dynamics of Structures

SOUSTELLE Michel
Handbook of Heterogenous Kinetics

2009

BERLIOZ Alain, TROMPETTE Philippe
Solid Mechanics using the Finite Element Method

LEMAIRE Maurice
Structural Reliability

2007

GIRARD Alain, ROY Nicolas
Structural Dynamics in Industry

GUINEBRETIÈRE René
X-ray Diffraction by Polycrystalline Materials

KRYSINSKI Tomasz, MALBURET François
Mechanical Vibrations

KUNDU Tribikram
Advanced Ultrasonic Methods for Material and Structure Inspection

SIH George C. *et al.*
Particle and Continuum Aspects of Mesomechanics

Printed and bound by CPI Group (UK) Ltd, Croydon, CR0 4YY

15/09/2026

14962362-0005